ALSO BY JOYCE E. CHAPLIN

Round About the Earth:
Circumnavigation from Magellan to Orbit

Benjamin Franklin's Political Arithmetic:
A Materialist View of Humanity

The First Scientific American:
Benjamin Franklin and the Pursuit of Genius

Subject Matter:
Technology, the Body, and Science on
the Anglo-American Frontier, 1500–1676

An Anxious Pursuit:
Agricultural Innovation and Modernity in the Lower South,
1730–1815

THE FRANKLIN STOVE

THE SECOND MODEL OF FIVE "STOVES"
INVENTED BY BENJAMIN FRANKLIN.
*(Mercer Museum Library of the Bucks County Historical Society,
accession no. 04519)*

The FRANKLIN STOVE

An Unintended American Revolution

JOYCE E. CHAPLIN

Farrar, Straus and Giroux
New York

Farrar, Straus and Giroux
120 Broadway, New York 10271

Copyright © 2025 by Joyce E. Chaplin
All rights reserved
Printed in the United States of America
First edition, 2025

Library of Congress Cataloging-in-Publication Data
Names: Chaplin, Joyce E., author.
Title: The Franklin stove : an unintended American revolution / Joyce E. Chaplin.
Description: First edition. | New York : Farrar, Straus and Giroux, 2025. |
 Includes bibliographical references and index
Identifiers: LCCN 2024035184 | ISBN 9780374613808 (hardcover)
Subjects: LCSH: Franklin, Benjamin, 1706–1790—Influence. | Stoves, Wood—
 Pennsylvania—History—18th century. | Climatic changes—History. | United
 States—History—18th century.
Classification: LCC TH7440.F73 C43 2025 | DDC 697/.1—dc23/eng/20241126
LC record available at https://lccn.loc.gov/2024035184

Designed by Patrice Sheridan

Our books may be purchased in bulk for promotional, educational, or business use. Please contact your local bookseller or the Macmillan Corporate and Premium Sales Department at 1-800-221-7945, extension 5442, or by email at MacmillanSpecialMarkets@macmillan.com.

www.fsgbooks.com
Follow us on social media at @fsgbooks

1 3 5 7 9 10 8 6 4 2

For Jack P. Greene

CONTENTS

INTRODUCTION: MAKE HASTE SLOWLY 3

1.
PROMETHEAN 19

2.
FORGING 50

3.
THE PENNSYLVANIAN FIREPLACE 85

4.
ATMOSPHERES 128

5.
GROW OR DIE 155

6.
FUELED BY FOSSILS 191

7.
THE AMERICAN REVOLUTION? 236

8.
NOVUS ORDO SECLORUM 268

9.
KEYSTONE STATE 303

CODA: GOD HELPS THEM WHO HELP THEMSELVES 345

NOTES 355

ACKNOWLEDGMENTS 403

INDEX 409

THE FRANKLIN STOVE

INTRODUCTION

Make haste slowly.
—POOR RICHARD

Yesterday's technology is trash, or maybe it gets recycled—unless it goes to a museum. Those were the fates of Benjamin Franklin's famous stoves, only one of which survives and now resides in the Mercer Museum in Pennsylvania (see frontispiece). It's roughly the size of a side table (30.75 by 27.5 by 35 inches) and, like many side tables today, originally came as a flat pack with instructions for assembly into what the museum describes as "a cast iron stove." But it wasn't just a physical thing, forged from iron. The stove was a hypothesis. Franklin was proposing that, armed with science, he and others could invent their way out of a climate crisis, creating comfortable indoor atmospheres for buyers of new stoves during a period of natural cooling (yes, the exact opposite of our problem) now called the Little Ice Age. He believed his stove could accomplish this in spite of another, related crisis: a widespread shortage of wood fuel. And he conceived of his invention as equal parts scientific instrument and home appliance, its many iterations identifying basic principles of atmospheric science that might allow him to figure out why the climate was changing in the first place. It may seem ironic that an old woodstove is now coddled by a museum's modern HVAC system, but it's a logical outcome. The search for better heating was part of a seismic shift in material life, the most important revolution of Franklin's era: the Industrial Revolution, with its historically unique rate of burning things, eventually resulting in the artificial heating of our entire atmosphere.

Given the high hopes today for inventing a way out of our own climate crisis, the history of the Franklin stove is useful, even if the physical object no longer is. Physical objects, I should say, because Franklin didn't invent *a* stove—he invented five of them. He kept reinventing them, over and over, because he and his contemporaries were worried (like us) that two grim trends—climate change and resource depletion—were converging on them, with the added threat (like us) that war might at any time further compromise access to resources; Franklin lived during what historians call the Second Hundred Years' War, beginning in 1688 with the War of the League of Augsburg (before his birth) and ending with the close of the Napoleonic Wars in 1815 (after he died), taking in the Seven Years' War and American War for Independence along the way.

With five different versions, "the" Franklin stove was never one single object. It was an ongoing project, preoccupying Franklin for more than half of his life, from 1738 to 1784, and eventually involving a great many other people, including the proprietors and workers at three Pennsylvania ironworks, a couple of Franklin's brothers, his London landlady and her daughter, several of the era's biggest scientific names, many aristocrats (from multiple countries) who liked buying the latest tech, two famous steam-engine inventors, and the chief of police in Paris.

It's from Franklin's era that we inherit our techno-optimism, hope that a single device might solve a huge problem—silver-bullet thinking. Faith in technology was part of an eighteenth-century rage for "improvement," the belief that planning and scientific knowledge could better the human condition. As an early incarnation of the charismatic scientific inventor who makes big promises, Franklin proposed a way to manage an emergency with a technical device whose virtues, he shrewdly hinted in his marketing materials, were uncountable: "&c. &c. &c." In his era, many improvements took the form of projects, requiring investment and expertise, expected to run for certain periods and to achieve specific goals, often involving new technology.[1]

We now expect even more from technology, maybe too much. Electronic consumer devices are our *friends*, glowing presences that make us feel autonomous and yet connected, an intoxicating combination. AI is our *rival*, a hypothetically self-governing power that might do us in and replace us. Other new technologies are our *saviors*—electric cars, solar panels, heat pumps, carbon capture, geoengineering, gigantic orbiting parasols—that are supposed to solve the climate crisis without our having to make any other changes to our society or economy.[2] But all three of these possibilities ignore how new inventions tend to remain tethered to their human creators and users. Devices depend on the social relations around them; their designers encode preferences into them, deliberately or inadvertently. We should ask: What gets encoded when technology is supposed to tackle climate change, as in Franklin's era?[3]

The Little Ice Age was the biggest change in climate before our own. This period of interglacial cooling in the North Atlantic lasted (roughly) from 1300 to 1850. It wasn't global in extent, nor caused by human beings, but its effects were sometimes devastating.[4] People disagreed as to whether they could mitigate it, that is, actually warm up the weather, but they knew they had to adapt to the cold.[5] And so, in their adaptations—including a growing reliance on science and technology—Franklin and his contemporaries conducted for us a kind of fire drill, one we can study in retrospect, seeing what worked or didn't.

If the stoves were hypotheses about inventing a way out of a climate crisis, their story shows that surviving climate change is not just a technical problem: there is no magic bullet. Climate change is a social and political problem, one with a much bigger history, though one early and important chapter of that story conveniently fits into a device the size of a side table.

The Franklin stove* is an icon and a mystery, obviously famous, though few people can explain why. It's always there on the lists of the celebrated man's celebrated inventions (lightning rod, bifocals, glass armonica, swimming fins, stove . . .), but there are few serious investigations of it. This isn't entirely surprising. Many details about material life in Franklin's era get brushed away, as if they're traces of some kind of backwardness. This is above all true of premodern energy use, which seems too embarrassingly carnal even for Hollywood to depict.[6]

Consider the movie most often praised for displaying the eighteenth century in living color, Stanley Kubrick's *Barry Lyndon* (1975). Famously, Kubrick shot footage in historic places, used only natural light or candlelight, and demanded period-correct costumes and makeup. But in the entire film, whatever the weather, there's not a single lit fireplace. True, in one scene, the title character morosely chops wood to make it totally clear he's a poor, rural nobody. It's as if getting fuel and keeping warm worry only the powerless rustics,

---

* Franklin invented his stoves to heat, not cook, and heating will therefore be the focus of this book. (Temperature is, after all, the central element of climate change.)

while the rich characters have hidden HVAC systems and keep their fireplaces just for show.

Kubrick began filming *Barry Lyndon* in spring 1973, right before the first energy crisis (which started later that year), but more recent representations of colonial America—many crises later—follow its odd logic about indoor heat. In *John Adams* (2008), John and Abigail show their sturdy origins by chopping and toting firewood at home in Massachusetts. And they share scenes with some fireplaces lit for specific tasks, like ironing clothes or burning government documents. (To protect historic interiors, many of these "fires" are fake.) But as in *Barry Lyndon*, the fires fade as the Adamses ascend the social ladder: no hearth gleams in the episodes set during Adams's diplomatic service in Europe. Things are better in the miniseries *Franklin* (2024), where the French foreign minister, at least, has a cheery fire in the winter of 1776 when Franklin arrives, though no one else does. A couple of winters later (episode four), everyone's chamber is aglow (did certain permissions for using historic interiors finally come through?), but only with the open, wood-burning fireplaces Franklin deplored and didn't depend on in France. The fireplaces are set design, signaling a past that's passé—except when they're not there and, instead, that amazing transhistorical HVAC is softly purring somewhere out of sight.

Beware any household fire that does burn in a historical film, however, because it's usually there to signal the wretchedness of premodern life. In *Sleepy Hollow* (1999), a chimney rarely appears unless venting the blackest smoke ever. The opening scene features a lit fireplace, the first of many. A snapped bit of kindling is a plot twist, a hearth's decorative detail is a clue—antiquated heating is inescapable. And then there's the doomed colonial family in *The Witch* (2015), whose passage to death and damnation is tracked by the growth of their woodpile, which in the end collapses onto the dying Puritan patriarch who built it up for a winter's warmth in which no one will live comfortably (let alone deliciously).

But *Sleepy Hollow* and *The Witch* don't add these details about early American domestic heating in good faith, to supply historic verisimilitude. No, the fires burn to highlight each movie's real form of energy: sorcery. The assumption seems to be that colonial people must have had some demonic equivalent to modern technology, the wood-burning kind being so patently ridiculous. We may wish we had a device *like* a magic bullet; maybe colonists had the real thing?[7]

To be sure, any tourist in Salem, Massachusetts, can see that early America is deeply and colorfully associated with witchcraft, but the real horror was not, in fact, having to chop your own wood, but having no wood to chop. Hollywood's disdain for material needs reflects a modern prejudice that stockpiling firewood was beneath anyone of consequence, reflecting how fossil fuels have, for a great many people, made energy use magically invisible.

Meanwhile, in the ivory tower, most historians also edit early America's fuel and fireplaces out of the picture to keep the spotlight on other developments, above all the American Revolution. Obviously, the origin story of the United States is important. But is it the most important transformation of its era? The events bundled under the label of Industrial Revolution, with its fateful turn to carbon-based energy, may have been at least as momentous. As this book goes off to the printer, plans are afoot to celebrate the 250th anniversary of the United States in 2026. But four years later, 2030, marks the date when, the Intergovernmental Panel on Climate Change has warned, CO_2 emissions must be halved to prevent the worst consequences of anthropogenic climate change.

The 2026 festivities may be eclipsed by awareness of the 2030 deadline for reversing trends—especially fossil fuel use—that the Industrial Revolution set into motion. (I dread chyrons along the lines of USA LIGHTS BIRTHDAY CANDLES AS PLANET BURNS.) Franklin, that famous American founder, may have had a greater impact on this other revolution whose consequences now overshadow everything else.

The technology that epitomizes the Industrial Revolution is probably the coal-burning steam engine, ancestor of anything powered by fossil fuel, including any electronic device that gets its spark, ultimately, from coal- or natural-gas-fired power plants. The steam engine's eighteenth-century inventors wanted to escape the limits of organic energy—sun, wind, water, wood, and food for humans and animals—by burning coal to generate power; they also wanted to make money and help other people to do the same: Industrialism is the most directly planet-eating part of modern capitalism. Few people in Franklin's era (including Franklin) foresaw what steam power would do, however. In their lifetimes, most engines pumped water out of mines—dirty, dangerous, and out-of-the-way places most people would never visit. Meanwhile, they thought new heating designs, distributed among too many households to count, might have huge impact. By the 1780s, improved fire grates would be one of the most commonly patented consumer items in Great Britain. And indeed, household consumption of coal, long before factory industrialization, was the gateway to fossil fuel dependence. Franklin designed his first stoves to consume wood—and conserve wood—but in the end he built heaters that burned coal, not his original intention.[8]

Really, what Franklin and his fellow stove inventors *thought* they were inventing was an artificial atmosphere big enough to live in. It was this new idea, eventually to converge with the relative novelty of burning coal, that was the first step toward central heating (and a regulated indoor climate) becoming the default in the industrialized world. Iron stoves were history's first consumer durable, ye olde ancestors of washers, dryers, dishwashers, and HVAC systems. Keeping warm under Little Ice Age conditions was a tall order, and yet the era's many heating innovators were promising indoor climate control *that saved fuel*. An artificial indoor climate was a snug hub of new confidence that nature could be controlled, defying conditions that might otherwise have caused fatalism.

Franklin's more distinctive ambition was to use indoor atmospheres to figure out the big one outside. His first two stoves generated his primordial insight: through convection, they heated the first atmosphere he investigated, a room in his family's home, which became his model for interpreting heat and the motion of air within the planet's atmosphere. Having turned the outside in, harvesting natural resources to control indoor climates, he then turned the inside back out. He conjectured that the Northern Hemisphere's climate was colder than it had been in the remote past. He described climatic phenomena (such as Atlantic storm systems and the Gulf Stream) as complex thermal systems that crossed multiple latitudes. He hypothesized that volcanic activity could cool the atmosphere.

Finally, Franklin examined the human impact on natural atmospheres. He questioned whether the deforestation of North America would result in a warmer climate, something his white contemporaries insisted was a benefit of their constant extraction of firewood, but which he worried might not work—or else might lead to catastrophic overheating. Relatedly, he observed what we now call microclimates, often the result of human modifications to dense urban spaces. Also, he refused to accept that atmospheric pollution from household fires was inevitable and tried to design heating to minimize it—a friend teased him for being a "universal Smoke Doctor." Franklin's anti-smoke campaign was an early attempt to scrub emissions from combustion (now most urgent in terms of carbon dioxide).[9]

Franklin's efforts, widely discussed, show that awareness of links between human and natural atmospheres began in the eighteenth century. True, no one proposed that collective, everyday fuel consumption might eventually transform Earth's entire atmosphere. But several eighteenth-century observations suggested that, if humans could modify atmospheres indoors, they might be forging a different one in the world beyond. Two (plus) centuries later, we see the truth of that and live with the consequences.

Like any successful revolution, the creation of modern industrial society produced deep cultural inertia, a thousand reasons not to change a thing. Most revealingly, Franklin's adopted colony of Pennsylvania was in the vanguard of this new materiality, which makes his fireplace (and fire drill) distinctively useful to consider. Once upon a time, the Industrial Revolution was assumed to have begun in Europe, especially Britain, motherland of coal mines and factories.[10] In fact, Europe's overseas territories supplied much of industrialization's raw material and capital, and some of its ideas—including Franklin's.[11] Above all, Franklin's colony was the most obvious prototype for North American industrialization.

Virginia has been proposed as the most quintessentially American of the original states, a mini United States, because it was the first English colony to invest in enslaved workers from Africa, the first to forcibly remove Native American populations, yet also a leader in creating the independent nation, given Virginia slaveholders' outsized role in the Revolution and early republic. This idea of Virginia as the United States in miniature does not acknowledge, however, the nation's eventual industrialization, which is usually associated with the northern colonies or states, especially New England and New York, and later with the Midwest, especially Chicago.[12]

But Pennsylvania contained all these elements. It would by the nineteenth century be called the Keystone State for its geographic position, midway among the thirteen original states, and for being the heart of the American Revolution, the place where both the Declaration of Independence and the US Constitution were written. It would also be the heart of US industrialization, the first American state to rely on free, waged labor for industrial production, and first to produce and burn coal on an industrial scale, then do the same with liquid petroleum. These outcomes had their genesis in the colonial iron industry that cast Franklin's stoves. The colonies were already, by the 1770s, producing one-seventh of the world's supply of iron, with Pennsylvania not just a follower in industrialization, but a leader.[13]

The colony's early Industrial Revolution was distinctive, as well, in exposing the limits of organic energy, the fuel and food produced photosynthetically by the sun. Long before Pennsylvania became a land of coal-burning, steel-forging factories, white settlers were burning through woodlands appropriated from Natives at a dangerous rate, and the colony's ironmasters were exploiting every kind of labor available, including underfed enslaved people. In his project, Franklin would validate the European-led transition to coal, but only by hiding the American aspects of his original stoves: their reliance on Indigenous removal and forced labor. This was cheating in any number of ways, but especially by foreshadowing the biggest cheat of all: industrialization's promise, through fossil fuel, to transcend the limits of organic energy sources like wood and human labor, and therefore triumph over natural limits entirely.

That premise bolted a glitch into industrialization because nature always has limits. Franklin both knew and denied this, much the way we know but deny it. He would hypothesize that plentiful natural resources guaranteed rapid population and economic growth—gloriously apparent in the British colonies—but admitted that even American abundance might falter, hence his fuel-conserving stoves. The sunnier half of his hypothesis became the core of growth economics, correlating human happiness with the development of natural resources.

This, too, was part of an age of improvement, when people were devising projects for profit on every possible scale. Adapting the Greek word for household, *oikos*, the English used "economy" to describe household management from the 1440s onward, then "political economy" to plot the wealth of entire nations—the origin of modern economics—starting in 1687, not twenty years before Franklin was born. Reflecting his era's confidence that wealth was easy to generate (and enjoy), he would describe human material desires with the most outrageous word possible: "infinite." No wonder he thought consumer discipline might be necessary; no wonder he didn't assume it would be a universal virtue. Only much later, in 1875, would "ecology" describe *nature's* households, belatedly recognizing a symbiosis of humans and nature, but without the assumption that nature can support infinite economic growth, an expectation, for better or worse, made in Pennsylvania.[14]

To promote economic growth, Franklin had to ignore two powerful Pennsylvanian counterexamples. First, Indigenous Americans rejected an ethics of human command over nature; they did not separate economy and ecology. For them, natural things were not passive assets. They burned fuel as part of diplomacy, for instance, in council fires that united people and even nonhuman species. Many other famous northeastern Native diplomatic practices—sharing tobacco, making speeches, presenting diplomatic gifts like belts woven with wampum (shell) beads—took place around such fires. It's no use thinking Franklin didn't know this. As official printer to the colony of Pennsylvania, he printed the treaty negotiations conducted around council fires. Those treaties, which now define Indigenous rights in relation to the United States, have always questioned the Industrial Revolution's split between humans and nature.[15]

Likewise, in suggesting that humans derive economic value from material things but are not raw materials themselves, Franklin ignored how enslavement violated this logic. Pennsylvania's Native Lenapes and its enslaved Black people protested this debasement decades before the state launched a formal program of emancipation that Franklin would in the end support, rectifying his earlier assertions that some people made economic value while others were made into it, including some of the ironworkers who produced his stoves.

In the end, however, Pennsylvania's history shows that industrialization was not—is not—an irresistible force. Some scholars (and pundits) have suggested that coal prompted abolition, with steam power making slavery seem unnecessary and backward. The corollary of this idea is that without fossil fuels, humanity will descend into a Hobbesian war of all against all. But Pennsylvania's history doesn't bear this out. Some people questioned and resisted enslavement long before the state's dependence on coal, let alone petroleum. Franklin, for example, who wanted to improve everyday life by crafting stoves made by enslaved workers, later realized that this amalgam of conservation and exploitation was not a step forward. That epiphany was an ethical shift for which he—and others—did not suggest coal as an easy fix. Liberation and societal uplift are possible because people make them possible, not because nature hands them the means to do it.[16]

As an adaptation to climate change, Franklin's stove project met only one of three desired criteria: it worked (certainly enough to be adopted and imitated), but it validated a short-term solution—fossil fuel energy—with its own colossal drawbacks, and, at least initially, it accepted social inequities rather than confronting them. The stove's story is complex—and must be told that way. A progress narrative might celebrate the stove for using applied science to improve material life and lay groundwork for climate science, with these good outcomes canceling out anything else. A tragic narrative would, conversely, stress the human costs as outweighing any benefits. I think the story is incomplete without all its elements. Cheerful, stove-heated parlors rarely feature alongside war zones, slave ships, and smoke-belching factories in American history. But if we are to understand energy, climate, and technology in their fullest historical impact, comfortable domestic spaces must feature within them. Franklin's project fits into optimistic and pessimistic visions of American technology; it was both the violent capitalist machine in an Edenic garden *and* far-out, new age, redeeming alt-tech.[17]

A complex history suits a complex challenge. Climate change is what ethicists call a wicked problem, something with so many variables, each with overlapping (and mutating) qualities, that solutions must themselves be multiple, overlapping, and quickly adaptable. Technologies never solve wicked problems unless they're linked to social transformations, because technology is always a system of things and people. If anything, inventions that tackle such problems risk embodying any bias that fueled the crisis in the first place. Today's solutions to climate and energy are unlikely to be free of the legacies of colonialism, for example, unless the inventors confront that history.[18]

A project leans forward in time—it's an experiment *projected* to run into the future. The Franklin stove, object and project, was supposed to solve two time-based dilemmas: historically atypical cold weather, and the threat of future deforestation and loss of fuel. As Franklin put it: "By the Help of this saving Invention, our Wood may grow as fast as we consume it, and our Posterity may warm themselves at a moderate Rate, without being oblig'd to fetch their Fuel over the Atlantick; as, if Pit-Coal should not be here discovered (which is an Uncertainty) they must necessarily do." Franklin believed human technology could mitigate climate change, but he knew this begged other temporal questions, especially in terms of growth economics, which were predicated on change over time and demanded new energy, as with that "Pit-Coal," which most definitely would be discovered in Pennsylvania just after his death.[19]

If a project runs into the future, we are Franklin's "Posterity," participant-observers in a continuing experiment in energy efficiency and resource conservation, sometimes challenged by warfare, always well-advised to consult the science of climate and atmosphere. Can climate-crisis technology operate on a suitable timescale? How fast can any adjustment be made, can it keep pace with population growth

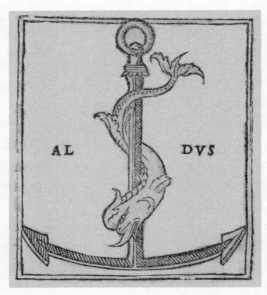

THE MOST FAMOUS *FESTINA LENTE* DEVICE, THE ONE FRANKLIN WOULD HAVE KNOWN.
(Aldo Manuzio, ed., Qvae hoc libro continentvr . . . *[Venice, 1449/50–1515], courtesy of the Grolier Club of New York)*

or consumer demand, and might it introduce a competing chronology in which resource depletion or toxic emissions undercut any benefit? The problem is choosing and implementing enough solutions in time. This is a particularly wicked feature of the wicked problem, the mismatch of timescales. Consumption of carbon-intense energy has, in a matter of centuries, damaged ecosystems that are thousands of years old and compromised an atmosphere millions of years in the making. Now, we need actions even swifter than those that brought us to this predicament—but without somehow making things worse.

For all his faults, Franklin had some good advice about this. His famous almanac's alter ego, Poor Richard, adapted a Latin proverb, *festina lente*, to advise that to get anything important done, you've got to "make haste," but, paradoxically, you'd better go "slowly." Since ancient times, people have translated the saying into visual emblems: a tortoise, proverbial for its steady progress, zips along with a sail rigged on its shell; a dolphin, a creature built for speed, is tangled around an anchor, which slows it down.

Each design suggests that technology can improve on nature, but that nature matters, too. Such images are reminders, as well, that "device" meant a visual emblem as early as 1375, only starting to mean a physical thing by 1570. Either way, humans are inventors. Left to their own devices, they can imagine many ways to do things with things—not with magic.[20]

Festina lente: both haste and resolve are necessary. To fix a wicked problem, people must get moving, but with intelligence, correcting errors both old and new as they go. The worse the problem, the greater the need for speed but also care. In any techno-optimistic rush to invent a way out of our current crisis, remedies require the swiftness of a sleek dolphin, even as awareness of errors committed in the past must be a steadying anchor, which I hope this history of Franklin and his stoves provides.

1

PROMETHEAN

Over his long and eventful life, Benjamin Franklin would win many splendid compliments but also some sharp rebukes, and, one time, a blistering fusion of the two. In 1756, philosopher Immanuel Kant declared Franklin the "Prometheus of the modern era" (*Prometheus der neuern Zeiten*). High praise, this comparison of the American colonist to the ancient Greek god who stole fire from heaven to assist hapless humans. It was a poetic summary of Franklin's scientific investigations of heat and electricity, including lightning, fire from heaven. But Kant was also warning that humans, and Prometheus their benefactor, had suffered Zeus's vengeance after the cosmic theft. "Promethean" describes titanic audacity, a dangerous and profane refusal to accept material limits.[1]

Kant was criticizing not just Franklin, but their era's science and technology, and indeed its confident ethic of improvement. When fuel supplies were shrinking, European investigators were promising to adjust heating systems to offset the loss, improving fire itself—Promethean, indeed. This was just one part of the bigger paradox, in which Europeans (and their colonizing cousins) expected an improved standard of living even as climatic conditions got worse and supplies of natural resources shrank. The Industrial Revolution, in its very earliest stages, was a firm refusal of physical limits.

All the bolder (if not just plain obnoxious) that Europeans exported these expectations to the Americas, where Indigenous ethics were, if anything, anti-Promethean. British settlers in Franklin's North America craved artificial heat in private dwellings and kept these warm atmospheres to themselves—whatever their collective toll on surrounding environments—in contrast to Native peoples' use of fires for collective and political purposes, not just personal comfort. These culturally different uses of fire had emerged after the very coldest atmosphere humans have ever experienced, a period ending a little over ten millennia ago, and they would further diverge during the second coldest, alongside the rising consumer expectations that underlaid the Industrial Revolution, a transition—defying conditions of cold and scarcity—that guided Franklin into his adulthood.

So this book must begin in earnest with what Franklin didn't invent: the conditions he inherited, the inspirations for his inventions. These began before his birth, though not too long before. On one of those long, long geological chronologies representing Earth's entire history, winding back into unfathomable millennia, we'll still be on the tiny bit representing not just human history, but very recent human history. Here we go.

Before the Little Ice Age, there was a big one. Earth's Last Glacial Period (LGP) took place between 115,000 to 11,700 years ago, fairly recent for a planet that's about 4.5 billion years old. During these chilly millennia, humans developed two technologies for keeping warm: fires to sit or lie beside in the coldest weather, and clothing sewn tight to retain body heat. The LGP happened long after ancient hominids had begun to deliberately set fires, at least 1 million years ago—they had one solution to the cold already. But the more intense cold encouraged greater technical sophistication: sharp, threaded needles (dating from some fifty thousand years ago in Siberia) to tailor animal skins into snug garments, thus creating the first artificial atmosphere right around the human body. Aided by the flame and the needle, humanity survived the coldest time in its past.[2]

Although these technologies, and the cold itself, are a shared human heritage, cultural differences in using natural resources to keep warm (or for other reasons) slowly emerged. By the time Franklin's English ancestors began to claim territory in North America, European and American use of material things had diverged significantly from long-standing global trends.

The peoples of northeastern North America were caretakers— consciously so—of the environments that were their homelands. Franklin would know best three extended groups: the Wabanakis of his native New England, the Lenapes in what would become Pennsylvania, and the Haudenosaunee (Iroquois Confederacy), stretching from western New York to the eastern Great Lakes, which would be the most powerful group of the three during Franklin's lifetime. If Europeans thought of themselves as existing in exile from the divine or heavenly, as in the Prometheus legend, or as in Christian scripture, with Adam and Eve's expulsion from paradise, these northeastern Natives thought the opposite.

With variations in the telling, they recounted a long-ago journey to their homelands, when a primordial traveler, a woman, fell and fell through a pitch-black sky, finally landing in a place where animals helped her reconstitute life, including human life. Indigenous anthropologist Roger C. Echo-Hawk (Pawnee) argues that these histories relate ancient migrations into North America during the cold, dark conditions of the LGP—testifying to the long duration of human history, extending back into geological time. In these Indigenous interpretations, the sky is a zone of danger, and Earth is where living creatures can look after each other; Sky Woman's fall is a deliverance, quite different from Adam and Eve's fall and alienation from God, or from Prometheus's trespass into the heavenly realm of the gods. If the earthly seemed second-best to Europeans, Native people in northeastern North America disagreed—the life around them was precious, to be used and preserved in thoughtful balance.[3]

The results were stunning. While Natives burned areas to help them plant or hunt, and gathered wood for construction and fuel, their actions actually protected forests by removing excess growth that might power devastating wildfires. On the Atlantic coast, the land now in New York State had the greatest extent of old-growth forest, at least 262,000 acres; Maine had the second-largest, with 36,000 acres, followed by Pennsylvania (27,000) and New Hampshire (15,000). A beech-hemlock forest covered 6 million acres in Pennsylvania's Allegheny Plateau alone. That expanse of trees took a colonial surveyor's breath away—he called it an "Ocean of Woods" and went away defeated, unable to chart it. At the time of European colonization, these northeastern forests had existed for ten to forty tree generations, thousands of years. This ancient biomass sustained multitudes of other species, plant and animal. Calling America's Indigenous peoples conservators of these spaces if anything underrates their commitment: they considered other forms of life their kin, inspirited beings who lived not above them like sky gods, but with them.[4]

Europe was different. There, during the LGP, hunters burned woodlands routinely to flush out prey. Their fires substantially reduced forest cover, with results that vie with East Asian irrigated rice fields as the first major anthropogenic transformation of Earth. And Europeans continued to harvest wood relentlessly, even when the cold receded during the Roman Warm Period (roughly 250 BCE to 400 CE) and the Medieval Climate Anomaly (around 950 to 1250).[5] In the Middle Ages, people cleared forests to cultivate land, to create visibility and security around settlements, to build everything from ships to cathedrals, but above all to get fuel for cooking, for crafts (forging metal, firing ceramics), and—always—for warm fires in cold weather.[6]

Wood and food were the most important forms of energy in the organic economy. When wood burns, the combustion reverses the chemical process that gave the plant its life, releasing the energy it captured from the sun. Human food was even more important, however, as the major source of energy before steam power. Animals did some work, though less than a quarter of the world's labor. Another tiny increment came from wind and water. All the rest, roughly three-quarters of the world's work, from building the pyramids to tending toddlers, was done by human beings. Fire helped those laboring bodies, keeping them warm and fed and arming them with forged or fired tools, and for all that fuel was essential and smoke a constant reminder of the sun-to-energy cycle.[7]

By 1500, Europeans began to worry about the wood running out. This didn't mean that Europe was being entirely deforested. But it hints that people were, for the first time, afraid it might be. People in all but heavily wooded places (like Scandinavia and Russia) began to complain about the cost of wood, even kindling. They also began to try to conserve surviving woodlands, sometimes to protect wood as a commodity, sometimes to preserve places that might yield other resources, like fish or game.[8] Although the scientific concept of an ecosystem was centuries away, rural people already knew that stripping a landscape of one thing might endanger others. Economic imperatives prompted choices between forest conservation versus extraction. And yet three different appetites for wood were rising at the time.[9]

The "age of sail" depended on wooden ships for trade, transport, and exploration, but especially for war—entire forests were launched into battle. A warship consumed 1,400 to 2,000 oak trees, each at least 100 years old, plus masts that might need to be 130 feet long, preferably made from tough but flexible conifers that would be several decades old. Such a ship would ravage wooded acreage of up to 0.2 square kilometers; a fleet of 100 warships depleted woodlands equivalent to nearly six Central Parks. And yet a single engagement could use hundreds of ships—the one-day Battle of Lepanto (1571) launched more than 400—and the rising militarism of the Second Hundred Years' War necessitated standing fleets, about 100 ships of the line and cruisers for the British Admiralty in a given year during the early eighteenth century, rising to 200 by midcentury. Any war might send this costly matériel to the bottom of the sea, and then replacement ships had to be built. Protection of trees for navies became standard policy. States established monopolies over certain forests and certain kinds of trees—no ruler wanted to be caught short, or foolishly dependent on a foreign supply, should war loom. The problem never got better. Just before the Battle of Trafalgar in 1805, two-fifths of Britain's ships of the line were out of commission, requiring a frenzy of building and repair, mostly using timber that came from outside Britain, which had long since run out of any Central Park–sized woodlands for the task.[10]

A second and rising demand for wood was as fuel for industrial use, making or refining everything from salt to iron. Iron was the most commonly used metal. It had to be melted from its ore at high heat; refining and shaping the metal required yet more heat. The English depended on ore- and tree-rich Sweden for much of their iron (and steel), though England's colonization of Ireland in the sixteenth century created a second source. But Irish landholders and iron manufacturers quickly began to worry about *their* trees and to enact policies for woodland conservation. Conflict between rural people and ironmakers over stocks of fuel would be a persistent problem, as eventually was the case in Franklin's Pennsylvania.[11]

The third and final reason for growing demand for wood—or any fuel—was that Europe's population was growing. After the Black Death of the early 1300s, Europe entered a long period of recovery and growth. By the early 1500s, Western Europe's population approached its pre-plague level (just under 80 million) and would rise to an unprecedented 120 million by 1700. Growing urban populations, above all in London, wanted wood for construction and everyday use. But the English worried about outgrowing their woodlands. Economic theories didn't yet celebrate growth, especially when stability was the goal of rulers; Renaissance commentators in Italy were the first to propose colonies as necessary outlets for populations that outgrew their nations' resources, whether those colonies yielded economic benefit or not. Although a larger population could increase a nation's productivity, it also consumed more, a risk within any organic economy.[12]

As elsewhere in Europe, English conservationists recommended preserving trees, not for their own sake, but for human advantage. In his landmark environmental work, *Sylva* (1664), John Evelyn defined forestry as something that had to persist over many generations and across the whole kingdom: "His *Majesty* must assert his *Power* . . . for the raising, and propagating of *Woods*, till the whole *Nation* were furnish'd for *posterity* . . . And which Work if his *Majesty* shall resolve to accomplish, he will leave such an everlasting *Obligation* on his *People*, and raise such a *Monument* to his *fame*, as the *Ages* for a *thousand* years to come." A thousand years! But only to benefit one nation—conservation began as a very exclusive business.[13]

Alternatives to wood, unsurprisingly, were highly attractive. In rural areas, dried animal dung was commonly burned. Processing wood into charcoal concentrated its energy-rich carbon into a hotter-burning fuel (necessary for forging iron) while making it lighter and more easily transported. Peat or turf, ancient vegetal material compressed in bogs, was excellent fuel once extracted and dried. But the big new energy alternative was coal, essentially older and pre-dried peat. Known since ancient times for its superb heat, coal had become an English substitute for firewood by the Middle Ages. It wasn't yet an industrial fuel. Before 1700, half the coal burned in Britain was for domestic use. Much of this was sea coal, a soft, easily ignited bituminous coal from deposits in oceans, retrieved from beaches where it washed up. A small but growing amount of coal was mined inland, from seams close to the surface.[14]

It was hard to ignore that smoke from coal was less pleasant than smoke from wood or peat, and that its soot clogged a chimney faster. English complaints about what would later be called "smog" are excellent evidence of the growing ubiquity of coal. In his landmark blast against pollution, *Fumifugium, or, The inconveniencie of the aer and smoak of London* (1661), John Evelyn (the same of *Sylva*) lamented all kinds of bad air in the busy city, but especially "that Hellish and dismall Cloud of SEA-COAL . . . perpetually imminent over her head." Evelyn proposed moving workshops outside the city, five or six miles lower on the Thames, where sea breezes would blow the murk off to somewhere (and someone) else. Another solution, at least within houses and workshops, were better grates, chimneys, and fireplaces, the start of the technical tinkering at which Franklin would excel. Coal in particular encouraged the tinkerers; it needs coaxing to ignite and, for domestic use, performs best in "basket" grates that, suspended above the hearth, expose the flame to air so it doesn't smother.[15]

None of these alternatives was yet a sure thing. Their distance from charcoal production, or their lack of coal mines or peat bogs, reminded many Europeans that they might need to conserve their forests. The problem grew acute as the weather got colder.[16]

Although thermometer readings of climatic variation didn't exist or weren't reliable before the nineteenth century, many people had plenty to say about the weather before then, and their comments corroborate proxy data of colder temperatures that scientists have extracted from tree rings, ice cores, and the like. Some of these observations, comparing conditions in Europe and the Americas, represent an origin point of the modern climate science Franklin would later develop.[17]

In Europe, the Little Ice Age's lower temperatures were only part of the problem. Wild swings in precipitation were equally menacing—first, too little to keep crops alive; then so much rain and hail that, along with damaging land and crops, it corroded buildings and bridges to the point of making them structurally unsound. Stretches of weather that sustained crops could be too short to do much good. Famine was a common result, and ignited societal and political conflicts, including witchcraft scares, with fears that the land was full of demons and all the larders empty.[18]

But one person's crisis can be another's opportunity. Several European states gained power precisely during the era's coldest winters, and some social groups—counterintuitively—somehow prospered. One way that Europeans coped was by going overseas, invading other people's territories, extracting their labor and resources, and lessening fears of overpopulation at home. This strategy would, for a time (and at an ethical price), ease the limits of Europe's organic economy, exacerbated by the bad weather. And the search for optimal regions for human life—preferably warm ones—would generate closer observations of climate.

Beginning in the 1500s, European accounts of the Americas defied received ideas about climate. As defined in ancient Greek geography, "climate" had been a synonym for "latitude," designating differently heated regions, hottest at the equator and cooling toward the poles. These climates were supposed to be globally uniform and unchanging.[19] But North America turned out to be distinctly cooler than the parts of Europe it faced, themselves colder during the Little Ice Age—oddness upon oddness. The transoceanic mismatch was the first to be amply recorded in writing, because the new transatlantic traffic elicited Europeans' repeated testimony (and disappointment) at finding cooler conditions over the water.[20]

But sometimes they did find heat—which they exploited. In the tropical climates of Brazil and the greater Caribbean, Europeans began to plant the distinctively destructive crop of sugarcane. That sweet plant decimated environments: entire islands were cleared and leveled, literally terraformed to grow sugarcane; processing the cane required mountains of firewood, often imported. Sugar also destroyed lives: Indigenous people were enslaved for the work, contributing to a death toll of an estimated 56 million Native inhabitants by 1600; starting in 1526, Europeans imported captives from sub-Saharan Africa, mostly for sugar cultivation, inaugurating the forcible removal of 12.5 million people across the Atlantic. The sugar industry was eventually the most capital-intensive in the Western Hemisphere and, in its financial and material demands, a model for the coming Industrial Revolution. And yet this was still a colonial economy. It exploited (and fetishized) warm climates overseas at a cold time in Europe.[21]

If deforestation and desertification of entire islands in the Caribbean were extreme forms of European colonization, was there a conservationist alternative? That wasn't an obvious goal. Europeans could better preserve their own forests if they had access to timber overseas, for example, which didn't encourage preservation of American trees.[22]

For the English, North American forests represented commercial opportunity. Starting in Virginia and New England, colonists turned trees into farmhouses, fences, and barns and grew crops on the cleared land. Naval stores, the pitch, tar, and turpentine that kept commercial vessels and warships afloat, came from coastal pine forests, prompting yet more extraction. The British Admiralty would claim a monopoly on mast-sized New England pine trees, though this was not intended to preserve forests in any real sense. And then there was settlers' constant demand for daily fuel. None of the early English colonies had easily accessible peat or coal. Wood was it. By 1637, two years after

Franklin's first New England ancestor had emigrated from across the Atlantic, colonists had stripped the Boston peninsula of trees.[23]

Rising material demands were part of an "industrious" or "consumer" revolution, one that powerfully shaped colonists' expectations of North America. The new consumerism is somewhat of a mystery, given that it began before the economic growth the Industrial Revolution sparked from the late 1700s onward, eliminating poverty as a near-default condition. Before that point, Europe had severe economic inequality (most wealth, in the form of land, belonged to a tiny elite) and no rise in economic output—there was little motion in the system, no real surplus, and hardly any capital or credit for anyone. (Our consumer credit would have seemed like something out of a fairy tale: caskets of gold delivered by elves.) And yet people began to buy stuff they didn't strictly need, seeking a different material life, with new levels of everyday comfort and abundance. The trend began a shift away from local and renewable resources, representing a divergence from cultures, including those of Indigenous Americans, with a moral imperative to respect natural constraints.[24]

Buying consumer goods because they made life more convenient, comfortable, or fun, or because they were eye-catching, in fashion, or (obviously) made someone else jealous, started to mark everyday life. New tastes, as for sugar, showed that human appetite represented a new engine of economic profit. Sugar fostered demand for related foods and goods, ones that carried yet more human and environmental costs: Asian tea, African coffee, and Mesoamerican chocolate, as well as Chinese porcelain and Spanish-American silver for serving these beverages, plus the sugar industry's by-products, molasses and rum. The new appetites began to descend down the social ranks, publicly displayed in what people wore. More people could buy shoes for cold and wet weather, even leather shoes instead of unforgiving wooden clogs that never broke in. Or they sought the silk, lace, or fur that had once indicated precise social ranks but could now be worn by whoever could afford them. "It is a hard Matter to know the Mistress from the Maid by their Dress," Daniel Defoe marveled in 1725—"nay very often the Maid shall be much the finer of the two."[25]

Among these new material expectations was a once-impossible bodily state: to feel comfortable indoors, in all seasons. This would require new technology as an everyday household presence, supplementing or substituting for the flame and the needle.[26]

Direct protection of the body had remained the primary way to keep warm. The ancient Romans had systems of underfloor heating (as did parts of Asia), but that technology had vanished from most of

WAYS OF KEEPING WARM, NOT ALWAYS WITH A FIREPLACE.
(Adriaen van Ostade, Dance in the Tavern *[1652], Harvard Art Museums/Fogg Museum, gift of Belinda L. Randall from the collection of John Witt Randall)*

Europe. Early modern buildings kept out the wind and the rain—period. To retain body heat, people wore layers of clothing during the day (even indoors), and in winter they retreated, early and often, to heavily blanketed and sometimes curtained beds (still wearing a fair amount of clothing, including headgear). True, they also kept fires indoors. But through the Middle Ages, most households had only one hearth, originally signifying the "heart" of the home (similarly, the Latin word *focus*, for a central hearth). Richer folks might keep one fire in a room or separate building for cooking and another in a large hall, the household's common area. No one expected a fire to heat all of a room, let alone an entire building, except in very wealthy households that might have a system of flues (quite rare) or else portable braziers to huddle over. In a 1652 illustration of a Dutch tavern, there's a nice big fireplace, but everyone (except maybe the cook) is heating themselves some other way—wearing warm layers, eating, drinking, dancing, making amorous contact.

Also, many dwellings had open hearths, not wall-set fireplaces with smoke-venting chimneys, a newfanglement that spread only after the twelfth century. In a great hall with a high ceiling, the smoke was at least diluted. But there were essentially only two options: being cold or getting at least lightly smoked.[27]

For men who headed households, a hearth announced prosperity and status. (Some English property taxes were calculated by the number of hearths or chimneys in a house.) Women might tend the hearth, an ancient role that signaled female domesticity and docility, with endless variations on Cinderella. But for a woman to preside over her own fire, in the absence of a household patriarch, was troubling. Early modern Germans considered a woman who lived alone, "under her own smoke," deeply suspect. Some accused witches, for example, were described that way—if a woman commanded fire, what else might she be capable of?[28]

By the late sixteenth century, a household's traditional source of warmth was newly tasked with creating an entire indoor zone of comfort, at first only for the wealthy, then for people in the middle ranks of society. This required installing fireplaces in rooms beyond the kitchen or the great hall (or the more up-to-date parlor) and using flues to heat rooms without their own chimneys. Networks of flues became more complex, zigging and zagging through walls, with average dimensions (width and height) of nine by fourteen inches, though they could shrink to six inches square. Chimneys and flues were adjusted to promote drafts to suck out the smoke; chimney stacks became bigger, to house multiple flues. People in Northern Europe had, by the late Middle Ages, also begun to use stoves, fireproof containers, typically metal or ceramic, vented to send smoke out of a house. Following German use of the word *stube*, the English used "stove" to describe a heated space, sometimes specifying the beneficiaries; "stove houses" protected plants, and "stove rooms" coddled people. Only gradually did "stove" mean just the heater itself. As coal replaced wood, its denser smoke was a particular plague indoors, an incentive to adopt flues with drawing power. Cleaning sticky coal soot from chimneys and flues required specialists, typically brush-wielding "climbing boys."[29]

This new trade was low and unenviable, marking a new divide between the haves and have-nots. Shakespeare's evocation of the equality of death—"Fear no more the heat o' th' sun, / Nor the furious winter's rages"—has a timeless quality. But his warning that "Golden lads and girls all must, / As chimney-sweepers, come to dust," documents a specific time, the early seventeenth century, when summer and winter seemed out of joint, and the multiplication of chimneys and adoption of coal were dusting cities with soot in a new effort to warm their inhabitants.[30]

The preferred indoor atmosphere was now a steadier state, with a temperature that varied less over the course of a year (or a day), and with winter's cold kept further away. At first, only the very rich could enjoy such spaces. Books on architecture, for instance, showed elaborate decorations—mirrors, marble, carvings, gilding—for large fireplaces destined for palatial residences. The grand decor, though beloved by Stanley Kubrick, wasn't practical. But the new ideal of being warm indoors gradually moved down the social ranks, and the heating systems slowly improved. Among other things, this permitted greater privacy. Someone who could afford it might now keep warm behind a closed door, shielded from anyone who, in another household, might catch them reading banned books, writing bad poems (until they finally wrote a good one), indulging in idleness, conceiving a child—or anything else they might prefer to do in peace and quiet. It was "alone in a stove-heated room" in November 1619 that René Descartes had his foundational philosophical idea: "I think, therefore I am."[31]

Other indoor comforts reinforced the trend for better heating. Greater use of window glass, mirrors, and candles made households brighter, day and night. Warmer rooms permitted lighter clothing and bedclothes, more commonly made of linen or, later, cotton. An increased manufacture of soap kept these fabrics clean. Several technologies (lighting, cleaning, heating) therefore existed in mutual support, with good chimneys diminishing the amount of soot that soap would have to remove from clothing (or people) even as better heating made life in winter less like semi-hibernation indoors. Notably, Franklin's father, Josiah, was a chandler, someone who made candles and soap from animal fat, sources of light and cleanliness, two new hallmarks of modern life whose production was now transplanted to Boston.[32]

These small revolutions in everyday experience had multiple costs. Consumers spent more to achieve material states that had once been luxuries but were now cheaper. Their collective spending affected the overall economy. People could choose to do more paid work, calculated against what they wanted to buy, and often (in a nice feedback loop) producing consumer items themselves, such as brewing beer to get money to buy soap, with a net gain in clean and merry citizens. Finally, the ballooning consumer demands had serious consequences for the natural environment. The collective fuel use was an obvious cost. Nearly everything being consumed had required fire for its production, and the people who consumed those products hoped to do so while kept nicely warm in winter, even the bitterest of the era's cold conditions.[33]

Long, cold winters were the American reality into which Benjamin Franklin was born. New England had proved a severe test of English commitment to "overwintering"—surviving year-round in a cold, snowy place. Franklin's ancestors settled in New England during some of the Little Ice Age's iciest intervals. His maternal great-grandfather, John Folger, relocated in 1635, in time for a 1650 temperature dip, the most apparent that scientists have identified in the seventeenth century. Franklin's father, Josiah, arrived in 1683, just before a cascade of complaints about the cold from (roughly) 1690 to 1710. The complaining might partly reflect how more people were putting their thoughts on paper, not that the weather was reaching a new nadir. But some winters were truly bad. Franklin would experience, as an eleven-year-old, the "Great Snow of 1717," when southern New England's February snowfall was three to four feet deep.[34]

Colonists had a simple strategy for such conditions: they stayed home. Members of Franklin's family—especially his pious father—probably knew the Reverend Cotton Mather's *Winter Meditations: Directions How to Employ the Liesure of the Winter for the Glory of God* (1693). Mather's advice, that colonists use winter for social removal and contemplation, snug beside their household fires, was culturally specific, importing a European preference that, as would become clear, made no sense to the region's Indigenous people.[35]

That was part of Mather's point. He reminded his readers that they were "New-Englanders," Christians who were superior to those back in ungodly old England and to the "pagans" whose lands they'd invaded. Winter was a war, but not against people—against irreligion. "There is the work of Watching and Fighting against our Invisible Adversaries; and this is to be done as much in the Winter, as at any Time whatever." In winter, "we can't *Go out* as at other times; Well, since we must *Sit still*, now let us more than ever, *Consider the Wondrous Works of God.*" Those works included snow, fog, wind, and ice, phenomena no less wonderful for being, to humans, uncomfortable.[36]

Christians should thank God for their protections against cold. "Let us also be Thankful for our FUEL," Mather continued. Although some "*Pagans*," including New England's Natives, worshipped fire as a god (no, they didn't), Christians knew it was a natural wonder "the True God" had created. And where was fire best kindled? "Let us be thankful for Our HOUSES too," Mather explained. "We are not left now to lodge abroad in the Cold, with none but the *Ground* for our Bed, the *Snow* for our Coverlid, and the *Sky* for our Canopy: nor are we obliged unto such Wretched *Wigwams* as were the best Habitations of the Barbarous Natives that were here before us." Finally, Mather said, "let us be Thankful for our TABLES. How many *Warm Dishes*, have we to cherish us, whereby we are strengthened against the *Cold* of the winter?" The short book validates a sheltered New England home life, with *Winter Meditations* good to read by the fire. Only Mather insists (of course) that the Bible is the best "*Fireside*" book; should any of his works threaten to oust scripture there, better "*that they were all thrown into the Fire.*"[37]

A sense of technological separation between the English and Natives would persist. Colonists worried that America—with its alien food, air, water, climate, soil—might alter their bodies, raising the (to them) horrifying possibility that this would make them equivalent to pagan Indians. Creolization was used throughout the Americas to describe that fate—"creolian Degeneracy," spat Cotton Mather (son of Increase), disgusted at the prospect. The unease prompted an anxious colonial literature insisting on the essential durability of the English body, its propensity to stay the same, even if uprooted and replanted elsewhere. But just in case, colonists were equally invested in consuming remnants of old-world culture, as with European wheat mixed into American maize in daily bread. A desire to live in permanent, European-style houses was part of this project—as was regulating conditions within them.[38]

Throughout the colonies, house design announced that sense of difference. Chimney fireplaces made of brick, stone, or clay—a design never present in Indigenous dwellings—were one important indicator, visible even outside the building. And considerable amounts of wood went into the fires within them. Proximity to remaining forests and woodlands was the astonishing reality that differentiated colonists from many Europeans. The trees would steadily diminish, but well into the nineteenth century, wood would be among the most common and valued raw commodities in the United States.[39]

An average colonial household burned somewhere between twenty to forty cords of wood each year. (The rich burned through more—they tend to do this.) A cord is a stack measuring four by four by eight feet. One thousand cords—for, say, a town of thirty households—is 128,000 cubic feet, with a BTU (British thermal unit) equivalent to a ton of coal. Each settler household was consuming, on average, more than an acre of woodlands each year, faster than woods can regrow if they've been clear-cut. (Oak, the wood that burns hottest, grows slowest.) The woodpiles translated into tangible, direct experiences with energy. As a colonist set kindling ablaze or fed a chunk of wood into the flames, they might remember chopping it into a usable fragment, or stacking it on the pile, or even first seeing it within the tree they were cutting down. Such was colonists' dependence on firewood that some Natives thought it was the reason they'd migrated in the first place. In Rhode Island, the clergyman Roger Williams said the Narragansett wondered "Why come the *Englishmen* hither?" They concluded, he said, "it is because you want *firing*," assuming that the wood-hungry English had burned through their homeland's forests. (Which isn't far from the truth.)[40]

The Natives thought pretty much everything the New Englanders did was strange, including their habit of staying home when snow began to fall, burning down their woodpiles and eating up their harvests, forgoing winter's best opportunity: hunting. Peoples in the Wabanaki Confederacy, which united the nations between the Gulf of Maine and the Gulf of St. Lawrence (the most powerful remaining in greater New England), considered winter—and not the autumn of agricultural societies—as the time of plenty. With cold and snow, nonhibernating animals were exposed as they sought food; hunting was at its best, and people grew fat and strong from the animals they consumed. Winter was also intensely sociable, a time to share the work of hunting and the food it yielded. To navigate snow, even the deep snow of the Little Ice Age, Wabanakis perfected snowshoes, all kinds of snowshoes, for different people and different winter conditions. A nineteenth-century Mi'kmaq dictionary (for one Wabanaki language) lists thirteen words for types of snow, but fifty for snowshoes—superb technology, though alien to Europeans.[41]

Likewise, many Indigenous people of northeastern North America had distinct designations and uses for fires: not just to warm an individual or family, but also to bring different people together. The Haudenosaunee, the Iroquoian peoples of the confederacy with which Franklin would negotiate several times, are an excellent example. In Haudenosaunee longhouses, families occupied opposite sides of the buildings but shared fires in the middle. A fire indicated membership, not in a single patriarchal household, as in Europe, but among extended kin defined by the mother's line. Beyond these households, council fires brought together present or possible allies. The confederated nations or houses within the Haudenosaunee oriented themselves in relation to a council fire: Mohawk and Seneca to the east, Oneida and Cayuga to the west, and Onondaga (as central arbiters) at center and to the north. The Onondaga were the seat of the council fire for diplomatic meetings with outsiders, including Franklin.[42]

The council fire as a tool, a technology of diplomacy, wasn't unique

A COUNCIL FIRE BURNS BETWEEN THE INDIGENOUS AND
BRITISH DELEGATES AT TREATY NEGOTIATIONS ON THE OHIO
RIVER WEST OF PENNSYLVANIA.
(Thomas Hutchins, A General Map of the Country on the Ohio and Muskingham . . .
[Philadelphia, 1765], detail. Library of Congress)

to the Haudenosaunee. Multiple Indigenous nations in northeastern North America kept such fires, including the Wabanaki of New England and the Lenape of Pennsylvania. For these nations, the fire emblematized the shared and delegated political power that characterized Native societies, unlike the centralized and hierarchical governments Europeans thought necessary for social order. The fires also designated kinship, including nonhuman kin, as when humans made decisions that might affect other living beings; in making plans to hunt animals (or not hunt them), Native men engaged in what can be described as interspecies diplomacy.[43] Above all, treaties that resulted from negotiations with colonists around council fires (physically present or invoked in oratory) established Indigenous sovereignty vis-à-vis European nations and, eventually, the United States. This was never perfectly congenial for colonists, whose ideas about fire were more overtly about control.[44]

In Europe and its colonies, for instance, outdoor fires often signaled violence and exclusion. The word "bonfire" derives from "bone fire," where human remains, or living humans, or dangerous human things (like books) were publicly burned, to instruct and intimidate. In England, executioners burned heretics and heretical tracts alike. Not unlike fireworks, bonfires might feature in public celebrations, marking out membership within a political community. In his Boston newspaper, for example, Franklin's brother James related news from Virginia in 1722 about how a former governor was welcomed back with "Bonfires, Illuminations, and other publick Marks of . . . Respect and Value."[45]

The bonfires Franklin and other colonists would know best were the ones that burned on Bonfire Night every fifth of November. This English holiday commemorates how a Catholic conspiracy to blow up the House of Lords was thwarted on November 5, 1605, with the arrest of Guy Fawkes, discovered in a gunpowder-packed cellar beneath Parliament. In 1606, Parliament declared Gunpowder Treason Day an annual holiday. Rituals included the burning of effigies of Fawkes, or of the pope, or of both, atop November 5th bonfires. New England's Protestant settlers took to the holiday with glee; "Pope Day" was celebrated as early as 1623 and, by the 1720s, when Franklin was a teenager, he could see the parades and bonfires that men from the lower social ranks organized. Their events blended ridicule of the local elite with patriotic anti-Catholicism, spiked with fear that French Catholics in Canada, with their Native allies, were a persistent threat to New England. Pope Day bonfires enacted group solidarity through implied violence, the complete opposite of the diplomatic function of Natives' council fires.[46]

The consumer revolution may have generated greater comfort, but also more garbage than ever before. Within an organic economy, however, even trash was precious and much of it carefully recycled. When Franklin, aged twelve, picked an occupation, he chose the part of the industrious revolution that turned rags and soot into some of the era's most sophisticated consumer goods: books. In 1718, Franklin was indentured to his brother James, an established printer. Using paper made from the tatters of other people's sheets and shirts and handkerchiefs, and ink boiled from soot scraped from lamps, he began to master a lifelong craft of putting into print what he thought readers, as consumers, might pay to read.

Printing was a printer's visible work; behind the scenes, and sometimes taking up even more time, they read *everything*—to find out what to print or reprint, to spy on rivals, to discover some profitable angle on a fast-changing market of all the ideas ink could put on paper. Among these were theories about nature and matter as well as recommendations for improving the human condition, and these would definitely catch Franklin's attention.

"Improvement" had come to refer, by the seventeenth century, to modifications to land that benefited people, gradually encompassing all kinds of schemes that yielded benefit. In a work Franklin greatly admired, Daniel Defoe's *Essay upon Projects* (1697), Defoe staked out a bold position: to be truly improving, projects had to stand the test of time. He defined his entire era as "The *Projecting Age*," with an ability to control the future, in contrast to "past Ages" when ordinary people couldn't aspire to that. Manipulation of future states was the goal of the insurance policies Defoe recommended, for instance, tied to the duration of human life, either of the insured person or another beneficiary.⁴⁷

Projectors gave science and technology new significance, as the means to improve material circumstances over the long term. Modern science was becoming part of cultural literacy, knowledge of nature that benefited everyday life. England's premier scientific organization, the Royal Society of London (founded 1660), announced its interest in discoveries likely to assist humanity and showcased the results in its journal, *Philosophical Transactions*. Many scientific findings became parts of the consumer revolution by being printed in cheaper formats—almanacs, newspapers, pamphlets—generating profits for printers while disseminating a world of improved understanding, a trend Franklin would happily exploit.

In science, as Franklin would discover, heat and combustion were central to the developing field of modern chemistry. Chemical experimenters relied on the concept of calor or caloric, a flowing substance colloquially known as heat, whose absence was perceived as cold, a phenomenon abundant at the time. In 1665, for example, chemical experimenter Robert Boyle published his punningly titled *New Experiments and Observations Touching Cold*. In his preface, he deadpanned that "Cold is so barren a subject"—or so it seemed. In fact, heat and cold "are the two grand Instruments by which Nature performs so many of her Operations." Boyle cited Francis Bacon on heat's being the right hand of nature and cold the left—in Latin, the *sinister* hand. While Boyle did not claim the era's colder weather as his inspiration, it definitely helped his experiments, as with his frequent use of snow to determine when and how different substances freeze, and in his descriptions of northern regions hardest hit by severe winters.⁴⁸

But Boyle warned that humans' sense of cold made them misunderstand it. Transitions between temperatures and levels of humidity often confused human perception. If only there were reliable "thermoscopes,"

Boyle lamented, instruments that showed precise "Degrees" of temperature, "preferable to those of our senses, because those Dead Engins are not in such cases obnoxious to the same Causes of uncertainty with our Living Bodies." Given that temperature extremes had serious implications for health, degrees on any thermometer could indeed be dangerously misleading. ("Wind chill" and "heat index" continue to warn people that other factors, like the movement of air and level of humidity, affect their bodies' response to temperature.) As Boyle put it in another work, thermometers gave temperatures relative to one another, in a kind of inter-instrument chatter, but "leave us in the dark as to the positive degree [of cold] thereof."[49]

Better comprehension of heat required understanding the materials it affected, especially air, which was what usually transmitted hot or cold to the human body. Every big name in science, and many smaller ones, was drawn to these phenomena. After Galileo Galilei defined air as having weight or pressure, Evangelista Torricelli constructed the mercury barometer (1643) to gauge levels of atmospheric pressure under varying conditions. Otto von Guericke made the first air pump (1650) and with it created the first artificial atmospheres: he evacuated the air within two closely fitted copper hemispheres, which teams of horses couldn't separate—a striking demonstration of atmospheric pressure, which humans (and horses) didn't entirely feel. Using an air pump, Boyle confirmed air's "spring" or pressure, also its compressibility and resistance. Boyle and Isaac Newton defined heat and air as fluids, varying according to the nature of the invisible particles within them; with less dense particulate structure, heat and air were "subtle" and "elastic." (Definition of gases according to their chemical compositions came later.) Following Newton's work on heat absorption (including the significance of color), Dutch chemical experimenter Herman Boerhaave investigated heat's ability to add volume (thus increasing the pressure of a gas, for example), and he advocated thermometers, using Daniel Gabriel Fahrenheit's mercury-filled devices, available after 1714. At first, thermometers and barometers were too expensive for everyday use. Over the eighteenth century, they became affordable consumer goods and redefined heat—and comfort; the instruments could finally speak to the humans.[50]

In the meantime, medical theorists were examining how humans were not just heat-seekers but heat-emitters. Was a body's innate heat the same as sensible heat, the exterior warmth the body sensed? And what caused it? Experimenters focused on circulation, respiration, and digestion, with various theories of how the heart's muscles or blood's motion might produce heat, or that respiration was akin to combustion (with breath "burning" some flammable element in air), or that body heat resulted from the stomach's grinding or fermenting of food. Theories about diet advised maintaining a temperate condition, neither too hot nor too cold. Alcohol, spice, and meat were thought to overheat the body, prompting some of the earliest calls for vegetarianism based on health, not religious piety.[51]

Inventors capitalized on all these developments. They became entrepreneurs, raising funds for production, commercializing devices that promised to solve some problem or other. Technology became part of government designs and it informed consumer choices. The sciences that investigated air and heat, for instance, guided the development of lamps, ventilators—and stoves. Governments encouraged such work by awarding patents. Steam power, later engine of the Industrial Revolution, emerged precisely from this context. In devices developed as early as the first decade of the 1600s, steam pressure drove machinery, with the first engines pumping water from mines, in Spain, by 1606. The first effective version was the Newcomen engine, which used steam pressure within a cylinder to operate a piston. This design was put into action (again in mines) with coal as its fuel by 1712, each such engine towering over the miners, who, because water was now extracted at a literally inhuman rate, could go ever deeper into the earth to extract its contents.[52]

AN EARLY STEAM ENGINE RAISES WATER WITH FIRE(POWER).
(J. T. Desaguliers, A Course of Experimental Philosophy, *2 vols. [London, 1744–1745], 2: plate XXXVII, call number EC7.D4508.B747p, Houghton Library, Harvard University)*

The efforts to use science and technology to improve the human condition intersected with emerging *social sciences* that calculated costs and benefits for entire populations. Foremost among these were the ancestors of economics and demography, under their earlier names of political economy (measuring national strength and human happiness in monetary units) and political arithmetic (defining and counting populations). The resulting calculations of energy used or saved, and populations kept industrious and healthy, informed many inventions and projects, as with heating technologies. Within Europe, and especially England, theories of population and commercial growth were beginning to define modern capitalism: quantifying humans and their activities, enshrining capital investment in schemes for improvement.[53]

It was all somewhat hypothetical, however. New demand for consumer comforts continued to provoke anxiety about the finitude of natural resources. Was there enough land and forest (or meat or bread) within Europe to supply a growing population that expected more of everything? Was it reasonable to stay warmer in winter than one's ancestors had managed, despite wood running out and winters becoming colder? In 1721, French naturalist René Antoine Ferchault de Réaumur (who invented a thermometer that briefly competed with Fahrenheit's) warned that consumers were destroying France's forests. "We build, adorn, and heat more apartments than our fathers did," he said, adding that the rising "number of forges, ironworks, glassworks" to supply consumer goods also took a toll.[54]

This is why the prospect of fresh woodlands in the Americas was so very welcome. Though, whether consumers back in Europe knew it or not, winter and wood were problems there as well. Stretched across the Atlantic, new expectations of prosperity, extensive and intensive, would be put to the test.

If one crisis is plenty for any generation in history, two are an emergency—and three a calamity. Franklin would come of age during just such a wealth of woes, when climate change, resource depletion, and warfare collided, not for the last time in his life. This unfortunate pileup occurred during New England's military confrontations with the Wabanakis, conflicts that had initially coalesced around the Indigenous insurgency called King Philip's War (1675–1678), on a per capita basis one of the deadliest conflicts in the history of what became the United States. As much as 5 percent of settlers and about 40 percent of the region's Natives died.[55]

Even as southern New England colonists claimed victory over the forces of Wampanoag leader "Philip" or Metacomet, the Wabanaki Confederacy continued the fight in northern New England, with the constant threat that they might move south into Massachusetts. These Anglo-Wabanaki wars (the most dramatic were in 1675–1678, 1688–1699, 1703–1713, and 1722–1725) were inseparable from winter conditions. During some of the era's worst winters, Wabanaki warriors at first had the advantage, using snowshoes to attack English towns. As James Franklin's Boston newspaper reported about the war zone around Kennebec (now in Maine), Native warriors were "well arm'd, had each of them two Pair of Snow Shoes." Colonial soldiers couldn't fight back until they literally stole the enemy's footgear. Meanwhile, multiple English accounts document how wartime displacements during extreme cold exposed both English and Natives to freezing conditions, sometimes fatally.[56]

This wasn't a war over abstract concepts of nationhood or identity—it was fought over land and resources, footholds for nations and identities, a struggle that continued even outside formal combat. Again, James Franklin reported this in his paper, sometimes using his younger brother's labor to set the type. In fall of 1721, for example, as peace was sliding back into war, *The New-England Courant* described, semi-comically, how the commander of the militia in Amesbury, Massachusetts, had ordered his men to shoulder "their Axes instead of their Firelocks." He marched them "into the woods where (with invincible Courage) they slew as many Trees as made 30 Cord of Wood." They returned to town with these spoils, destined for "the Relief of the Poor" in advance of winter.[57]

In managing cold by cutting down trees, New England settlers were, among other things, testing a novel hypothesis about climate: humans could change it. This new theory began as a theological debate over the flood God had sent to punish a sinful humanity, with Noah and his family riding out the Deluge on an ark stocked with pairs of terrestrial creatures. After the waters receded, life on Earth continued, but under harsher conditions. One geological analysis at the time proposed that the flood had wrecked Earth's originally smooth surface, with jagged mountains and valleys as reminders of humanity's second lapse, after the original expulsion from Eden. It also conjectured that the massive displacement of water, by altering the planet's position relative to the sun, had changed the climate as well. Without a definite sense, yet, of the longer duration of geological time, these hypotheses identified anthropogenic changes to the climate and environment—human sin made the planet harder to live on. But maybe humans could reverse the damage, atoning for past transgressions? That was something Catholic thinkers had suggested in the Renaissance, though Protestants, including New England Puritans, would also discuss the possibility, making a unique contribution to the discourse about climate change in Franklin's birthplace, starting just before his birth.[58]

It was precisely the Little Ice Age winters of the late seventeenth century that gave New England colonists a sense of their possible role in improving the Earth, both to benefit themselves and to atone for ancient sin. And so, they defined a theory that would inspire white settlement well into the nineteenth century: by cutting down forests and breaking open land for cultivation, colonists might make the climate milder and more predictable. After all, forests shaded land; also, Europeans thought they emitted noxious miasma. It made sense, given its wooded state, that North America was colder and had more precipitation than the parts of Europe opposite it. Get rid of the trees and conditions would improve. At first, some warmer intervals in the seventeenth century seemed to support the hypothesis. Phalanxes of soldiers bearing axes were not just defeating trees but also the bad winters—plus the Natives who'd simply accepted North America's fallen state, war by many means. New Englanders' thesis about anthropogenic climate change would become an important part of modern climate science, though—stay tuned—Franklin would be a notable skeptic.[59]

In the meantime, the uncut forests were prizes of war. Beyond the Wabanakis, other Native nations were keeping an eye on the confident, aggressive New Englanders. The Haudenosaunee and the Lenapes were, for the moment, spared the presence of land-hungry colonists. The Mohawks, Haudenosaunee's eastern house, were the main contact for Dutch fur traders at Fort Orange (later Albany), as Lenapes were for both the Dutch in New Amsterdam and the Swedes in what would become New Jersey and Pennsylvania. In what would later be called Canada, French migrants (small in number) developed relations with the Haudenosaunee, Wabanakis, and other nations, based on trade and mobile Catholic missions, not on land cessions—not at first.

The Wabanakis had made an alliance with the Mohawks in 1700, consolidating their own confederacy and representing themselves, as early as 1701, within a larger Federation of Seven Fires, council fires of the seven constituent groups. Wabanakis' simultaneous Mohawk and French affiliations cut across English political alliances, which were with the Haudenosaunee but not with France. That was complicated. The competing loyalties fueled more than a century of conflict over sovereignty and resources within North America.[60]

Quite often, the fur trade was the liveliest source of conflict. Using animal pelts for warmth long preceded the Little Ice Age, and the industrious revolution independently increased demand for furs once restricted to an elite. But frigid conditions made wearing lush fur more possible. In North America, hunters sought otters, foxes, bears, moose, martens, raccoons, and above all beavers, creatures whose pelts the cold made denser, who were perfect for winter hunting and conversion into winterwear. The longer the winter, the more time to hunt, and rivalry over hunting zones was keen. The Beaver Wars preoccupied Indigenous, French, Dutch, and English people through most of the seventeenth century, the tensions also inflecting the Anglo-Wabanaki Wars. Successful in war, trade, and diplomacy, the Haudenosaunee pushed west and south, and eventually positioned themselves along the Susquehanna River, close to the Lenape homelands that Pennsylvania colonists would claim.[61]

Whether he had followed all the news from Pennsylvania, Franklin certainly knew about the Haudenosaunee presence there. In August of 1721, he may have set the type for another of *The New-England Courant*'s updates, this one on Pennsylvania's recent diplomatic negotiations with Haudenosaunee delegates. Two years later, Franklin didn't need to follow such events from afar. He'd run away from Boston and ended up in Philadelphia, where he'd make himself into the Prometheus of the modern age.[62]

2

FORGING

It rained so hard on the autumn day he was walking to catch what he hoped would be a final ferry to Pennsylvania that Franklin was "thoroughly soak'd" and so miserable, he later confessed, that he began "to wish I had never left home." And yet Franklin had judged Philadelphia to be his best prospect, and he turned out to be right; good thing for him he kept going, on through the chilly rain. He chose the city based on its potential as a market for printed media, little realizing what else the region would offer him. With direct consequences for his stoves, he found unexpected opportunities in the region's iron industry and its relations with Native nations.[1]

These schemes in fact competed. Pennsylvania's founder, the Quaker pietist William Penn, wanted everything for his colony but war, not comprehending that the material plenty he expected from his new territory might make conflict inevitable. On top of the timbering, farming, and commercial trade that dominated New England (and other colonies), Penn predicted that Pennsylvania would also produce iron, one of the most fuel-intensive industries of the era—yet another Promethean expectation. Forging iron would test every bond between colonists and Lenapes, setting up the tension over use of resources that Franklin's stoves were supposed to resolve, preserving consumer comfort while conserving American trees, and keeping the peace over the land. If that were possible. As it turned out, Franklin's work as a printer would include a great deal about the weather, foundation for his later works on heating and climate, and a constant reminder that atmospheric conditions challenged any easy definition of a good life.

Even before the English showed up in Penn's colony, Lenapes had prepared themselves for the worst. Subject to mounting colonialist pressure, aware of what was happening to Native groups elsewhere, the Lenape nation made an important decision: they announced that they—all of them—were women. This was a distinctive ethical claim. Indigenous women were empowered to make peace and Lenapes had excellent reasons to prefer that state of things.

Within their homeland of Lenapehoking (Lënapehòkink), stretching from the Hudson River to the Delaware River, Lenapes were divided between mutually intelligible dialects: northern (Munsee and Unalachtigo) and southern (Unami), the latter dominating the coast of what would become New Jersey and Pennsylvania. Like most northeastern Natives, they grew crops in the warm seasons, then hunted from late fall through winter. When farming, it seems that they, like the Haudenosaunee, lived in longhouses—hall-like buildings occupied by families sharing matrilineal descent—then built smaller dwellings for tighter family units when hunting. Living on the coast and along great rivers (Hudson, Delaware, and Schuylkill, Europeans would name them), they also relied on fish and shellfish, with great respect for water and its creatures. And they appreciated fire, using it to generate community ties, as in sweathouses and with council fires.[2]

By the middle of the seventeenth century, Lenapehoking was in danger of invasion by the Dutch to the north, by the Iroquoian-speaking Susquehannock and Haudenosaunee to the south and west, and by ragtag bands of Swedes and English along the mid-Atlantic coast. This was why, to keep the peace, Unami Lenapes declared themselves a nation of women, a pledge that they no longer had any fighting men among them. Lenape women, as in other northeastern tribal nations, never engaged directly in warfare, and often fostered peace by intermarrying, withholding provisions from warriors, and giving hospitality to outsiders. Alongside the declaration that they would henceforth only perform these roles, Lenapes brokered an alliance with the Haudenosaunee, naming them as their maternal uncles, a way to define affectionate and respectful relations within matrilineal cultures.[3]

Peace was the promise, but never a guarantee. Earlier, Lenapes had fought European attempts to take their land. A 1631 attack on one Dutch settlement, for instance, may have been a response to a tin sign Lenapes thought the Dutch had tacked up to claim territory. But this was likely only one rebuke of Europeans' claiming land for agriculture and to extract commodities, including fish or whales, without permission. Lenapes' tolerance of the settlers of Pennsylvania would depend on proof that their sovereignty was not being ignored.[4]

White accounts of William Penn's 1682 initial meeting and treaty with the Lenapes state that it took place under an elm tree on the Delaware River in what is now Philadelphia. There, Penn met Tamanend, a man he took to be leader of all Lenapes within his land title (given by Charles II, not any Lenape), and they pledged peaceful relations. Images of the meeting, including the massive "Treaty Elm," became settler icons. Likewise, Charles II named the colony for its new owner and its trees: Penn's *silvae*, Latin for "woods," origin, too, of the word describing people who lived among trees, "savage," which was definitely the king's implication about Indigenous people. But Lenapes didn't just live among trees—they considered themselves to be, with the trees, co-beings in Lenapehoking. Penn's own instrumental view of his new land, which assumed its resources could be turned into property, made no sense to them.[5]

Unaware that Indigenous people couldn't imagine alienating land from themselves (and that repeated interaction with Europeans had made them wary about just that peril), Penn assumed they would transfer territory to settlers if paid. He said he would record and compensate sales of land, but, even when he followed through, he totally failed to honor Native ways of being. And so, if the Dutch "bought" Manhattan for a legendary $24, Penn "bought" Philadelphia for an actual £1,200—a greater sum, worth at least £242,600 ($302,000) in purchasing power as of 2023.[6]

Whatever its monetary worth, Penn's payment wasn't something Lenapes could have agreed to. They likely considered it part of an ongoing process of cohabitation, not unlike rent, or as a diplomatic gift to accompany the treaty. Moreover, Penn couldn't determine which Indigenous people might have sovereignty over land to the west, in the Susquehanna Valley, where multiple nations shared a complex landscape. He would assume a 1701 payment to Conestogas, with an Onondaga leader as witness, constituted a sale. In fact, this engendered rivalries over the region that lasted until the end of the century.[7]

But another part of Penn's plan did match Lenape ethics. As a Quaker, a "Friend" in the Religious Society of Friends, Penn was a pacifist. He abhorred the wars that had become endemic to colonization; his colony didn't even have a defensive militia. Lenapes respected Penn's commitment to keeping the peace, though in praising him for it, they unwittingly endorsed an English myth that the Quaker man was the region's major force for peace. Colonists took this as truth, ignoring all the Lenape work in doing that first. Pennsylvania became legendary for its unusual tranquility. The French philosopher Voltaire judged Penn's accord with Tamanend to be "the only treaty between

[American Natives] and the Christians that was not ratified by an oath," because Quakers thought that swearing of any kind was coercive and impious.⁸

Still, the fundamental way that English property law was never acceptable to Lenapes was a time bomb. The colony's profits were going to depend on economic activities done by a large number of non-Native migrants. If those pursuits did not benefit Indigenous people, or damaged their communities, there was no guarantee the colony would avoid the kind of violent conflict that was prevailing elsewhere.

Penn wanted to pack Pennsylvania with settlers, confident there could be no scarcity in America, and because he wanted revenue from land sales. He took advice on this from a friend, William Petty; both men owned land in colonized Ireland and in Pennsylvania. Petty was an early expert in political arithmetic, political economy's twin field, each of which suggested that while humans were natural beings, they were also economic ones; this made a hypothetical distinction between economics and what would later be called ecology. Petty estimated an optimal number of settlers (and proportion of women of childbearing age) for the best rate of natural increase. He assumed this population would clear land to farm, forming a landscape where Indigenous Americans were a minority. He did not recommend any kind of conservation program, nor justify the political (and ethical) costs of displacing the original people.⁹

This is surprising, given that Petty worried about trees and fuel in Ireland. The island had an alternative fuel in peat, but Ireland's emerging iron industry required charcoal made from wood. Plus, as Petty noted, people needed timber for construction, for cooling shade, and for everyday wooden items. Only with replanting, cultivation, and judicious use could Ireland (or England) retain woodlands for posterity. In this, Petty revealed his key assumption: England, somewhat, and Ireland and Pennsylvania, definitely, were underpopulated, lacking workers to develop their economies. Any conservation must foster colonists, not nature itself—and not the original populations, either Irish or Lenape.[10]

Penn defined his colony according to the European idea of wilderness, as a kind of blank slate for colonization. Indigenous conservation efforts were invisible to him. "The *Trees* of most note," he wrote, were "*black Walnut, Cedar, Cypress, Chestnut, Poplar, Gumwood, Hickery, Sassafrax, Ash, Beech* and *Oak* of divers sorts . . . of *All* which there is plenty for the use of man." And he rejoiced that woodlands near the coast were "not one quarter so thick as I expected." (Colonists saw the interior's denser forests only after Penn's death in 1718.) Like most Europeans, Penn feared that woodlands emitted unhealthy "Mists and Vapours." But "as the *Woods* come by numbers of People to be more *clear'd*," he said, the air "will Refine."[11]

Cutting down everything wasn't desirable, however. A 1681 law, passed during Penn's own government, stated that for every five acres cleared, one should be left wooded, especially in oak and mulberries, trees that, respectively, could build ships and feed silkworms. (A projected silk industry never took off.) Although a 1700 act of the Provincial Assembly, requiring householders in Philadelphia and two other cities to plant and maintain "shady and wholesome trees before the door of his, her or their house," was repealed by 1704, another provincial law of 1700 regulated rural lumbering by specifying that some lands had to be untouched. The lumbering law was restated in 1735. And again in 1794. This was a recurring reminder of colonialism's warring impulses, growth of population and economic output versus conservation. Much of England's landscape was already showing this was a doubtful balancing act.[12]

And yet Philadelphia was—is—mapped in trees. Penn said he wanted a "greene Country Towne, [that] will never be burnt, and allways be wholsome." He stated that wish in the wake of two urban catastrophes: the Great Plague of London (1665–1666) and the Great Fire of London (1666). Like others, Penn hoped that better-planned cities might be healthier and safer. Philadelphia's original layout, on a rectilinear grid, reordered natural space, but Penn mandated that the streets in the grid be named for "things that grow . . . and are native." The oldest east-west streets include Spruce, Sassafras, Mulberry, Chestnut, Walnut, Pine, and Cedar. More than 150 Philadelphia streets are now named for native plants and animals, from Acorn to Verbena, Deer to Sandpiper. Philadelphians eventually ran through their state's growing things and began to name streets for local minerals, from fancy Agate to basic Rock.[13]

Because he founded his colony on religious toleration, Penn called it a "Holy Experiment." But in his attempts to control nature—renaming thousands of acres of trees for himself, paying off people he thought were "savages," gridding wild things firmly into place in a cityscape, imagining endless fuel for industrial fires—he was just slightly Promethean, a tiny bit profane in thinking humans could tame all of God's creation. It was a bold vision of a new-world colony. That made it attractive to a great many people—including Benjamin Franklin.

In late September 1723, Franklin had run away from home, breaking the legal terms of his apprenticeship to his brother. He paid for ship's passage from Boston to New York, then continued to Philadelphia, first in open boats, finally on foot, increasingly exposed to worsening weather as September turned into October, when he wished he hadn't set out in the first place. On the last night of the journey, on Cooper's Creek just north of Philadelphia, he and some other travelers scavenged "an old Fence . . . the Rails of which we made a Fire, the Night being cold, in October." (If the fence, despite its age, was someone's property, turning it into fuel was in fact theft.) The next day, October 6, Franklin entered Philadelphia, aged seventeen, a fugitive from the law, exhausted, and "very hungry." He found a bakery and bought some bread, which he ate while wandering down Chestnut and then up Walnut.[14]

His choice of food, three "puffy rolls" (made from the colony's main commodity, wheat), was an early hint of his beliefs about heating the body from within. Medical theories inherited from classical antiquity had recommended a balance of material states—including heat and cold—in a constant interplay between body and environment, with different foods able to supply warm or cool qualities as needed. It is notable that, at this cold time, some medical practitioners began to recommend a "cool diet," one that didn't overheat or otherwise overstimulate the body. Following some of the early theories of animal heat, Franklin considered grain the optimal fuel for humans, superior to products processed, basically, from grain, including the flesh of herbivores, or distilled or fermented alcohol. Franklin was especially influenced by the dietetic theories of Thomas Tryon, a champion of "Temperance," which he (and others) defined as maintaining the body in moderation, "a cool airy pleasantness," without the fever that meat or alcohol—"Riotous" living—could generate.[15]

Franklin had read Tryon in Boston and adopted his vegetarian diet, not least because it saved money. (Tryon may have been the first author he read who discussed improvements in indoor heating and cooking.) Franklin later modified his diet to include fish, confessing that, on seeing fish eat other fish, he decided: "I don't see why we mayn't eat you." Fish were in fact associated with dietary moderation and with the coolness of water, antithesis of hot-blooded land animals.[16]

Franklin remained a critic of drinking alcohol, however, and hypothesized a kind of energy theory about why it overheated the body. When he briefly left Philadelphia for the first of three intervals in London where, from 1724 to 1726, he honed his craft in two printshops, he fought an unwinnable battle against his fellow printers' habit of drinking beer—when they might have been eating bread. He tried to convince one coworker that "the Bodily Strength afforded by Beer could only be in proportion to the Grain or Flour of the Barley dissolved in the Water of which it was made," which would always be cheaper as actual bread and water. Some of the printers used the same energy-based reasoning, though, to object that they needed beer whenever they felt "*their Light*, as they phras'd it, *being out*."[17]

It was just that spark that Franklin thought needed diligent attention, in body and soul. When he returned to Philadelphia, and "conceiv'd the bold and arduous Project of arriving at moral Perfection," he composed a list of thirteen essential virtues he wanted to cultivate. Significantly, "TEMPERANCE" topped the list, with sub-instructions to "Eat not to Dulness. Drink not to Elevation." This virtue commanded the other twelve, Franklin explained, because it "tends to procure that Coolness and Clearness of Head" necessary to support all other virtues. Staying cool, avoiding anything that shocked or seared the body, or dulled one's ability to reason, was the foundation of a good life. Borrowing from a widely circulated set of comic descriptions of drunkenness, Franklin reproduced, twice, how intoxication made a body seem "*almost froze, or feavourish*," in a state so extreme it could feel like its opposite, another warning that actual heat and human sensation of it were not identical and required care to regulate.[18]

Prudent use of material resources would continue to concern Franklin, leading him to theorize human population increase as the key to economic growth. As a young worker, still in his twenties, he became convinced that human labor was the main source of economic value. This opinion endorsed analyses that recognized how, within a preindustrial economy, labor was indeed a significant form of energy, though these statements also reinforced an econometric theory of the human condition as dependent on matter yet master over it, somehow.

When he printed a pamphlet in 1729 recommending a paper currency for Pennsylvania, Franklin argued that "the Riches of a Country are to be valued by the Quantity of Labour its Inhabitants are able to purchase," not the gold or silver they possessed: "the Value of all Things is... most justly measured by Labour." The next step was to try to calculate the number of people in Pennsylvania. Unlike most colonies in British America, it never had a population census, but Franklin identified two proxy measures. Philadelphia kept count of its recorded burials and its arriving ships, so Franklin compared these to totals for cities (in Europe and the colonies) that had censuses, to estimate the colony's size.[19]

By either measure, Pennsylvania's settlers were multiplying rapidly, both because of natural increase and in-migration. Franklin was one data point in a relentless stream of newcomers. Tens of thousands were arriving from Britain, Ireland, Africa, and the German-speaking lands, plus other British colonies, with notable acceleration in the 1720s and 1730s. For the moment, abundant food and lack of warfare kept the birth rate high and mortality low, compared to other colonized zones, including Franklin's native New England.[20]

Pennsylvania's economy was also growing in qualitative terms, developing what was, compared to the other British colonies, the most mixed economy, combining agriculture, shipping, manufacturing, and extractive industries, including mining. This blend stood in contrast to the predominantly agricultural economies of southern or Caribbean colonies, where enslaved Black workers, in some places the majority, produced tobacco, rice, and sugar. It was also different from colonies, like those in New England, more dependent on shipping and mercantile activities. Most importantly, Pennsylvania's per capita output was increasing at what was, for the time, an impressive rate; European economies were also beginning to grow (on a tiny scale) in the eighteenth century, but the British colonies' growth rate was at least twice that of Britain. For the moment, Pennsylvania derived most of its revenue and wealth from grain, the colony's main export, which depended on appropriation of Indigenous land for farming, a pattern particularly evident between 1720 and 1760 with the constant removal of trees to create farmland, which Indigenous cultivators had not done.[21]

Throughout the colonial period, Lenapes would resist this constant encroachment, remaining consistent critics of all forms of domination, over land or over people. They knew most settlers had no such qualms: to the north, New Englanders had enslaved thousands of Natives they defeated in war; to the south, settlers in Carolina made war on Natives expressly to enslave people; in both regions, colonists also imported captive Black people. While Pennsylvania Quakers are often praised for their eventual criticism of slavery, that was a belated outcome. Early in the eighteenth century, they profited from both the Atlantic slave trade and enslaved labor. It was the Lenapes who were the first unbending critics of enslavement, just as they, not Quakers, were the first agents of peace in Pennsylvania.[22]

For a while, Lenape antislavery ethics influenced policy. In 1706, the Assembly passed a law banning imports of enslaved Natives, given that it "hath been observed to give the Indians of this province some umbrage for suspicion and dissatisfaction." Six years later, the Assembly added a steep duty of £20 on any imported captive, either Indian or Black. And in 1719, the Philadelphia Yearly Meeting of Quakers prohibited any sale of enslaved Natives, again responding to Lenape horror at the practice. But this would only temporarily slow colonists' rush for profits, using whatever labor yielded gain.[23]

Alongside market-oriented agriculture and trade, Pennsylvanians were beginning to mine iron ore and smelt it into crude iron, generating the industry that would eventually forge Franklin's stove. By the early eighteenth century, Pennsylvania had eclipsed New England, the first English colony to produce iron, and—as Penn had predicted—joined New Jersey and Maryland in creating the first North American industry based on mineral resources. By 1775, the British mainland colonies were the world's third-largest producer of unfinished iron, trailing Russia and Sweden, but accounting for 15 percent of the total global output. Most colonial iron came from the Chesapeake, primarily Virginia. Not until the War for Independence would Pennsylvania begin to outstrip Virginia and Maryland. But the foundations for this lay in the colony's mixed economy, with seventy-three ironworks opened from 1716 to 1775.[24]

Aside from investment in ships and shipping, iron production was Pennsylvania's most heavily capitalized activity. It required such high investment that individual proprietorships were rare; nearly all successful operations were held jointly among investors who pooled capital. And each operation had to attempt a kind of vertical integration: everything from mining ore; maintaining a wood supply; burning the wood down into charcoal; farming food crops; constructing buildings, roads, and bridges; and even sewing clothing and cooking for workers. Despite its high capital investment, the industry grew steadily, from 1,500 tons of crude and wrought iron in 1700 to 30,000 tons in 1775.[25]

Three types of iron enterprises existed. At bloomery forges, workers (usually assisted by water-powered equipment) hammered the slag out of heated ore to isolate raw iron. Blast furnaces used greater heat—2,800 degrees Fahrenheit—to smelt the iron from the ore. These furnaces were large, square stone chimneys, about twenty-five to thirty feet tall and twenty-five feet wide at the base, tapering to around eleven feet at the top, their fires fanned by water-powered bellows at the bottom, delivering unheated air, in what's called a "cold blast" method. They needed tall platforms around them, or were nested into the sides of hills with a bridge linking the hilltop to the maw of the stack, into which workers emptied baskets of ore, charcoal, and a fluxing agent (limestone or oyster shells) that flushed impurities from the ore and produced a runoff of liquidized crude iron; remelted, this could be cast in molds, as for stove plates. Refineries used yet more heat and pounding to purify crude iron into more durable wrought iron. Refining was the most skill-intensive operation, blast furnaces the most taxing. Once lit, the furnaces had to be constantly tended, with workers adding ore, charcoal, and fluxing agent and monitoring the stream of iron through a channel and into the molds of "pigs" (portable bars) embedded in sand.[26]

It was a conspicuous industry. A blast furnace was visible for miles around, and its fire lit up the sky at night. Its stack towered over countryside that might have few tall buildings and was a match for the newer if equally imposing technology of the steam engine, also about thirty feet in height. In Britain, such engines might pump water from a mine producing coal to blast iron ore into usability, the engine and the furnace like twin chained trolls huffing and glowering at each other.[27]

Although a great deal of the region's iron was turned into consumer products sold locally, starting in 1717, war was a great incentive to produce iron for export. Conflict between Sweden and Britain in that year cut off British manufacturers' main supply. Rising demand encouraged colonial iron production, and iron proprietorships proliferated. By the 1720s, most Pennsylvania iron resulted from the split production of blast furnaces and refinery forges. The former became even more visible, with properties extending for miles, well beyond the central workspace, sometimes connecting mines, woodlots, furnaces, and refineries into a landscape not fully industrial, but getting closer. In his newspaper in 1744, Franklin ran an advertisement for a not particularly extraordinary property:

> To be LET, A FURNACE, Sawmill and Forge, within 13 Miles of the City of Lancaster, for 20 Years . . . all of them being almost new, with good Water and Timber, with an unquestionable Quantity of good Iron Ore . . . a Quantity of Coals [charcoal] housed, some Wood cut: She may be put into Blast early this Summer if required. There are 80 Acres of Land within Fence, 20 Acres of Meadow cleared, 50 more may be easily made, 7000 [fence] Rails ready mauled, with other Conveniences, to accommodate an Iron-Work.

By 1756, the middle colonies and Chesapeake had at least twenty-eight blast furnaces and forty-six refinery forges. Like the one described above, many were mini-settlements, seated on waterways for transport or power, surrounded by villages of workers and their families.[28]

The intense exploitation of natural resources fed on human exploitation. Ironworks were first called plantations when that word still implied colonial settlements, plantings of English people into America. But by the eighteenth century, "plantation" had acquired its current connotation, describing highly capitalized business enterprises, as with the large agricultural estates (especially in the Caribbean) that specifically exploited the labor of enslaved Black people. Both old and new meanings of the term applied to iron plantations and their extended operations. Virginia's ironworks had, by the 1720s, begun to shift from free white workers to enslaved Black people; by the end of the century, some Chesapeake ironworks had mostly if not entirely enslaved workforces. Most of Pennsylvania's enslaved Black people were, however, concentrated in Philadelphia and other towns, as domestic workers within white households or as urban tradespeople, doing everything from heavy construction to craftwork.[29]

But Black men and women were present in the iron industry and were therefore actors in early America's emerging industrial economies. Iron production was one of the few Pennsylvania activities that paid profits high enough to enable the purchase of coerced labor, which rose after 1725. In the Pennsylvania countryside, enslaved Black people increasingly did everything related to iron: they mined ore and limestone, cut and produced charcoal from wood, and made the metal itself.[30]

West Africans' expertise with iron may have been relevant. That skill was apparent in political and religious life in the African diaspora elsewhere in the Americas. One hint of a self-governing expertise among Black ironworkers comes from the region's labor regime: iron plantations tended to operate according to production goals—tasks rather than supervised labor—a system that assumed workers' autonomous skills. At least one establishment in Maryland had an exclusively Black workforce, men and women. Moreover, the workers' ability to craft iron made white enslavers' common resort to metal restraints questionable. The industry produced these objects, as with a surviving harness that locked around a person's neck and forehead, with an attached bell to betray location and discourage flight. But one enslaver who advertised a man's escape from an ironworks in 1775, describing an "iron collar round his neck," admitted "it is likely he will soon get it off."[31]

The iron industry's reliance on enslavement by no means discouraged white investors, even Quakers. As early as 1716, at least one forge, a bloomery forge, belonged to English Quaker Thomas Rutter. The colony's first successful ironworks and blast furnace were, respectively, Quaker Samuel Nutt Sr.'s Coventry (1718) and Rutter's Colebrookdale (1720). All three of these operations were in a part of Chester County that would later be hived off into Berks County. (Nutt would by 1732 establish Pennsylvania's first steel furnace.) Slightly later, the Potts family (also Quakers) in the Schuylkill Valley set up ironworks, including John Potts's (the founder's) Mount Joy Forge on Valley Creek. Similar businesses spread into the Delaware Valley, the Lancaster area, the Susquehanna Valley, even into Philadelphia itself, which among other urban industrial metal operations had, by 1747, a steelworks. In the Cornwall area of what would become Lebanon County, settlers developed what would be, until the early nineteenth century, North America's largest mined iron-ore deposit.[32]

The successful proprietors, above all from the Rutter, Nutt, and Potts families, became prominent citizens. The Potts family, in particular, was amazingly extensive, with rich and powerful Pottses popping up everywhere. Men from these families served in the Pennsylvania Assembly and as judges; many were noteworthy in the politics leading into and through the American Revolution. Several iron proprietors were active in cultural life, including the Library Company of

Philadelphia (1731), North America's first public library. ("Public" meant it was available by subscription.) It is notable that women of these families were significant entrepreneurs themselves. The larger and successful iron families intermarried, and the daughters who forged these connections expected to be active business partners. Three daughters of Thomas Potts, the English immigrant who had founded the family's iron-making concerns, married sons of other ironmasters; most of their children continued in the industry. Given the need to pool capital to make iron, intermarriage was an excellent economic strategy: on the eve of the Revolution, the Potts clan owned or had shares in seven forges and two blast furnaces.[33]

At first, the iron industry coexisted, however uneasily, with Natives who stayed near the coast. And when ironworks were established west of Philadelphia, there was still a significant Indigenous presence there. Thomas Rutter evidently paid Native men to cut trees and clear ground for his Colebrookdale Furnace. But in 1728, the iron region was the site of the first significant violence between Natives and Pennsylvania settlers; Miami warriors attacked Colebrookdale. Nor was it clear that some ironworks had any right to be on Indigenous land. The proprietors of Durham Furnace (founded a decade earlier) didn't offer formal payment to the Lenapes until 1737.[34]

Above all, the iron industry's insatiable need for fuel was as powerful a threat to Indigenous sovereignty as colonial farming. Not until the nineteenth century, well after Franklin's death, would Pennsylvania forges burn coal. Before then, iron proprietors depended on two extractive economies: mining beneath the earth and timbering above it. Smelting one ton of pig iron required charcoal from 800 cubic feet of wood; making this into one ton of wrought iron took another 960 cubic feet. (As a reminder, an average colonial household burned 2,560 to 5,120 cubic feet of wood per year.) Depending on the quality of the charcoal and the ore, it required about 500 acres of woodland to make 1,000 tons of pig iron, but this could soar up to 1,500 acres if both charcoal and ore were poor. One of the largest Pennsylvania operations, Elizabeth Furnace, was therefore seated on 10,124 acres of land, far more space than the immediate iron production needed. On such an estate, the most time-consuming work was felling trees, cutting them into pieces, and then slowly burning the chunks into charcoal. Iron forges and furnaces without woodland had to purchase fuel from elsewhere.[35]

By 1775, there would be at least eighty-two iron furnaces in the British colonies, mostly in the mid-Atlantic region, a great number of them in Pennsylvania—all of them fired by charcoal. In that year, the 30,000 tons of iron that Pennsylvania exported had burned through the equivalent of 13,750 acres of woodland, giving a sense of the thousands of acres already (or about to be) lost. Vast environmental damage was inevitable.[36]

The results of all this industriousness, however ravaging to the land, were highly profitable, as with the greater economic growth rate apparent in colonial regions compared to Britain (or other parts of Europe). And Pennsylvania's mixed economy generated what was, for the colonies, a fairly advanced consumer economy. That was why Franklin, as a young printer, had made his way there.

The wisdom of his choice is apparent from the records of his flourishing printshop, where he and his wife, Deborah Read Rogers Franklin, sold what they printed plus a range of stationery: papers, inks, pens, slates, pencils, sealing wax, pasteboard, parchment, memoranda books with thin sheets of erasable ivory. Franklin did a brisk trade in printed forms, also standard reading matter printed by others, such as Bibles. He ordered books on request, as with an edition of Afro-Roman author Terence, whose famous quotation, "I am human, [therefore] I think nothing human alien to me," somewhat describes the Franklins' consumer wares. These included esoteric items for people of high rank or for family members. Franklin got what was at the time an extremely rare fountain pen for a member of the prominent Biddle family, galoshes and billiard balls for another high roller, ice skates for a brother-in-law, and a telescope for his mother-in-law. He edged into the region's fuel economy by becoming a middleman for firewood. One of his firewood suppliers bought books from him.[37]

While profiting from his new home's prosperity, Franklin oriented himself, and others, in relation to its iron production. In his very first *Poor Richard* almanac, for 1733, he noted distances between major landmarks, including Maryland's Principio Iron Works. In 1738, he would supply a "Mr Baxter at the Ironwoorks" with a copy of Eliza Smith's *The Compleat Housewife*, one of the first cookbooks printed in the British colonies, complete with medicinal recipes. Franklin's printing press depended on repairs done by metalworkers. His associated work with papermakers generated further work with forges, as when he ordered paper molds in 1737. Iron proprietors towered over him on all sides. The prominent Pennsylvania political figure James Logan, whom Franklin would appreciate for his intellect and voracious reading, had founded Durham Furnace; Sir William Keith, governor when Franklin arrived, created Keith's Furnace on the Christina River, now in Delaware.[38]

The iron manufacturers Franklin would know best were Robert Grace and his wife, the previously widowed Rebecca Savage Nutt, whose mother-in-law, Anna Nutt, had established the Warwick Furnace that Grace would obtain through marriage—a rare example of an ironworks that a woman possessed in her own right and passed to another woman before it was co-acquired by a man. All these Nutts and Graces were customers at Franklin's printshop. There, Anna Nutt purchased a copy of *The Pilgrim's Progress*, one of Franklin's favorite books as a child. His most enduring connection with the Nutt-Graces was a shared interest in Philadelphia's life of the mind. He and Grace were two original shareholders in the Library Company.[39]

WARWICK IRON FURNACE, CHESTER COUNTY, PENNSYLVANIA, WHERE MOST OF FRANKLIN'S PENNSYLVANIAN FIREPLACES WERE PRODUCED.
(Benjamin Henry Latrobe, watercolor on paper [1803], Maryland Center for History and Culture, Object ID 1960.108.1.6.10)

It's not too much to say, not at all, that without the region's iron industry, well established and profitable by the time Franklin arrived, the Franklin stoves might never have existed, or indeed the famous Franklin we know today. It was only by acclimating himself to his adopted region's mixed economy, its profits and its economic incentives, that Franklin became a success, initially by becoming the colony's foremost printer and source of information, the foundation of his later and better-known work on science and with iron.

It was as a printer that Franklin first turned to questions about the climate and atmosphere, tracking winter conditions in his newspaper and almanac in an era often overlooked by contemporary climate scientists and historians today. Most studies of the Little Ice Age in America have focused on the late seventeenth century, particularly in New England, where conditions during the "Maunder Minimum" (1645–1715), especially the 1690s, elicited comment—and distress. The interval dashed hopes that, by removing forests and cultivating land, colonists would warm the climate—the early hypothesis about anthropogenic climate change. But during the eighteenth century, possibly because of some warmer spells, settlers and Europeans began to predict again that colonization might moderate the climate. Optimistic statements about this *possibility* may have misled historians into thinking that the eighteenth century was warmer than it actually was.[40]

But climate anxiety was in fact reemerging in the eighteenth century, as printed sources (including Franklin's) make clear. Printing was the important new element, one often overlooked in early climate history. While it's possible that warmer spells were longer and more frequent during the eighteenth century, that may have made colonizers more impatient when the trends didn't last—and they had greater scope, in printed media, to record their exasperation. By the 1700s, newspapers and almanacs were common vehicles of information, including facts about the weather. Some basic text mining reveals settlers' ongoing worry about winter conditions. The keyword-searchable Early American Imprints, Series I (1639–1800), a digitized compendium of printed material from the British colonies, shows a distinct seasonal tilt: 8,042 hits for the word "winter" but only 3,202 for "summer." The disparity is all the more striking given that summer was by many measures the more eventful season. This was still an agricultural society, dependent on warm conditions for life itself. And news from abroad, or even from distant colonies, required optimal sailing conditions, which winter rarely supplied.[41]

In his own newspaper and almanacs, Franklin diligently tracked the cold months. At his printshop, he published *The Pennsylvania Gazette* from 1729 to 1747 (it would thereafter continue under other printers) and his *Poor Richard's Almanacks* from 1732 to 1757. These time spans encompassed his formative work on heat and the climate—also a series of colder spells in the region, which he recorded.

He stressed the human and material costs of hard winters, sometimes hearkening back to his icy boyhood home. An entry in *The Pennsylvania Gazette* in February 1730 said that Philadelphia's unusually deep snow hinted things were worse in New England, so much "as to prevent all Travelling on the Roads," including transport of any mail or newspapers. For want of fresh news, Franklin had delayed printing the *Gazette* for two days. In December, Franklin again warned that winter conditions meant the post only arrived every two weeks. So, "this Paper will be published on a Tuesday" and only as a half sheet, given that "Winter generally Occasions a Scarcity of News in these Parts." There was more bad news about the lack of news in early 1732. "We have no Entries this Week," Franklin confessed on January 4, "the River being full of Ice" and blocked to ships. In late November, the Delaware River (which included the city's port) froze again. "No Vessels can go up or down, a Thing rarely happening so early in the Year." In addition, "Wood is risen to an excessive Price," so people had less heat as well as reading material.[42]

His paper's careful reporting of winter conditions contrasted with Franklin's reluctant performance of his task, as an almanac maker, to forecast the weather. Prognostications were nearly universal in almanacs but dismaying for readers who might take them too literally. In his *Poor Richard* for 1737, Franklin promised never to include useless platitudes like "*Snow here or in New England,—Rain here or in South-Carolina, —Cold to the Northward.*" He vowed to do forecasts only for the specific region and for seasonal trends, begging forgiveness for any predictions that erred by only a day or two.[43]

Franklin's more serious discussions of winter included early thoughts on fuel and heat. He printed someone's verses on "THE RAPE OF FEWEL. *A Cold weather POEM*," in *The Pennsylvania Gazette* in early November 1732, just as "The bleak Norwest [wind] begins his dreaded Reign." The bees in their hives, ants in their nests, and better-off humans in their houses are snug and warm. But the poorest settlers, "made desp'rate by their Wants," go "shud'ring forth in early Hours and late" to steal firewood. In the dark, the thieves pry fuel from other people's sheds and fences, then sneak home "Loaden with interdicted Posts and Rails," another day's heat stolen "to save a wretched Life." The poet doesn't blame the foragers. For the desperately cold "to wrong their richer Neighbours" was forgivable. People with pillaged fences should compare their losses to the hazard of floods: Ultimately, nature was to blame, and why expect justice from nature?[44]

Household heat also features in the first edition of *Poor Richard*, prepared around the same time, in late 1732, with Franklin referring to fireplaces three times. His almanac's alter ego, the feckless mathematician and astronomer Richard Saunders, confides that his wife threatens to "burn all my Books and Rattling-Traps (as she calls my Instruments)" unless he makes money by them. (Hence the almanac, which made money for the nonfictional Mr. and Mrs. Franklin.) And Franklin inserted, for the cold months of January and February, two Poor Richard maxims about wives and fires, though with divergent sentiments about them:

> January: *A house without woman and Firelight, is like a body without soul or sprite.*
> February: *Ne'er take a wife till thou hast a house (& a fire) to put her in.*

To this mixed advice Franklin added sayings that more clearly reflected his personal beliefs. Very often, these were about food as bodily fuel: "Eat to live, and not live to eat" and "To lengthen thy Life, lessen thy Meals." He kept issuing this warning. In *Poor Richard, 1736*, he wrote, "I saw few die of hunger, of Eating 100000"; and the next year,

"Three good meals a day is bad living." These admonitions assumed food was so easy to get, it could easily become an overindulgence, a comfortable assumption, hinting that the poorest Pennsylvanians didn't buy *Poor Richard*, whose scoldings they didn't need anyway.[45]

If too much fuel threatened health from the inside, might cold damage it from the outside? In his newspaper for late November 1732, Franklin reported widespread "Colds" in the colony and neighboring Maryland, plus anxiety over where they came from. Some people thought their coughs and sniffles came from "the sudden Change of Weather into hard Frost," the ambient cold triggering the human colds. But others "believe it contagious . . . communicated by infected Air, after somewhat the same Manner as the Small-pox." Franklin, who had a policy of avoiding unnecessary controversy, helpfully cited a scientific article from the *Philosophical Transactions* of the Royal Society of London about outbreaks of colds in Ireland that would "favour both Opinions."[46]

Keeping home fires burning presented other dangers. White Pennsylvanians liked to pile up fuel before the snows came and stick close to home until spring. Their decision was part of the ongoing colonial differentiation from Natives, but it introduced the hazard of urban fire, which already plagued towns and cities in Europe, as if Prometheus was avenging himself on the humans for whom he'd suffered so much. The risk of even a small stray ember was such that the Pennsylvania government levied a fine on anyone caught smoking tobacco "in the streets of Philadelphia, either by day or night."[47]

News items from Franklin's *Pennsylvania Gazette* repeatedly warned that domestic fires could quickly get out of hand. One April, the paper reported a fire in a blacksmith's shop that consumed it and four other workshops and threatened the city's prison. A November fire was swiftly suppressed but considered newsworthy because a visiting sheik from Beirut happened to be staying where it happened. (Shedid Allhazar, a man of learning who toured the British colonies, arrived in Philadelphia with a letter of introduction to James Logan.) In November of another year, Franklin reported a fire in a dancing school. The gloomiest event was possibly an April fire that destroyed several semiconstructed buildings that "would have been a great Ornament" to the town but now might never be completed.[48]

To protect the "green country town" Penn had wanted to fireproof, the solution was not fewer or smaller fires, but organized firefighting. Just as London's Great Fire had publicized the hazard of bigger, denser cities with proliferating indoor heat sources, so the city pioneered social coordination to manage those hazards. Indeed, the source of the problem paid for its solution: a duty on coal subsidized London's rebuilding and its new firefighting system, in which each quarter and parish of the city had to buy and maintain pumps, buckets, and tools to demolish buildings right in the fire's path that were doomed anyway, mere tinder for spreading the disaster. There were also twice-yearly inspections of hearths, ovens, and stoves, and all wooden chimneys had to be rebuilt in brick.[49]

Along those lines, Franklin wrote an essay about fire safety for the civic-minded Junto, a club he cofounded in 1727 as a kind of mutual improvement society for young working men. The essay advised regular chimney cleaning, and greater use of incombustible building materials—"an Ounce of Prevention is worth a Pound of Cure"—but also an urban fire force. This last was a public agency, requiring investment in pumping "Engines" to spray water onto fires, and organization of squads of white male citizens to operate the engines and to wield axes and buckets. Franklin published the Junto essay in his newspaper in February 1735. That was also the year the Assembly stipulated that the area around its State House should "remain a public open green and walks forever," a nice urban amenity that buffered the center of government from anything flammable.[50]

One year later, Franklin helped organize Philadelphia's Union Fire Company, possibly modeled on the Boston Fire Society in his hometown. The inaugural Philadelphia company, of twenty men, signed articles in December 1736. What they'd formed was essentially an insurance scheme: each man pledged to buy two leather buckets for carrying water and two sturdy bags for transporting goods to safety, and to stand ready against the dreaded call of "fire!" knowing he'd get help should his own house or shop catch fire. The association fined anyone who shirked his duties and promised to assist widows who survived members. Again, the overall thinking was that, given how fires had to be kept in households and workshops, their hazards should be acknowledged and planned for.[51]

Firefighting was just the first step in Franklin's mastery of domestic heating. If spiritual tracts had been fireside reading for his Puritan ancestors, he seems to have settled before his fireplace with books on science. Let's peek over his shoulder at one volume he also decided to offer for sale in early 1734—this one because it dealt with climate science. In his *Telluris Theoria Sacra, or Sacred Theory of the Earth* (1681–1689), English clergyman Thomas Burnet refined the theory that the scriptural flood, by swiftly transferring subterranean waters onto the Earth's surface, had permanently changed the planet for the worse by scarring its once-smooth surface and altering its tilt relative to the sun, changing climates and seasons, ruining any prospect of steadily good conditions. Human action—specifically sin—was therefore a distant reason for the bad weather in the seventeenth and eighteenth centuries. Many people accepted Burnet's theory, on either scriptural or scientific grounds, or both. It would be background for some of Franklin's thinking, even as he also read practical texts on how to fight the cold the ancient Deluge might have caused.[52]

Three authors in particular would influence his thoughts: J. T. Desaguliers, Nicolas Gauger, and Martin Clare. Each was not only an author but an inventor, and their devices championed the new desire to stay warm indoors while minimizing use of fuel and exposure to smoke. From their works, Franklin would define his efforts to do the same.[53]

Desaguliers, all but forgotten now, was hard to miss in Franklin's era as a premier science communicator, the Carl Sagan or Neil deGrasse Tyson of his day. He was one of Isaac Newton's protégés, famous for his public lectures on scientific discoveries. He was also an important consult for anyone trying to improve heating or ventilation; his published lectures describe his devices to refresh air in the House of Commons and to heat the House of Lords. He may have encouraged in Franklin a distrust of thermometers by noting they lacked a universal system of measurement, guaranteeing confusion "until every body agrees to graduate their Thermometers in the same Way," which we still haven't done.[54]

Desaguliers (French by birth) translated Franklin's next reading on heat, Gauger's *La Mecanique du feu, ou l'art d'en augmenter les effets, et d'en diminuer la dépense* (1713), on the mechanics of fire, including reduced fuel costs. But Desaguliers's edition, first published in 1715, and then with an appendix in 1736, was titled *Fires Improv'd: Being a New Method of Building Chimneys, So as to prevent their Smoking: In which a Small Fire, shall warm a Room better than a much Larger made the Common Way.* (This was like having Neil deGrasse Tyson endorse your book—but also retitle it and explain what was wrong with it.) The rechristening translated the natural phenomena even more explicitly into human goals. Climatic conditions, precisely, prompted such thinking; Gauger wrote in the wake of a very bad winter, 1708–1709, when Venice's lagoon froze solid.[55]

Desaguliers further emphasized that he'd improved *Fires Improv'd* by omitting stuff he thought "superfluous" and adding practical advice of his own. The deleted content included attempts to fix smoky chimneys that Gauger had admitted didn't really work; the additions were about coal and peat, domestic fuels that dominated in Britain and

Ireland, unlike France. But Desaguliers faithfully showcased Gauger's main point: a fireplace's power could be increased by installing into its lower chimney a set of metal baffles, rectangular passages that restrained airflow, preventing loss of heat up the chimney and sending more into the room. Gauger took thermometer readings near ceiling and floor to verify his design's greater heating capacity—even better than the stoves used elsewhere in Europe, where people marveled at British tolerance for shivering while watching wood burn, or as Desaguliers put it, being "*roasted* before, or *frozen* behind." With better fireplaces, Britons, too, might bask in a state of equal warmth.[56]

Plus, fireplace improvement might reduce smoke, not least to preserve health. Establishing a line of inquiry Franklin would find crucial, Gauger wondered if smoke could be reburned, generating a cleaner emission. He cited Herman Boerhaave on the debatable distinctions between fuel, soot, smoke, and ash, the Dutch chemical experimenter having explained that he could ignite the residues, proving they were also fuel, if in diluted form. Gauger further extolled the experiments of André Dalesme, reported in the *Journal des Sçavans* for 1686, reprinted in the Royal Society's *Philosophical Transactions*. Dalesme's device caused much excitement: it was an early downdraft model that reburned its smoke (a detail that particularly intrigued Franklin). It was essentially a metal tube with a fixed grate, positioned so that smoke from a fire would (because it was cooler than the heated air) sink back down into the flame and be reburned, reducing smoke and soot. "Tho' we may not at first feel the Inconveniency," Gauger warned, "yet at long run we shall find the pernicious Effects of drawing Smoke into our Lungs," or even "Air impregnated with Smoke."[57]

Martin Clare's *The Motion of Fluids, Natural and Artificial* . . . (1735), the third major source on heat Franklin would read, examined how air responded to different conditions, including temperature. Like Desaguliers and Gauger, Clare used thermometers, but he complained about them. If anything, the best data came from "the Old and Infirm,

[who] are so many sensitive Barometers," given their bodies' vulnerability to atmospheric conditions. He noted that even healthy youths shrank from sharp drafts (findable by holding a lit candle to crevices and keyholes). An optimal fireplace heated a whole room by convection, warm air's tendency to rise—visible if someone put a set of revolving metal vanes in front of the fire—which could be augmented by installing "hollow iron Plates, disposed about in the Back and Sides of the Chimney," the system of capturing and radiating heat that Gauger had recommended. Again, and despite the era's worsening cold, heating was supposed to be getting better—the books said so.[58]

Not that Franklin lingered with a book by the fire if there was something else to do. From his printshop, he built an impressive network, cultivating those above him, with power and connections he might hook himself into. His growing authority in reporting public affairs led to his selection in 1730 as official printer to the colonial government and in 1736 as clerk of the Pennsylvania Assembly. One year later, he was appointed postmaster at Philadelphia, a British imperial post. His government work at both colonial and imperial levels gave him a firsthand view into the colony's Indian relations as they deteriorated, a story he told as official printer of the treaties that documented the decline of Native sovereignty and rise of colonial control of natural resources.

William Penn's policy of negotiating land sales with Natives had helped keep the peace, just about, even amid mounting expectations of the wealth the colony might yield. As land transfers proceeded, and settlers made their way into more and more of "their" colony, they were astonished at its wooded state—and Natives were astonished at their unfamiliarity with it. Inexperienced colonists would return from treks with hair and clothes studded with twigs. Lenapes couldn't believe their lack of outdoor common sense: sleeping under trees that could shed branches (or creatures), trying to burn green, fresh-cut wood, taking forever to accept that canoes were often the fastest way to travel. Also, such was their quickness to cut down tree after tree that the Seneca word for Europeans, *Asseroni* (axe makers), obviously applied to the white Pennsylvanians, too.[59]

Certainly, settlers were bringing the forests down quickly. If an average colonial household was burning roughly thirty cords of wood extracted each year from an acre of woodland, that acre could be cleared in a mere three days. One able worker could then segment, split, and stack the wood at a minimum rate of a cord a day, with an average household's annual fuel needs met, therefore, within thirty-three working days. This was good work for winter, when farmers had more time and when cold, dry weather made timber easier to cut and split. Burning the wood for heating itself took little time and labor; the indoor work related to fire that took the most effort (even more than woodcutting) would have been preparing food.[60]

Erosion of Native land tenure continued through Penn's lifetime and after his death in 1718, though amity between Lenapes and settlers held through the governorship of his widow and fellow proprietor, Hannah Callowhill Penn. But Penn's will and associated legal documents didn't make clear who inherited authority to buy, sell, or even survey land. Most of what's now Pennsylvania remained sovereign to Indigenous people, though with little regulation of white colonists' behavior beyond areas of dense settlement. In 1722, four years after Penn's death, a murder on land near the border with Maryland threatened the colony's peace. In this incident, one of two white traders, brothers, struck a Seneca man, Sawantaeny, on his head; he died the next day. Multiple Indigenous groups were outraged, not just the Haudenosaunee (of which the Seneca were members), and controversy continued until colonial and Haudenosaunee leaders met at Albany at the end of the year. There, condolence ceremonies, apologies, and atonement led to forgiveness and a "Great Treaty of 1722," reaffirming amity with the Haudenosaunee. It's now the longest continuously honored treaty in US history. But the circumstances that prompted it seemed unresolved and likely to spin into violence again.[61]

When Hannah Penn died in 1726, the Penns' sons and heirs, John, Richard, and Thomas, wanted clear title—and legal jurisdiction—over as much land as possible. At first, they made gestures to their parents' pacifist legacy. And they capitalized on their father's decision to accept Haudenosaunee claims over land in the upper Susquehanna Valley, which other Native nations, including the Lenape, disputed. As Franklin reported in *The Pennsylvania Gazette* in late 1736, the Haudenosaunee had confirmed William Penn's 1701 purchase of land west from the mouth of the Susquehanna River. This formal assent honored how the "late Honourable Proprietor" had "made it his principal Care, to cultivate and maintain a good Understanding with all the Indians," which the "present Honourable Proprietors" defended in the peaceable release of the Susquehanna lands. As the colony's printer, Franklin published the *Treaty of Friendship held with the Chiefs of the Six Nations, at Philadelphia* . . . , the first of several such treaties he would print.[62]

These documents record how Haudenosaunee diplomats chose from several political devices, depending on who they were dealing with. When meeting with colonial or British officials, they often used the originally English concept of a Covenant Chain binding the Haudenosaunee to the British colonies, starting with New York's tie with the Mohawks. But with each other, the Haudenosaunee emphasized a shared council fire at Onondaga, the central house of the original five Iroquois nations. Finally, Haudenosaunee diplomats would speak of the road or path that could lead to diplomatic meetings, or else be blocked with fallen trees or human bodies, or fouled with blood. The images reflected Haudenosaunee belief that they and trees were similar beings, equally native to the same place, part of their interspecies ethic. A chain, in contrast, implied the work of the human hand, especially the manufactures white people made. By invoking a council fire, Natives extended to white settlers their very different assumption that nonhuman things did powerful diplomatic work.[63]

In the speeches opening the 1736 negotiations at Stenton, the acting governor James Logan's house outside Philadelphia, the Haudenosaunee delegates stated: "*They were now come, after a full Consultation with all their Chiefs at their great Fire or Place of Counsel, to return an Answer to the Treaty, that some of them had held with us four Years since, at* Philadelphia: *That they intended to . . . proceed to the Fire kept for them at* Philadelphia; *where after two Nights more, they would at that Fire give their full Answer.*" Told there were cases of smallpox in the city, they clarified that "*tho'* Philadelphia *is the Place where their Fire is kept for them, yet it may upon Occasion be brought out hither; and they are free either to proceed to* Philadelphia *to treat at the Fire there, or to stay here.*"[64]

Colonial officials arranged to keep the Native delegates safely housed in the city, far from any smallpox cases. Proceedings reopened at the Quakers' Great Meeting House, with Kanickhungo (Seneca) recalling the 1732 Haudenosaunee-Pennsylvania meeting: "As you received us kindly, and at that Treaty undertook to provide and keep for us a Fire in this great City, we are now come to warm ourselves thereat, and we desire and hope it will ever continue bright and burning to the End of the World." Thomas Penn acknowledged, among his colony's ties to the Six Nations, that very "Fire." This wasn't a physical entity, given the extreme unlikelihood that the meetinghouse had a heat source; churches rarely did, beyond flickering candles and fervent worshippers. But Penn was recognizing the concept as essential to Indigenous diplomacy, the sharing of light and warmth.[65]

Franklin's printing of these treaty negotiations is thus one of the first documented accounts, in English, to show settler acceptance of the Indigenous concept of a council fire, signifying a shared and egalitarian political task. In notable contrast, the "Great Treaty" of 1722, adjudicating Sawantaeny's murder, had mostly cited the chain and never the council fire. In 1736, for the Natives present, oral statements about a Philadelphia council fire were crucial, defining the essence of the meeting; for the colonists, putting those declarations in print was likewise politically salient, if perhaps more dutiful than heartfelt.[66]

But Pennsylvania's concord with the Haudenosaunee did not benefit Lenapes. In 1736, the same year as the treaty negotiations with the Six Nations, the younger Penns orchestrated their colony's most nefarious land grab, the so-called Walking Purchase, within a decade exposed as a fraud. The Penns claimed to have excavated a never-executed 1686 deed of sale from the Lenapes. In this, they said, their father was promised lands stretching from the meeting of the Lehigh and Delaware Rivers to a point as far west as a man could walk in a day and a half. This was territory that William Penn and Lenapes had evidently assumed would be co-inhabited, not ceded entirely, and surveyed by walking it out together.

The younger Penns announced their intention to finalize payment for the land if Lenape leaders would confirm the purchase. It was Governor Logan, who'd been William Penn's secretary, who claimed to have found the stray document. Logan had a private interest in the land cession because Durham Furnace, in which he was a partner, would benefit from greater access to timber that could provide charcoal. The proprietors showed the Lenapes a map of the intended sale with misleading landmarks, making it seem as if more of the affected land had already been ceded anyway than was the case. The Lenapes reluctantly agreed to the terms, not realizing the trick on the map, but suspecting (rightly) that the 1686 "deed," mysteriously mislaid for half a century, was a forgery.[67]

And if the Lenapes thought they'd consented to a land sale based on a "walk" through wooded land, the proprietors cheated them yet again: they cleared a trail in advance and engaged three fast runners, promising 500 acres to the swiftest. Only one man staggered to the end of the thirty-six-hour marathon. He had covered 70 miles, his track demarcating one side of a triangle that yielded a catastrophic concession of 1,200,932 acres, roughly the size of Rhode Island.[68]

Coming on top of that year's confirmation of western land from the Haudenosaunee, the Walking Purchase enlarged "Penn's Woods" dramatically. Long settled on large rivers near the ocean, many Lenapes were forced into inland exile, the first of several banishments to landscapes alien to them, deprived of their nonhuman kin, orphaned. Around the Susquehanna River, they joined Onondagas positioned to hold down Haudenosaunee claims to the area, plus refugees from other Native groups. Franklin's edition of the 1736 treaty of friendship with the Haudenosaunee appeared on September 22, 1737, two days after the conclusion of the infamous walk, documenting that colonists knew of more honest ways to negotiate with Native people.[69]

From this point on, monitoring what they assumed would be a shrinking Native presence, Pennsylvania's leaders, increasingly unlikely to be Quaker, would prioritize treaty relations with the Lenapes and the Haudenosaunee, while hoping the latter would disempower the former. Until 1762, Franklin printed texts of the treaties, including summaries of the key diplomatic meetings, recording colonial acceptance of the reality of Native political autonomy, but also the limits of that acceptance.

These treaties, extending over two decades, would frequently refer to the council fire: in 1742, 1744, and 1747, again in 1753, 1756, 1757, 1758, and finally in 1761 and 1762, always recorded in print. Franklin attended the proceedings, becoming—unexpectedly (not least to him)—one of the most important sources of published information about Indigenous political beliefs. In these negotiations, he himself used language about council fires at least twice. In November 1756, he served on a commission that drafted speeches for the governor, William Denny, to give at a conference with Lenapes in Easton. The first of these refers to "our Council Fire, kindled by us both"; the second, in Franklin's handwriting, promises to exchange diplomatic gifts for white hostages Lenapes had taken during the French and Indian War, offering to do so "at a Council Fire to be rekindled at Philadelphia for you and us, or here, as you shall chuse."[70]

Even as he quoted Native diplomatic concepts, Franklin knew that fires had different political associations for the English. This was obvious in his descriptions, in *The Pennsylvania Gazette*, of bonfires lit to celebrate British military actions. In spring 1740, at the outbreak of the War of Jenkins' Ear, the *Gazette* related that Philadelphia's festivities included "a Bonfire on the Hill." In 1745, when the French surrendered their main Atlantic Maritimes fort, Louisbourg, to a joint British–New England force, Philadelphians were even more exuberant: "upwards of 20 Bonfires were immediately lighted in the Streets."[71]

Colonial erosion of Lenape land rights continued alongside these British triumphs. Land from the Walking Purchase was opened to settlers during one of Pennsylvania's fastest phases of growth, from the 1740s into the 1770s. By 1750, the colonial population of the greater region (Pennsylvania, New York, New Jersey) would reach about three hundred thousand. Pennsylvania now had all the foreboding conditions that William and Hannah Penn had hoped to avoid, lest they incite violence. Lenapes would never forget how the younger Penns had treated them so contemptuously. And the Penns were risking the warfare their parents had shunned by pushing the colony's boundaries toward powerful competitors: the Haudenosaunee and the French (who claimed territory at Pennsylvania's western edge, running toward the coveted Mississippi River valley and up to the Great Lakes).[72]

In essence, homelands were being chopped into homesteads, consigned to private investors as individual sites of production and consumption. William Penn had realized that Lenapes and other Natives didn't live or think this way. He'd said that "their pleasure feeds them," reflecting a European assumption that hunting and fishing were rural pleasures that happened to yield food, and for Indigenous people, "this table is spread everywhere." Only that natural abundance was no longer available "everywhere."[73]

Franklin joined this relentless takeover. He seems to have begun speculating in western land as early as the 1740s, exactly when he, as a successful business proprietor, had the wealth to do so—printing could produce that kind of prosperity. Indeed, Pennsylvania's mixed economy generated multiple incentives for white settlers to go west, including the quest for trees to burn into charcoal to forge iron. That endless erosion of forests violated the "thousand-year" scale of European conservation plans, as John Evelyn had hoped might prevail in England. In Pennsylvania, English colonists were convinced they could create a combined agricultural *and* industrial economy, using ever-mounting quantities of fuel (at a time of unusual cold) without running out of the necessary resources, land, trees, and ore. This projected economic growth, taking advantage of new-world abundance extracted from the Lenapes, was the impetus for Franklin's first published work in science, in which he warned that at least a little conservation might be needed to keep the whole project running.[74]

3

THE PENNSYLVANIAN FIREPLACE

Maybe some odd drift in the smoke above the chimney gave it away. But otherwise, a historic laboratory in a private home at 131 Market Street, Philadelphia, was invisible from the outside. The building is long gone, no plaque commemorates what was discovered there, but this was where Franklin investigated his first atmosphere. Later portrayed as a godlike master of nature—thunder, lightning, ocean currents—he began it all with experiments in heating his family's sitting room, starting in the late 1730s, trying to minimize both fuel use and emission of smoke, right as the weather got colder (yet again) in the 1740s. The result was Franklin's first invention, the Pennsylvanian fireplace, a metal insert for a hearth with a chimney, starting with the one in his parlor on Market Street.[1]

Franklin's fireplace may have conserved wood, but it used a lot of paper. Printed accounts of the worsening winters and of the newly invented fireplace made paper media part of ongoing adaptation to climate change, no less than the stoves. Franklin did so much reporting about cold weather that his work as a printer was, intentionally or not, the foundation for his new heating system. He established himself as a weather expert, a reliable source of information about how bad the cold might be, and yet how it wasn't as bad as elsewhere, all of which made him a reassuring guide to some of Pennsylvania's coldest intervals, for which he proposed a scientific solution.

His project was conservationist, an energy regime to slow the loss of woodlands that already troubled Britain and was beginning to plague Philadelphia and its environs. The connections to greater atmospheric phenomena would come later—Pennsylvania's social and material atmosphere was the immediate context for Franklin's initial work. The parlor on Market Street was the microcosm, the colony the macrocosm; functionally interdependent, neither could have existed without the other, with everyday heating the measure of Pennsylvania's opportunities, inequities, comforts—starting with the discomforts of yet more bad winters.

In *The Pennsylvania Gazette* for March 5, 1741, Franklin related how "Accounts from all Parts of the Country are fill'd with Complaints of the Severity of the Winter, no one remembering the like. The Cattle are daily dying for want of Fodder; many Deer are found dead in the Woods, and some come tamely to the Plantations and feed on Hay with other Creatures." By spring, the weather was better but the damage more apparent. On April 9, the paper reported that some colonists had resorted to eating the dead deer. And, Franklin said, Natives warned that wild animals might be "scarce for many Years." This Indigenous testimony, from past experience, identified an unexpected hazard of clearing away natural resources. *Poor Richard's Almanack* for 1742 further noted, in January, that, with the freezing of waterways, "Foot, Horse, and Waggons, now cross Rivers, dry, / And

Ships unmov'd, the boistrous Winds defy." Everything was strange, "by Snow disguis'd." Bad news for everyone unless the trend reversed. Which it did not.[2]

The winter of late 1740 to early 1741, with its deadly white landscape, was a turning point. It stood out, not as the worst winter of the eighteenth century, but as the memorable first of a sequence, all occurring within colonists' living memory and documented in multiple accounts. A rival Philadelphia printer, Franklin's former employer Andrew Bradford, gave an extended account of the winter in his *American Magazine*, North America's first magazine, launched on February 13, 1741—three days before Franklin's own attempt, *The General Magazine*. Bradford's magazine lasted only three issues, and Franklin's, six, but while they appeared, they gave the city's snowbound readers very welcome reading material. *The American Magazine* offered a weather diary for "this last hard winter"—snow, ice, frozen roads, more snow, more ice, roads thawing into icy mud—from October 27, 1740, to April 19, 1741. The account concludes that it would be "a great Apple Year . . . but few Peaches," foreign trees marking the colony's natural fortunes. Scooped by Bradford, Franklin frostily ignored the ongoing weather in his 1741 magazine, but commemorated it later, in his almanac for 1749. In the calendar notes for March, he reminded readers that, "on the 13th of this month, 1741, the river Delaware became navigable again, having been fast froze up to that day, from the 19th of December in the preceding year." So far, he said, 1740–1741 had offered "the longest and hardest winter remembered here."[3]

That winter was also the first time Philadelphia's officials considered subsidizing firewood for poor white people who, in this wealth- and credit-starved era, struggled to scrape together money for anything. Franklin indicated his support with some foreign news: as in Pennsylvania, "the Winter has been excessively severe in England, beyond any thing in the Memory of Man, by which the Poor have suffer'd extreamly, notwithstanding vast Sums of Money, and Quantities of Coals, Bread, &c. distributed among them in Charity." One way or another, in harsh winters, someone had to pay for heat.[4]

This particular winter, icy midwife to the Franklin stove, raises questions about how people experience and respond to climate change. Historians have studied the winter of 1740–1741 in Europe, where its shortening of the growing season caused famine in several places. In Ireland, as much as one-fifth of the population died, a toll possibly higher than during the better-known Great Famine of 1845–1852; the disaster triggered notable out-migration (including to Pennsylvania). And yet climate scientists cannot identify proxy data for a significant temperature variation. This indicates weather's complex relation to other material circumstances (in Europe, the cold interacted with drought and disease), and that human accounts of weather events may document not just natural conditions but cultural expectations. The new demand for greater indoor comfort may have made colonists into more sensitive weather instruments—and into pickier consumers with a greater sense of entitlement. Toughing out a bad winter was less acceptable. By the 1740s, people expected to be able to make artificial atmospheres indoors, as Franklin did.[5]

In case a little schadenfreude cheered up anyone in Pennsylvania, things were even worse elsewhere in North America. In his *Poor Richard* of 1748, under January, Franklin would note that "we complain sometimes of hard Winters in this Country; but our Winters will appear as Summers" compared to those in "Hudson's Bay." Franklin supplied several detailed pages about this place where, in winter, everything existed in various states of ice. "And now, my tender Reader, thou that shudderest when the Wind blows a little at N-West, and criest '*Tis extrrrrrream cohohold!*' . . . what dost thou think of removing to that delightful Country?"[6]

New England remained Franklin's main point of comparison for any harsh weather. On February 11, 1752, *The Pennsylvania Gazette* reported that, on January 6 in Boston, it had "been observed by Gentlemen who keep Thermometers, that for several Days last Week, it was colder by 7 or 8 Degrees, than it has been for many Years past." Mail from England was delayed. Meanwhile, Boston Harbor froze so hard that, by sled or on foot, an "abundance of People pass daily from the Town to Castle William, and the Islands." Only with much heroic chopping of ice from the harbor's channel could ships depart.[7]

Even after Franklin retired from his printing business in 1748, his printer contacts and his newspaper (edited by his business partner) would monitor the weather. The winter of 1764–1765 set more records, as *The Pennsylvania Gazette* reported multiple times: three feet of snow between December 25, 1764, and January 14, 1765; a two-day blizzard in March that deposited another two and a half feet of snow; and the freezing over of the Delaware River, which closed Philadelphia's port and delayed all traffic, including postal service. In January, Franklin's protégé James Parker, public printer in New York and eventual postmaster in New Haven, sent word to Franklin (at that point in London) of the dire conditions. Snow had begun to fall hard on Christmas Day and didn't stop until it lay three to six feet deep; the Delaware and Brunswick Rivers froze and were closed to navigation. "It is thought by most," Parker wrote on January 14, "that from the 25th Decem. to this Day, the Weather has exceeded the hard Winter of 1740."[8]

The solution, winter after winter, was to try to keep warm at home, a running joke (at best) on the human condition in the northern colonies. A Massachusetts almanac for 1777 would admit, for example, that fireplaces could smoke up a house, making people cough and dredging their furnishings and clothing in soot. "How to hinder a House from smoaking?" The obvious answer was to "KEEP no fire in it; if a fire is already made, throw a sufficient quantity of water on to quench it, and the smoak will soon depart." (This is indeed foolproof.) Another Boston almanac for the same year joked about finding anything to burn in the first place. Almanacs often supplied recipes and this one obliged with a "Recipe to keep one's self warm a whole Winter with a single billet of Wood," meaning one large, reusable chunk. Just carry this remarkable source of heat upstairs, at top speed, throw it out the window into the yard below, then race downstairs to fetch it up again. "This renew as often as Occasion shall require." Again, there's an undeniable logic to it.[9]

Logical, except the solution simply swaps one fuel for another, food in the belly for wood on the hearth, giving a person the strength to douse a smoky fire or thunder up and down the stairs. Food wasn't necessarily cheaper than firewood. And, therefore, all jokes aside, the real problem with fireplaces was securing wood to burn in them in the first place, rather than being cold while listening to the wind moan in the chimney.

The cost of keeping warm fell unevenly, with the richest suffering the least—a transhistorical constant. For James Logan, who built Pennsylvania's grandest house, Stenton (completed in 1730), money was no object: smack in the front hall he had a huge open fireplace, the most ostentatious and wasteful heating choice. This despite a complex ventilation system in Stenton's dining room, indicating Logan's familiarity with Nicolas Gauger, whose book on heat and fuel efficiency he had in his grand and expensive library. He knew about minimizing fuel use; he didn't need to bother with it, except to show off his book learning about it.[10]

But for an ordinary colonial family in the early eighteenth century, fuel represented 9 percent of a typical household budget—in a typical year. (In comparison: the average energy burden for non-low-income US households in 2024 was 2 percent, though for low-income households it was 6 percent.) In the Little Ice Age, unusual cold persistently raised the price of keeping warm. For poorer people, that meant decisions that affected health if not life. Things were even worse for enslaved people, dependent on provisions of clothing and food that were barely adequate in any winter, and, if they lived outside a town, often tasked with getting their own firewood. All this was exacerbated by rising fuel costs. By the 1730s, seaport towns in British America had cut down their nearest woods. They had to import fuel from farther away, transport costs adding to the expense. In Philadelphia, a cord of wood ran from 8 to 10 shillings between 1700 and 1735, then from 10 to 13 shillings between 1736 and 1750.[11]

Within a scant two generations, English colonists along the coast of North America had replicated conditions that, in Europe, dated back to the Middle Ages. The "new world" turned out not to be a wondrously refilling cornucopia. Its resources were finite, too. Unless some burnable treasure lay buried within it? In his *Winter Meditations*, Cotton Mather had noted that God's creation included "Numberless *Fossils* in the Bowels of the Globe, which probably contains above Ten Thousand Millions of Cubic German Leagues." *Probably.* If true, then huge amounts of mineral wealth, including coal, remained to be discovered, probably. Only, in the meantime, it was not just probable but obvious that North America's trees were vanishing.[12]

If a pun is forgivable, Franklin had warmed up to his work with stoves by pondering several questions about heat, fire, and smoke. He did his first investigation of thermal variation after a deep snowfall in the winter of 1729–1730. He'd duly reported extreme conditions in his newspaper (noting, as usual, that things were even worse in Boston) and then described using a Philadelphia snowbank, on one pleasantly sunny day, to demonstrate that cloth patches of different colors sank into the snow at different rates, with darker swatches, made hotter by sunlight, descending deepest. (Franklin might have learned about this demonstration, devised by Dutch experimenter Herman Boerhaave, based on Isaac Newton's work with heat and color, from a copy of Boerhaave's *Elementa Chemiae* in James Logan's library.) Around the same time, Franklin's bookish club, the Junto, was debating various questions in the sciences of heating and hydrostatics, including "the phenomena of vapors," why a candle's flame ascended "in a spire," and "how may smoky chimneys best be cured?"[13]

Frigid winters had already prompted colonists to adopt European technology that augmented indoor heating. The most basic was the fireback, an iron plate set into the back of a fireplace to radiate heat out into a room. The earliest known colonial firebacks date from the 1660s. Next, in the half century from circa 1726 to 1773, householders who could afford them bought five-plate stoves, iron boxes with one side opening into another room, often the kitchen, where they received fuel (and leaked smoke), with the closed box on the other side of the wall, radiating smokeless heat into a parlor or sitting room. This technology was mostly restricted to Pennsylvania, where (partly due to the busy iron industry) it became commonplace. But the scholars who described these technologies didn't notice that they followed changes in the weather: firebacks proliferated after the late-seventeenth-century temperature dip, and five-plate stoves multiplied through succeeding bad winters, from the early to middle 1700s.[14]

It was at this point, in early 1738, that Franklin seems to have begun his work to improve heating systems. That year didn't bring a particularly bad winter, but it was embedded within a longer experience of cold—one of the ironworks Franklin would collaborate with reported in February 1737 that they'd had to buy ore instead of using their own "heap," which was blocked by ice. In January of 1738, Franklin purchased a stove (likely a five-plate model) from William Branson, a merchant and investor in Reading Furnace. Later the same month, he bought 28¾ pounds of steel from Branson, probably plates and other components (lighter and easier to shift around than iron), measured against his fireplace to fashion a prototype. In June, he bought another 10½ pounds of steel and, in November, another stove. Thus armed, Franklin went into another winter prepared for a second round of tinkering. Maybe it adds to the charm of his experiment that it might have been a family affair. His son William, who would later help him fly a kite in a thunderstorm, was at home in 1738, when he turned eight.[15]

Two years later, Benjamin Franklin entered the stove trade himself, as a middleman for an unnamed ironworks, selling stoves of someone else's design. Amid the alarming news of the freezing cold in February 1741, Franklin's newspaper advertised "Very good Iron STOVES to be sold by the Printer hereof." There is also a hint that the cold was driving up the price of heating fuel. Anthony Morris Jr. advertised coal in the same February issue of *The Pennsylvania Gazette*, explaining that "at 18 pence a Bushel" it was "cheaper Fewel . . . than Wood at the present Rates."[16]

And then, on December 3, 1741, Franklin printed the first reference to his own invention: "To be SOLD at the Post-Office Philadelphia, The New Invented Iron Fire-Places." (Franklin was postmaster; he ran the post office, which means he was using an imperial position to promote a private project.) The *Gazette* also announced that one of the fireplaces was currently "in Use" at the Post Office, where anyone could go to observe it and maybe buy one. On January 20, 1742, "a fresh Parcel of IRON FIRE PLACES" had arrived "from the Furnace," maybe because the previous stock had sold out. It may seem anticlimactic that these brief newspaper entries constitute the birth announcements of the famous stove. But at the least, they testify to Frankin's businesslike view of his science and its ultimate benefits; also to his cannily interconnected activities—printing, postal service, industrial investment, and inventing—through which he made himself into an inescapable Philadelphia presence (and phenomenon).[17]

Although the ads didn't name the stove's manufacturer, it probably wasn't Branson's Reading Furnace but rather the neighboring Coventry Forge, owned by Thomas and John Potts (father and son), also in Chester County. Branson had been, until 1736, a partner in this forge. It's possible that Franklin had formed a connection with Coventry through Stephen Potts, a bookbinder he'd known since 1727 and a fellow founder of the Junto. It was a good connection for both men given that, at the time, printed material was sold unbound—binding cost extra. In the early 1730s, Potts rented a workroom and then the garret at the top of the Franklins' combined shop and home, paying for meals and firewood, indicating he may have lived and worked there; Franklin's accounts record that in 1731, Potts was busy binding one thousand of James Logan's books.[18]

The Pottses' ironworks was a logical choice, anyway, because it already produced stoves. Franklin's business records show exchanges with various Pottses as early as 1735. Although they must have been producing his devices for a year, given Franklin's announcement in his paper, the earliest that Coventry's records document his invention is September 22, 1742; the Pottses shipped "Seven Small New fashioned fire places," billed at £3, 6 shillings, and 2 pence apiece, to two Philadelphia merchants, Israel Pemberton and William Hyme, who acted as middlemen. (This description also indicates that the devices, from an early date, came in different sizes.) If these, like the fireplaces advertised earlier, were priced for sale at £5, each represented a 34 percent markup, presumably divided among the forge, the designated agent of sale, and Franklin (maybe), covering their respective expenses for transportation and advertising, plus a profit. It's possible that the fireplaces Franklin advertised earlier were done speculatively, for him alone, and if at Coventry, recorded in letters not included in the forge's ledgers.[19]

The "fresh Parcel" of fireplaces in early 1742 hints of sales of them in 1741, but later events indicate these had been disappointing, prompting reorganization. (In fall 1742, Coventry launched a "new fashioned stove" evidently not designed by Franklin; maybe that one sold better.) One of the Pottses' business partners, Robert Grace, placed another order for fireplaces, this time from Mount Pleasant Furnace, in November 1742. But then Grace decided to make the fireplaces himself at Warwick Furnace, which had come to him through his marriage to Rebecca Savage Nutt. This decision was likely amicable: the Potts and Nutt families were intermarried; Coventry Forge was, in 1744, described as the joint property of "Grace & Potts"; and all three of the ironworks Franklin would work with—Reading, Coventry, and Warwick—were right next to one another in upper Chester County. (On the map below, find the big "Coventry" at center, then drop down to the left to see "Rob.t Grace," "Coventry Forge," "Warwick Forge," and "Reading Forge," all within walking distance.) Warwick Furnace and Coventry Forge seem to have run a split production, the former making iron from ore that both operations could then fashion into consumer goods with some managerial cooperation and shared labor.[20]

Franklin and Grace, as Philadelphia neighbors, could talk in person about their collaboration, leaving no record of how they reached their accord. It may not be irrelevant that, to set up his printshop in 1728, Franklin had borrowed money from Grace, who was also, perhaps as early as 1739, his landlord. Grace owned the house at 131 Market Street, and nearby property in Church Alley for the Franklins' printshop–post office.[21]

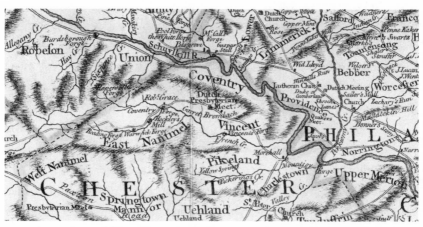

(William Scull, A Map of Pennsylvania *[London, 1775], detail, Library of Congress)*

The house was the largest space Franklin had yet occupied as he moved up in the world, but at the cost of being more expensive to heat. If he was acknowledging a debt to Grace by collaborating on a heating system invented there, the arrangement reflected a real household's real heating problem.

We may only know what this new heating system was because Franklin's modifications made it more expensive—he needed to sell more units than he'd managed through post-office foot traffic and newspaper ads, which, in this era, rarely boosted sales of branded products. Meanwhile, his household and its expenses had grown in 1743 when, after the death of their four-year-old son Francis, he and Deborah had a second child, Sarah. So Franklin in 1744 printed a work, *An Account of the New Invented Pennsylvanian Fire-Places*, to explain his device, deftly merging commercial promotion with scientific experimentation. It's the only account we have of his experiments; if there were earlier notes or diagrams, or rough drafts of his pamphlet, they may have ended up as kindling for the household's morning fire. Franklin seems to have burned documents (everyone did), and in the case of his fuel-frugal stove, this would have been apt and thrifty, if deeply annoying for the historian. Franklin may have cleared his other debts to Grace, because he charged him for all of the pamphlet's printing costs, including its engravings. And he monetized every angle of the project, advertising the pamphlet in his newspaper and publicizing Warwick Furnace on the pamphlet's flyleaf. The pamphlet cost much less than the stove: one shilling, the same as an imported bar of soap. But this was, for Franklin's printshop, comparatively expensive. In 1740, a *Poor Richard's Almanack* cost three and a half pence. A *Pennsylvania Gazette* subscription for the same year cost ten shillings, but this yielded months of information, cheap at the price.[22]

The price of the pamphlet indicates that Franklin and Grace weren't assuming stove sales would subsidize their printing expenses. Rather, Franklin was selling one idea in two media, paper and iron.

The iron version seemed the simpler of the two, a tangible thing (in three handy sizes: small, medium, and large). The essay was more cerebral, describing the stove's genesis in a series of scientific investigations, disciplined examinations of heat, complete with scholarly citations—all this being an advertisement not just of the stove, but of Franklin as a man of learning. To illustrate the device's operation, the pamphlet had two engraved plates, the main reason for its relatively high cost. A Pennsylvania draftsman and cartographer, Lewis Evans, had drawn the diagrams, which were engraved in Boston. Franklin later said the "Pamphlet had a good Effect," suggesting its embedded science was a selling point, as he must have hoped it would be.[23]

Both Franklin and Grace were interested in science. Grace had studied works in metallurgy, donating to the Library Company his copy of Nicolas Gauger's work on heat and heating systems, in its second English edition (1736) as Desaguliers's translation, *Fires Improv'd*. Franklin may have consulted this very copy, though he also knew at least some of the original French work, probably from James Logan's copy, an edition printed in the Netherlands.[24] In any case, Franklin put his *Pennsylvanian Fire-Places* in dialogue with the three texts he'd studied, by Gauger, Desaguliers, and Clare. To show his fluency in the sciences of fluid substances, including heat and air, Franklin credited these works. But he had new points to make and, in the end, diverged significantly from their interpretations of heat and climate, especially their assumption that the climate didn't change over time.[25]

At the very start of *Pennsylvanian Fire-Places*, he explains that "in these Northern Colonies the Inhabitants keep FIRES to sit by, generally *Seven Months* in the Year; that is, from the Beginning of October to the End of April; and in some Winters near *Eight Months*, by taking in part of September and May." He followed this warning about a possible new normal with a prediction about the long-term consequences of heating. "WOOD, our common Fewel, which within these 100 Years might be had at every Man's Door, must now be fetch'd near 100 Miles to some Towns, and makes a very considerable Article in the Expence of Families." The mile-per-year calculation was ominous. Philadelphia was about 100 miles from the northern part of the Walking Purchase, on the Delaware River. Franklin was warning that "Penn's" woods might not keep up with colonial settlement, especially given the threat of long, cold winters.[26]

Note here that Franklin was questioning the hypothesis that cutting down forests would warm North America's climate. Against an overwhelming colonial tendency to chop away woodlands, and despite suspicion they fostered a colder climate, Franklin recommended conserving trees, not least for future fuel, especially in a region with high energy demands. His time-space estimate reveals an acute awareness of resource scarcity. Forest conservation would become more common in the late nineteenth century, but Franklin's fireplace pamphlet was one of several publications in which he began to warn settlers, much earlier, about the wood supply. "As therefore so much of the Comfort and Conveniency of our Lives, for so great a part of the Year, depends on the Article of FIRE; since Fuel is become so expensive, and (as the Country is more clear'd and settled) will of course grow scarcer and dearer," he said, "any new Proposal for Saving the Wood . . . may at least be thought worth Consideration."[27]

Burning less fuel was the obvious solution, but if that left people shivering, it wasn't optimal, not in an era when consumers expected improvements. Franklin was aware of the other variable in the equation: all those recurring bad winters. He was so fixated on those that he gave the wrong publication date for Gauger's *Fires Improv'd*, saying it was 1709. The book only appeared in 1713, but Gauger had said the sharp winter in 1709 (when the Venice lagoon froze) had inspired his research—that's the part Franklin remembered. But how could adequate heat be created if winters were becoming worse and lasting longer? Some mechanism was needed, obviously, that would increase a fire's ability to generate heat within a closed space. That was Franklin's real goal: making an entire room warm—but with less fuel.[28]

Franklin did his foundational investigation into that problem in the big house on Market Street that cost more to heat. Most likely, his lab was the ground-floor sitting room or parlor, probably running to most of the building's full interior width and half its depth. Insurance records reveal that the house's exterior was 17 feet wide by 45 feet deep. If its floor plan was typical of other houses of this era, the ground floor might have had a hallway to the side. So, the main sitting room (at the front) is plausibly estimated as 12 by 20 feet, allowing for wall girth and the hallway, a total of 240 square feet—an approximation, but a reasonable one.[29]

Franklin began with a basic question: How does a fireplace heat a room? That required explaining two essential physical components, "AIR and FIRE," the latter a contemporary synonym for heat. Franklin assumed the room was full of air (one hopes so), which he thought was an elastic substance of loosely distributed material particles (now most commonly called a "gas") that, if heated, expands. A room felt warm if it had warm air throughout it, not just near the fire; in other words, the room needed some movement of the air, both to circulate heat and to keep the fire going and the people respiring as they enjoyed the warmth.[30]

Feeling was everything. The lack of any description of scientific instruments in Franklin's essay is striking. In contrast to Gauger, he never uses the word "thermometer" (or "barometer," for that matter). This is a contrast, as well, with Martin Clare's analysis of fluid substances, including air—Clare had a whole chapter on thermometers. Instead, Franklin tells the reader to use common household items to visualize how air expands when heated and contracts as it cools. "Take any clear Glass Bottle," put it "before the Fire, and as the Air within is warm'd and rarified, part of it will be driven out of the Bottle." To see that happen, the heated bottle could be turned upside down and its mouth lowered into a container of water. "As the Air within cools and contracts you will see the Water rise in the Neck of the Bottle, supplying the Place of just so much Air as was driven out." A similar device could be rigged up from an animal bladder that, partly inflated, firmly tied, and placed near a fire, will blow up tight as its air expands.[31]

Franklin proceeds from those small containers to a whole "Room," one supplied with a fireplace, a chimney, and windows and doors that open and close. This is not a hypothetical space but a real one, plagued with chinks and drafts and therefore complexity. That had the advantage, at least, of revealing the interaction of cold "fresh Air coming in thro' Doors and Windows, or, if they be shut, thro' every Crevice . . . as may be seen by holding a Candle to a Keyhole." Those drafts would continually cool the air in a room, but by replacing any hot air that sought the mouth of the fireplace, they also sent the smoke gathering there up the chimney; without those cooler eddies, "the Smoke being no longer driven up [by the draft] must come into the Room."[32]

Oh, that smoke. Fire's irritating by-product would end up fascinating Franklin nearly as much as heat. He would always distinguish smoke from heat and from air, even though it existed because of them. Any fire would emit heat, but also light and smoke. Heat and light, he thought, moved quickly and in straight lines. This was in contrast to smoke, heavier because it "is but just separated from the Fuel," and only moving, in rippling streams, if carried by the heated air. Otherwise, smoke would linger, even around the fire, which it could "stifle." In this way, Franklin considered smoke a suspension of matter "just separated" from whatever was burning, interacting with air but no longer just air, not even the heated air that was the essential part of any domestic warming system.[33]

These physical patterns could be verified simply by feeling the warmth of the (non-smoky) air, which was of medical benefit. In using his parlor and his body as the units of verification, Franklin made an important statement: science is not removed from lived reality—it describes phenomena relevant to human life. The human body was the best thermometric gauge, after all, if the whole point of indoor heating was furnishing comfort, particularly during cold snaps. As Franklin was careful to add, his new fireplace had been "experienced now three Winters by a great Number of Families in Pennsylvania," a description that again indicates the spread of Franklin's first model in 1741.[34]

Inexpensive, everyday comfort, despite the bad winters, was part of Franklin's idea of improvement, with fuel conservation an integral part of a better life for colonial communities. To justify his own invention, he described six older designs, dwelling on their drawbacks—above all, their inability to heat much of anything unless fed absurd amounts of combustible stuff. To make that point as clear as possible, Franklin listed the competitor devices in terms of their fuel efficiency, least to greatest, worst to best.

First and most wasteful were the "large open Fire-places used in the Days of our Fathers," still common in kitchens but also in the "Country," anyplace near woodland and therefore cheap firewood. In their classic form, these fireplaces were like miniature rooms within rooms (called "inglenooks" in Scotland), providing a space where two people could sit, each taking an outside corner, or for a cook to have ample elbow room while tending roasts, pots, and pans over the fire. In practice, such fireplaces were often too hot for comfort; meanwhile, the surrounding room remained dismally frigid. Plus, this kind of hearth needed a large chimney funnel to remove all the smoke, though this would also evacuate much of the heat. To dilute any lingering smoke, a door had to be kept at least slightly open, exposing the room to renewed blasts of cold. Screens or settles with tall backs might protect people from drafts, but darkened the room and made it into an obstacle course. "I suppose our Ancestors never thought of warming Rooms to sit in," Franklin concluded; "all they purpos'd was to have a Place to make a Fire in," to go warm themselves at, or cook over, coughing in the smoke, banging into furniture in the half-light.[35]

The second kind of fireplace Franklin assessed was better because it was smaller—less like a miniature room. The open mouth of the fireplace was both lower and narrower, with upright jambs inserted on each side. This was a design so successful that, though recent, it had become the standard, with consequent reduction in the use of bricks. Franklin spent the most time on this model, mostly because he found it such an exasperatingly small step in the right direction. But he clearly knew he was describing a kind of fireplace most Pennsylvanians were familiar with. True, this new type used less wood. And it sent more smoke up the chimney, though only by generating a vacuum that sucked cold air into the room from any chink or crevice. The local heat these fireplaces supplied was a dubious gain—"a Man is scorch'd before, while he's froze behind." All manner of drafts came "Whistling or Howling" into rooms, which was uncomfortable, and a sign that fuel was wasted, its heat constantly diminished.[36]

The drafts were also, Franklin thought, likely to make people ill. In a footnote, he admitted that he was "neither Physician nor Philosopher," so offered the reader "the Authority of some that are so," from Britain to China. (Note: none of these authorities were from the Americas.) He began with his fellow British subject, Clare, citing him on the bad practice of sitting near a window or door in a room with an ordinary fire (or even a great many candles), because the heated air's rising motion would pull cold drafts against their bodies. He also cited an Italian physician on how, during a cold and wet winter in Venice during the 1600s, many people became sick. This was in contrast to the German-speaking lands, where people kept their entire bodies warm by luxuriating in "stove rooms" heated by their greater abundance of wood compared to Venice. Meanwhile, "the Spaniards have a Proverbial Saying, *If the Wind blows on you thro' a Hole, Make your Will, and take Care of your Soul.*" Colds could severely affect the teeth and jaws, Franklin thought. Drafts had "destroy'd early many a fine Set of Teeth in these Northern Colonies."[37]

Above all, this kind of fireplace was wasteful. "The greatest Part of the Heat from the Fire is lost" straight up the chimney, and some of the rest absorbed by the masonry of the hearth and jambs, which had little ability to radiate heat. "Thus Five Sixths at least of the Heat (and consequently of the Fewel) is wasted," Franklin scolded, "and contributes nothing towards warming the Room." Meanwhile, the pointlessly roasting temperature right in front of the fire, so tempting to anyone in the otherwise chilly room, would damage their eyes and shrivel their skin, even as the cold drafts sneaking up on them from behind delivered other miseries.[38]

Hence the obvious appeal of the third kind of warming technology Franklin discussed, an early attempt at forced-air heat. This was Gauger's design, which inserted a set of iron plates into an existing chimney. These rectangular baffles slowed leakage of heat up the chimney, and collected it over an extended, folded-up surface area that then kept heating air, which, as it expanded, would push through vents into the room. This was an excellent idea, Franklin allowed, though the necessary ironwork was expensive to make and install, discouraging its adoption. (Case in point: this was the very system wealthy James Logan had at Stenton.) Plus, Gauger's design still let a lot of heat drift up the chimney.[39]

The fourth heating design was much older, being the closed iron stove traditionally used in the Netherlands. (They were common elsewhere in Northern Europe, though Franklin didn't note that.) He praised the device's ability to make "a Room all over warm," because this design, with its closed box and connected metal flue, prevented drafts from entering the room while expelling the smoke. Also, heat issued from all six iron plates of the stove. The only drawback was the stove's front door, intended to be shut for optimal operation. This prevented any "Sight of the Fire, which is in itself a pleasant Thing," and without which people didn't know when to add fuel. Also, because the fireplace had a narrow top, and functioned best with its door closed, its heat could not be used for other purposes, such as warming food or drink. The Dutch stove had none of the drawbacks of the old open fireplaces, but none of their cheery advantages, either, particularly for "the English (who love the Sight of the Fire)."[40]

Fifth, and with little patience, Franklin assessed "the German Stove," describing it as "like a Box, one Side wanting." It was made of five iron plates "scru'd together" and mounted against a hole in a wall, with fuel fed into a fire on the other side. "'Tis a kind of Oven revers'd," Franklin explained, and "certainly warms a Room very speedily and thoroughly with little Fuel." With its open mouth in another room, it neither emitted smoke nor tempted cold drafts through cracks and crevices. On the other hand, its dancing flames were as invisible as those of Dutch stoves, making it tricky to keep fueled, and it did not allow any circulation of air. People shut up with such a stove breathed "the same unchang'd Air continually, mix'd with the Breath and Perspiration from one anothers Bodies," which was unpleasant and unhealthy.[41]

Finally, Franklin described how simple "Charcoal Fires, in Pots," were sometimes useful, mostly in workshops that might lack fireplaces and chimneys. These portable braziers could warm a reasonably sized room with a small amount of fuel, but without the necessary draft a chimney created, the air did not circulate, and the accumulating fumes could be unpleasant, unhealthy, and sometimes—"when the Door is long kept shut"—fatal. Of the six designs Franklin listed, these cheap pots of charcoal were the most efficient in terms of fuel consumption, if potentially lethal.[42]

Franklin omitted one heating improvement, so well-known in Pennsylvania it's obvious he'd decided to ignore it: the fireback. Iron firebacks were certainly cheaper than full stoves and much easier to install than anything like Gauger's baffles. Cast with moralizing images (Cain smiting Abel, sinfully; England's heraldic lion and unicorn rearing up, patriotically), they were very common, as their frequent survival in museums indicates. Were they notably efficient, compared with the other designs Franklin surveyed? He refused to say. His silence on a piece of technology associated with Germany matched his claim that the disappointing two-room fireplace, fueled on one side of a wall, was "a German stove." It was used elsewhere, as Franklin knew; it appears in the English book he confessed was a childhood favorite, *The Pilgrim's Progress*. And firebacks had in fact been first used in England and France, in the fifteenth century, before moving into the German-speaking lands.[43]

Franklin's motive, of course, was to alert his readers to salient features (and problems) of indoor heating, not as a full survey, but selectively, to sell his fireplaces. This was somewhat awkward, given that the Potts and Grace ironworks, Coventry and Warwick, produced several of the very devices Franklin disparaged. As well as firebacks, the Graces' Warwick Furnace was, by the 1740s, producing stoves in three sizes. Their design was unspecified, but in 1744 Grace was maker or middleman for a "Dutch" stove and in 1745 paid "for two Dutch Stove Moulds," wooden templates for sand-cast stoves; if "Dutch" meant Pennsylvania-Deutsche, these were presumably the German designs Franklin disdained in favor of his own.[44]

His "PENNSYLVANIAN FIRE-PLACE" consisted instead of eight iron plates that, installed in an ordinary brick fireplace and chimney, would avoid the drawbacks but "retain all the Advantages of other Fire-places" he'd evaluated. The two principles of his fireplace were convection, the tendency of fluid substances to rise when heated and sink when cooled (which could channel smoke out of a room), and the expansion of a stove's heating surface using baffles. These would support the main goals for all the stoves Franklin invented: maximizing heat and minimizing smoke.[45]

His first step was to explain how the eight plates fit together, providing the kind of diagram that anyone who's assembled an IKEA bookcase might find familiar. But this item's assembly required a specialist workman. Three of the thirty-five pages of Franklin's essay are "Directions to the Bricklayer," and the second-person address and chatty detail express confidence that this person is intelligent and literate, a trowel-and-chisel-wielding amalgam of science and workmanship. (In 1746, Franklin would sell a copy of Boerhaave to a bricklayer.) The bricklayer would use the iron plates as a template to reconfigure the bricks or tiles of the hearth and chimney, then be on hand to mortar the assembled ironwork into the correct position. If you ever fret over the instructions for assembling the contents of a flat pack, be grateful that at the least, your guide and diagrams don't deliver the abrupt news that, by the way, you need to hire a specialist, and that they don't come with footnotes* citing scientific literature.[46]

Should you ever need to do it, here's how to set up a Pennsylvanian fireplace. You are building two iron boxes that together give warmth and fresh air to any room in need of both.

"To put this machine to work," first, you need a firebox, formed from six of the iron plates, to sit within an existing fireplace. These six pieces handily slot into each other along their cast ledges or grooves: bottom (i), top (vi), back (ii), two sides (two pieces marked iii), and a front piece (v) that doesn't fully cover the mouth but extends partway

---

* At the point where he is scolding the reader about overheating the front of their body in front of an ordinary fireplace, Franklin inserts a long footnote with an extended passage in Latin, et cetera.

THE PENNSYLVANIAN FIREPLACE

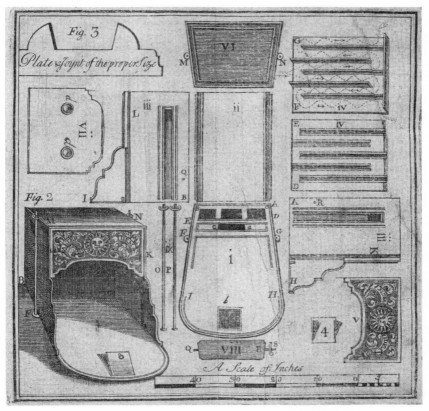

INSTRUCTIONS FOR ASSEMBLY. DIAGRAM FROM BENJAMIN FRANKLIN, *AN ACCOUNT OF THE NEW INVENTED PENNSYLVANIAN FIRE-PLACES*... (PHILADELPHIA, 1744). THIS COPY BELONGED TO FRANKLIN'S BROTHER JOHN.

(The Morgan Library and Museum. PML 64766. Purchased as the gift of Henry S. Morgan in honor of the fiftieth anniversary of the museum's becoming a public institution, 1974)

down it. (The assembled whole is pictured at bottom left of the diagram.) The bottom plate is much deeper than the top one, pouting a curved lip out into the room. The side pieces have slopes, descending from back to front, to connect top and bottom plates of the same width but different depths. The design is such that, with the plates correctly positioned, "a Pair of slender Rods with Screws, are sufficient to bind the Whole very firmly together." (See again the frontispiece in this book—a butterfly screw tops the rod visible at the fireplace's back corner.) Mortar can then seal the edges. Franklin would later explain that this "Firebox" was his original design, as manufactured by Potts and Potts, before the switch to Grace's forge and the addition of a second component, one that added cost but also radiated additional heat.[47]

Franklin's second design had, in addition to the firebox described above, a separate "Air-Box" formed by the two final plates of the "Machine" described in his pamphlet. Each of these plates (the two pieces marked iv at top right of diagram) has baffles or channels cast into it, two inches deep; helpfully, the baffles match up when the plates correctly face one another and can then be cemented with a heat-resistant glue around their sides—not, however, around top and bottom. As with Gauger's (suspiciously) similar design, the baffles extend the metal's surface area so it absorbs more heat. The airbox's warmed air enters the room through vents at the firebox's side plates, requiring no additional energy to force it out, just pressure from the air's expansion. A simple design, though one requiring greater outlay for the two extra plates that go into the chimney. It also requires a chimney of sufficient size and shape—two more reasons for Franklin's detailed, printed explanation.[48]

In addition to the heavy iron pieces, the flat pack includes two thinner plates that could move within the device. One is a "Shutter" (vii), what we might call a damper. Equipped with brass knobs as handles, the shutter slides on grooves along the side plates, either to open the fireplace for use or shut it down for the night. Finally, another light iron plate or register (viii), this one to fit the space between the back plate and the airbox behind it, can be turned with iron rods (also attractively topped with brass, if preferred) to narrow or widen the passage from fireplace to airbox.[49]

To house the assembled firebox and airbox, a fireplace needed two modifications. The bricklayer had to construct a false back and a channel under the hearth, both designed to evacuate the smoke with minimal loss of heated air. Franklin provided an engraving of the chimney and fireplace components in profile, showing the bricklayer the configurations for "a false Back of four Inch- (or, in shallow small Chimneys, two Inch-) Brick-work." That required a wall of bricks (B) built four inches from the actual back of the chimney (E), tall enough to extend above where the hearth opened into the chimney, with a sloping segment of brickwork closing off this new space. Bricks were

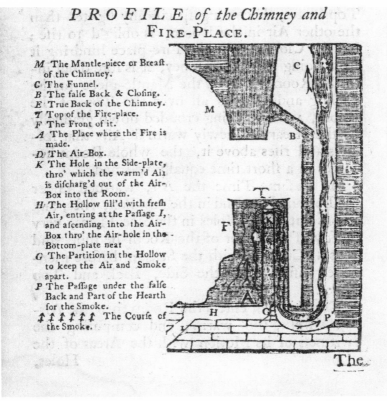

INSTRUCTIONS FOR INSTALLATION. DIAGRAM FROM BENJAMIN FRANKLIN, *AN ACCOUNT OF THE NEW INVENTED PENNSYLVANIAN FIRE-PLACES*... (PHILADELPHIA, 1744). THIS IS ALSO FROM THE COPY THAT BELONGED TO FRANKLIN'S BROTHER.

(The Morgan Library and Museum. PML 64766. Purchased as the gift of Henry S. Morgan in honor of the fiftieth anniversary of the museum's becoming a public institution, 1974)

to be removed from the hearth to form both a space under the bottom plate (starting at I) and a channel up the passage between the chimney's false and actual walls. A thin partition (G) separated the area under the hearth from what was now the flue for smoke.[50]

This profile image of Franklin's fireplace technology shows how he visualized convection and air pressure. This is why his pamphlet, describing the second version of his device, emphasizes that two kinds of air, merely warmed or also smoky, are channeled through distinct parts of the apparatus, including holes in the bottom plate of the firebox, which Franklin calls the "Hearth-Piece." Its curved front edge has a ridge at the front to act as a low fender. At the back of this plate is a long slot where fresh air can enter the airbox once it is heated. Behind this slot are three more, meant for the smoke. The airbox is placed upright over the single slot (see how the lower edges of the two baffled pieces are marked D, E, F, G, corresponding to their positions on the bottom plate). This creates a heat-absorbing barrier. The bottom plate also has a trapdoor (b) with lid (4) that, if raised, sends a draft to blow up the fire, if needed.[51]

Once wood is laid and lit on the hearth plate, its flame and smoke rise faster than any significant amount of air can be warmed. The smoke's increasing pressure will send it, first, down the back of the airbox, then into the three slots in the bottom plate, where it next transits the space between the chimney's false and true backs, and finally, up, and out of the house—good riddance. Meanwhile, warmed air is accumulating and begins to flow into its slot in front of the rising smoke (which helpfully blocks it), entering the airbox. This metal chamber absorbs heat in two ways: from the fire itself, of course, but also from the hot air most fireplaces send straight up the chimney.[52]

The firebox also augments the heat. Its protruding hearth plate lets the fire burn farther out into a room, and all six of its plates collect heat to warm the air, even after the fire dies down. As this warm air becomes "specifically lighter than the other Air in the room, [it] is oblig'd to rise." The hot cloud expands and "rises by the Mantle-piece to the Cieling and spreads all over the Top of the Room." Any further heated air will, being hottest, rise above and push down the earlier layers. "The whole Room becomes in a short time equally warmed," Franklin promised. This is so easily achieved that "when the Plates are no hotter than that one may just bear the Hand on them, the Room will generally be as warm as you desire it." The efficiency prevented mishap. If a lively child or unsteady adult bumped into the insert, they were less likely to be burned than if they collided with an entirely closed and very hot stove.[53]

With his airbox design, Franklin offered heat plus ventilation. In his device, he said, the rising column of smoke sends any currents of warm air seeking the chimney into the slot leading into the airbox. Once enough hot air accumulates there and is further heated, it expands out of "Holes in the Side-Plates" that align with the airbox. This works best if "the Door of the Room be shut," fostering draft-free circulation. Also, a small hole in the firebox's bottom plate generates pressure into the airbox from a rising draft of cooler air, pushing the warmed air out. "By this Means, the Air in the Room is continually changed," he explained, "and kept at the same Time sweet and warm." He calculated that nearly 10 barrels of such air emerged every hour. A barrel held about 36 imperial gallons, so 10 barrels of air was equivalent to about 57.8 cubic feet. A parlor 12 by 20 feet, with maybe a 10-foot ceiling, contained 2,400 cubic feet of air, meaning the 10 barrels' worth of air would refresh 2.5 percent of the room's air per hour. This did not fully ventilate a sealed space, but it regulated temperature and evacuated smoke. Did it work? Yes: the big crack on the base plate of the surviving stove was most likely the iron's reaction to having cool air beneath it and hot above; the design outperformed its medium.[54]

If the main problems of other heating systems were that they let too much heat escape up the chimney, or sucked cold drafts into the room, or kept the room too smoky—or some wretched mix of the three—Franklin believed his device avoided all of these hazards, generating a healthful atmosphere, comfortingly warm, without drafts or smoke, and with abundant fresh air. "Let the Room be made as tight as conveniently it may be," he instructed, so that most air had to enter via the airbox, "by which Means it will not come cold to your Backs"—ventilation and heating combined in what may be the very first statement about making a "tight room" or "tight house," now an industry standard for indoor climate control. Franklin's debts to previous designs were obvious. He had built on the trend to modernize fireplaces with smaller hearths, openings, and chimneys, combined with Gauger's chimney metalwork, therefore proposing a solution that would seem familiar to many consumers, with easily understood improvements.[55]

To his overall design, Franklin added three smaller tweaks. First, his adopters should maintain a square opening for a trapdoor in the chimney for the sweep to use, with the trap angled so that, when opened, it dumped the accumulated soot forward onto the hearth (where it could be swept away) rather than behind the false back, where it would be irretrievable without dismantling the brickwork. Second, for rooms where tobacco smokers congregated, Franklin suggested cutting a hole, five to six inches square, where the top of the chimney formed its exiting funnel. The resulting small draft above the smokers' heads would remove their emissions "and keep the Room clear," also handy for cooling the room should it become too hot—suggesting this possibility was a clever way for Franklin to restate his confidence in his invention. Third, Franklin suggested rigging up the whole room as a heating machine, by mounting its main door with "Screw Hinges, a Spring, or a Pulley" so it would shut automatically, without people having to keep getting up to close it. Also, if the fireplace had any gaps around it, small mesh grates would thwart heat-seeking mice.[56]

A "machine," to us, implies moving parts and action that's at least semiautomatic. That's what Franklin meant, too. He was careful to explain how his heating machine needed to be fed its proper fuel, wood, but also the fuel's necessary medium for combustion, air. With correct infusions of each, the fireplace could be "set," configured to as near a thermostatic optimum as was possible in this pre-electronic age. After all, humans were themselves respiring machines. In winter, they and their fireplaces had to breathe air and consume fuel (food or wood), with Franklin offering a way for them to optimally coexist.

The wood for his fireplace had to be cut to fit its metal hearth. (This seems to have been unusual, indicating regional uniformity in the dimensions of commercial firewood intended for the bigger hearths Franklin wanted to make obsolete.) Dry hickory, ash, or other wood that burned with a clear flame was preferable. The morning's first ignition should be dry "Brush-wood" (twigs) that would catch and burn quickly—"its sudden Blaze heats the Plates and warms the Room" faster than any larger and possibly damper pieces of kindling could do. Once the fire burned briskly, someone could adjust the firebox's sliding "Shutter" and register, watching the direction and quantity of smoke, readjusting to corral the smoke while maintaining a draft to replenish the air bound for the airbox. Together or individually, the two moveable partitions could also adjust airflow to slow the burning of the wood, to damp down the fire for the night, or to ignite it in the morning.[57]

The regulated movement of air through the room, keeping the area both warm and ventilated, evoked what was happening to the human bodies in the room. Again, Franklin never offered thermometrical measures of his fireplace's output, nor any instrumental measurements other than those ten barrels of fresh air per hour. He seems to have owned a thermometer by about this time; in 1751, he would complain that he owned three that never agreed, an expensive way to be misinformed. At his printshop, Franklin sold a thermometer to his pamphlet's illustrator, Lewis Evans, priced at £1 and 4 shillings, about a quarter of the price of a fireplace insert. But Franklin couldn't assume everyone had a thermometer to monitor their heating. No matter, as they could use their bodies to do that.[58]

The first of his fireplace's "Advantages," Franklin claimed, was "that your whole Room is equally warmed." People no longer needed to "croud" around the fire but could sit where they liked, near the window, for instance, and thus during the day "have the Benefit of the Light for Reading, Writing, Needle-work, &c.," with fingers warm enough to write or sew or turn pages. This pervading "Comfort" was especially welcome in large families that couldn't fit around one fireplace. And the diminished draftiness protected anyone anywhere in the room from the maladies—"Coughs, Catarrhs, Tooth-Achs, Fevers, Pleurisies"—that a "sharp Draught of cold Air playing on you" might cause. Candles burned evenly and didn't blow out unexpectedly. Nor was there the persistent smoke that in other heated rooms damaged "both the Eyes and Furniture." The warmth could last all night, diminishing any morning chill, and the fireplace could be made so secure while the household slept "that not one Spark can fly out into the Room to do Damage."[59]

Franklin stressed the healthfulness of the very iron of his fireplaces. Allegations that iron stoves generated foul air were simply false. Citing Desaguliers's experiments with metal and air, Franklin said iron was "one of the sweetest of Metals." A clean stove was "as sweet as an Ironing-Box," a charcoal-filled iron for pressing clothes, "which, tho' ever so hot, never offends the Smell of the nicest Lady." What offended people was not the stoves themselves but the grease from hands rubbed over them, the wax from candles placed on them, and above all the spit too often launched at them as an "inconsiderate, filthy unmannerly" way "to try how hot they are." The burn-off from these substances could indeed stink "nauseously." But, working with the "perfectly wholesome" material itself, ironworkers were renowned for their health, and the enriched water from iron mines was recommended as a medical remedy.[60]

Franklin emphasized the aesthetics of his design as well. With a Pennsylvanian fireplace, "you do not lose the pleasant Sight nor Use of the Fire, as in the Dutch Stoves." And also, unlike those heaters' narrower projections from the wall, his fireplace's deep top plate supplies a handy surface to "boil the Tea-Kettle, warm the Flat-Irons, heat [portable] Heaters, keep warm a Dish of Victuals by setting it on the Top, &c. &c. &c."—too many marvels to list. Once assembled, the new fireplace should be washed, warmed, brushed all over with a solution of rum, water, and powdered black lead, let dry, then rubbed to a "Gloss" so the joints were disguised and the metalwork made to "shine like new Iron." That, plus plastering the false back white and keeping the red brickwork clean, "will make a pretty Appearance."[61]

Nor was Franklin convinced that comfortable rooms made people *"tender and apt to catch Cold."* His family and friends were, if anything, *"actually hardened"* against illness by sitting in "warm Rooms for these four Winters past," as if they were oozy green wood slowly drying out. Franklin's dating here is also helpful, confirming that his experiments with steel prototypes had begun in the winter of 1740–1741 (and that his friends and family were guinea pigs). He denied any heat-induced tenderness on the grounds that a body could adjust to a sudden change of temperature; it was constant exposure to *both* extremes that damaged health, as he, following Desaguliers, had described for anyone baked in front of a fire while slammed with cold drafts from behind. Besides, the Swedes, Danes, and Russians spent their intense winters in rooms "as hot as Ovens" yet were stupendous winter soldiers, marching through snow, sleeping on ice.[62]

Franklin's most important measure of his fireplace was the amount of fuel it used, meaning not much. "As very little of the Heat is lost . . . *much less Wood* will serve you, which is a considerable Advantage where wood is dear." Franklin said people reported saving anywhere from three-quarters to five-sixths of the wood they had once burned; it all depended on how bad their old fireplace had been. His family's "common Room," he said, "is made twice as warm as it used to be, with a quarter of the Wood I formerly consum'd there." He noted that people in the cold northern nations who used stoves instead of fireplaces also used less wood: "Tho' those Countries have been well inhabited for many Ages, Wood is still their Fuel, and yet at no very great Price; which could not have been the case if they had not universally used Stoves, but consum'd it as we do, in great Quantities by Open Fires."[63]

Franklin betrayed only one doubt: his device might, just might, set the chimney on fire. This was a hazard for any fireplace. Rising smoke carried soot, which accumulated in the chimney, requiring removal by a sweep at least once a year. Between cleanings, the sediment (especially where the chimney might narrow into a funnel) could reignite if flames leapt up from the hearth. In an unintendedly comic run-on sentence, with a double set of conditional parentheses nestled inside it, Franklin acknowledged the danger while trying to deny it, extolling his fireplace's mazelike interior, its need for less smoke-producing wood in the first place, and its fail-safe damper and register:

> And if the Chimney should happen to take Fire (which indeed there is very little Danger of, if the preceding Direction be observ'd in making Fires, and it be well swept once a Year; for, much less Wood being burnt, less Soot is proportionably made; and the Fuel being soon blown into Flame by the Shutter (or the Trap-door Bellows) there is consequently less Smoke from the Fuel to make Soot; then, tho' the Funnel should be foul, yet the Sparks have such a crooked up and down round-about Way to go, that they are out before they get at it) I say, if it should ever be on fire, a Turn of the Register shuts all close . . . and so the Fire may be easily stifled and mastered.[64]

Did he convince even himself? Obviously not, as his continuing annoyance over smoke and bad chimneys would reveal.

Franklin's trump card was the "considerable *Publick Advantage*" of his fireplace. "By the Help of this saving Invention," he praised himself, "our Wood may grow as fast as we consume it, and our Posterity may warm themselves at a moderate Rate." That "saving" would extend into the future not only genealogically, but materially; posterity would not be "oblig'd to fetch their Fuel over the Atlantick; as, if Pit-Coal should not be here discovered (which is an Uncertainty) they must necessarily do." Here again was the hope, however alarming to us, that coal saved trees, at least in parts of the British Empire that had coal.[65]

In case these geographic and historical measures weren't enough, Franklin summoned modern experts who calculated public benefits for designated populations. "We leave it to the *Political Arithmetician*," he said, "to compute, how much Money will be sav'd to a Country, by its spending two thirds less of Fuel; how much Labour sav'd in Cutting and Carriage of it," and how much more labor could be shifted to cultivate the cleared land. It was also far more economical, he argued, to generate a warm workshop where people could labor steadily, no longer "oblig'd to run frequently to the Fire to warm themselves." Confident they would agree with him, he left it "to Physicians to say, how much healthier thick-built Towns and Cities will be, now half suffocated with sulphury Smoke, when so much less of that Smoke shall be made and the Air breath'd by the Inhabitants be consequently so much purer."[66]

That scenario, in conjunction with Franklin's goal of making an entire room warm yet unsmoky, underscores his belief that spaces filled with air might be controllable by humans. Here is where his emphasis on health and comfort established a significant metric. Franklin defined the human body and human experience as optimal measures of material conditions. True, by arguing his stove's effectiveness, he accepted that technology could mediate between humans and the rest of nature, one sign of modern confidence that humans can live in controlled conditions. But in an *early* modern era, when any division between people and the natural world was still partial and porous, Franklin defined the human body as a perfect sensor; technology worked based on how it made a person feel—an elemental measure of resilience during climate change, defining human control over an interior space and artificial climate as compared to the overall state of nature.

But surrounding that indoor microcosm, standing between it and the rest of the universe, was the colony of Pennsylvania—the place, its peoples, and their relations with each other, all of which went into forging the Pennsylvanian fireplace. Franklin doesn't say much about that macrocosm. When he mentions the humans warmed by indoor fireplaces, he is describing the white body—not all bodies. Colonists' unease over the prospect of their physical adaptation to American environments had not gone away. For white people born in the colonies, that anxiety was reinforced by concern that Europeans might assume them to be physically similar to Indigenous people or to the enslaved Black people who had also adapted to new-world places. All the more important, therefore, to construct homes that resembled European buildings and in which at least some atmospheric conditions might be manipulated. Strewn over an expanding landscape of settlement, such homes (even rustic cabins) were—like other consumer choices that aligned the colonies with European material life—protective pods where colonists could control their physical surroundings, separated from ambient conditions that Indigenous people, for instance, did not bother to alter.[67]

Moreover, as with many a flat pack and set of instructions, a consumer who bought a Pennsylvanian fireplace learned little about the making of the thing they're left to assemble. The records of the forges Franklin worked with supply that backstory. Behind his short, printed booklet are the manuscript ledgers and daybooks that document who, exactly, made the fireplace inserts, and how. What those records reveal is the iron industry's promiscuous exploitation of every available kind of labor, with strategies that either erased or controlled the Indigenous and Black people who also lived in Pennsylvania.

The ironworks Franklin collaborated with record a receding frontier economy, with a diminishing Native presence. The 1727–1730 ledger for Coventry has a list of about a dozen named Native people who kept accounts at what was essentially its company store. (These stores functioned as rural equivalents to shops in towns.) Two men, Nenockeman and "Daniell," for example, bought tobacco, molasses, gunpowder, rum, a blanket, and a knife and fork, paying in venison and white peas. Another Native man, Josiah, in 1727 exchanged deerskins and venison, also a crafted broom, for a blanket, tobacco, pipes, gunflints, shot, salt, and cloth; his ledger entry specifies his wife and "Boy," maybe so they could buy on his account.[68]

The frontier nature of ironmaking in these early decades is summarized in Thomas Potts's shipment to England, in 1734, of both pig iron and buckskins. A year later, Potts paid an "Indian" for dressing more skins and, in 1737, for the services of an "Indian Doctor." Slightly later, the frontier commodities that arrived at ironworks came from settlers, not Natives. In 1751, John Kent sold two bearskins to Coventry. Entries for deerskins in the Warwick Furnace ledger simply give totals, as in September 1751, one month before Franklin received the plates of his two-box model from Warwick. Steadily depleted of deer, bears, and native trees, ironworks were Europeanized landscapes, restocked with foreign, cultivated plants. Thomas Potts bought thirty young apple trees for Coventry in 1737, for instance, then two hundred more in 1738.[69]

As Indigenous people were displaced, making iron depended on white and Black workers of many types—waged workers, skilled and contracted services, casual labor, indentured servants, convicts, and enslaved people—with some blurring of these categories. Iron proprietors had to pool multiple sources of capital; similarly, they drew on all labor options. In fact, ironworks' records reveal a unifying principle by translating all human effort into energy consumption—and overall consumer behavior—in both ways prefiguring a high-energy, modern consumer economy, though still operating within the constraints of an organic system.

Like Natives, many white people made use of ironworks' company stores, their labor represented only indirectly in the goods they exchanged. Other people appear in terms of specific commodities or tasks. In 1742 and 1743, John Gibbons was a tailor for Coventry Forge, Jacob Wence was a waggoner, David Jones a carter, and Thomas Ford a collier (charcoal burner). John Davis was Coventry's "pork man" in 1745, supplying meat. Robert Ellis was their "lime burner" that same year. Even more fleeting were Barbara Hildebrand and Margaret Enters, hired for the 1742 harvest at Coventry, where some foodstuffs were evidently grown. William Campbell is described as "Th.s Pott's Man," maybe a personal servant doing errands; Jonathan Hackett is listed as a "Servant" for 1744 and Israel Willson as a "Labourer" that year. In such cases, white men received a daily rate for their work, except in the case of indentured servants, as Campbell and Hackett may have been, who were paid only if doing something beyond contracted expectations. Other establishments likewise used indentured servants for ironwork, and convicts as well.[70]

Conversely, some white men received wages or even salaries. Samuel Meridith, for example, was a "Manager" circa 1744. Some skilled employees listed family members on their accounts, typically their wives, acknowledging their higher level of economic value, also their status as heads of households. In 1743, Martin Kingelo was not just an occasional transporter of goods but "our Carter," and paid in cash—his wife came to collect that. Other men's accounts more passively registered the families they supported. A collier at Warwick in 1742 who may have been supplying both that estate and Coventry with charcoal around the time Franklin's fireplaces were made, deducted family-sized units of food from his payments—four bushels of wheat, a side of beef, a peck of salt—plus sundries such as pins, thread, stockings, and shoes for a man and a boy.[71] John Mills was entered into a Coventry ledger as hammerman, with wages at roughly £14 per month, for a total annual value of £168, recognizing the skill necessary to operate or supervise the forge's machinery. About a decade later, Franklin's rent in Philadelphia (paid to Robert Grace) would be £55 per year, which gives a sense of Mills's purchasing power. His income supported a wife and possibly other members of his family; his account has deductions from his wages for buying an almanac, some ribbon, an ivory comb, and women's stockings and shoes.[72]

Listed down to the halfpenny, the energy exchanges of these transactions were often explicit. Certainly, the Warwick collier demonstrated this by swapping charcoal for human fuel—wheat and beef. In 1744, a white man, Jacob Johnston, was charged for "6 Days Diet When not Worked," indicating that he was free, and therefore responsible for his own subsistence when not producing value for Coventry; he was also charged for shoes, cloth (including fancy damask), rum, a knife, and other items. Against those charges on his account, Johnston received extra pay for processing iron ore at night, maybe when getting the product ready for a buyer was critical, with Johnston paid more to buy more food, or damask, or whatever he pleased.[73]

Despite the iron industry's use of waged or contracted labor, it expanded precisely with the colony's reliance on enslavement. From 1725 to 1750, up to half of the people at some Pennsylvania ironworks were enslaved. When Grace's mother-in-law, Anna Nutt, advertised a quarter share in her ironworks in 1744, it included a quarter share in the estate's "two able Negroes." Her example in particular shows that the industry's economic opportunities for white women depended on enslavement. When he died in 1752, Thomas Potts, who had cast the Pennsylvanian fireplace's first model, owned eleven enslaved people.

(His son, John Potts, would also list eleven enslaved workers in his will of 1768.)[74]

The number of slaves in the iron industry in fact surpassed those listed in the ledgers. Some lived elsewhere because they were enslaved by other white people who collected payment for their work. Colebrookdale Furnace, another Potts property, listed nine Black people there, but also hired Ishmael, deducting provisions each day he worked.[75] Coventry Forge recorded six Black people, all male, with George and "Little George" possibly father and son, which implied the existence of at least one woman (partner and mother) who may not have worked for the forge—many Black people had to find partners at some distance. In another case, a man named Will, described as the property of a white man, was hired as a carter for Coventry Forge, where his wife bought rum and butter on account.[76]

Those consumer items are but one example of how ironworks' proprietors assigned enslaved Black people two competing identities: chattel and freelancer. Colonial law and industry practice both facilitated this. Enslaved people could, in Pennsylvania, own property (and also learn to read), which embedded into their lives some dimensions of apparent choice. And ironworks tended not to have fixed assignments for workers, who were multiskilled and multitasked. Whenever advantageous to them, iron proprietors solicited semiautonomous work, and sometimes treated enslaved people, like Will and his wife, as consumers comparable to free people. These were choices only in the absence of any law that regulated contingent work. Most workers, including enslaved people, did tasks tangential to production of iron—in essence coerced gig work, an extremely precarious economic condition.[77]

Such arrangements show a complex division of labor across the ironworks landscape. John Potts paid Jonathan Robinson for "Negro Dick Driving Team," doing deliveries and hauling. In this instance, Potts also referred to the driver as "Negro Dick Privateer," as if Dick negotiated his terms of employment, raising the question of whether

Robinson split Potts's fee with him. Warwick subcontracted the work of getting wood as well as charcoal, paying John Lack, a white woodcutter, in 1746 for one delivery. Likewise, Robert Grace paid a white man "for ye Negroes Washing," presumably because the ironworks lacked the water or the labor to do laundry. Meanwhile, Ben at Coventry Forge earned income by making and mending clothing that might ultimately get washed off-site. Altogether, the decisions assumed some Black autonomy, though with lesser rewards than for white men who could contract for larger units of effort.[78]

And yet Black people did significant semiautonomous labor. At Coventry, Anna Nutt noted in 1749: "Cash gave the Negros." By definition, this could not represent wages; Coventry's ledgers describe everyone as either "Negroes" or "Servants," as if they were distinct categories. But paying enslaved Black people in cash hints that they earned money not only through informal tasks but in recurring functions, most often by cutting wood. This was definitely a Little Ice Age adaptation, in which fuel for everyday life and for burning into charcoal was a constant presence on ironworks. Moreover, the records indicate that enslaved people, like white workers, purchased food and clothing above whatever was supplied—the imperative to keep fed and warm in cold weather affected everyone. At Warwick Furnace, items listed under Black men's accounts include sugar, flour, dried beef, and liquor, indicating that their subsistence did not include these items, or not enough, either for the men or for their families.[79]

The clothing Warwick's enslaved men bought followed a similar pattern. In 1744, both Cesar and Streaphon purchased (with money or credit accrued through extra work) osnaburg cloth, a coarse fabric for work clothes. It's not clear whether this was for themselves or for family members who also needed tough garments. (The accounts for 1744 list both an old Cesar and a young one, "Junior," possibly father and son.) They also bought shoes, caps, and "coating," woolen cloth that could be tailored. Again, the record does not state whether this supplemented what their enslavers provided or was for dependents who received nothing directly. Either way, the layered clothes made from multiple fabrics shows concern to keep warm in this era of bitter winters. So did Cesar's purchase of "thread and Mohair" to repair clothing, as did a charge to Streaphon for a tailor "Making a pair of Trousers."[80]

Evidence of consumer choice is most apparent in the silk handkerchiefs Cesar and Streaphon bought. Whether for themselves or loved ones, these were not workwear. True, a previous year's handkerchief could mask against dust or smoke, wipe away sweat, protect the head, or add a layer in cold weather. But such kerchiefs primarily announced social rank and respectability. A cravat tied around the neck, like a necktie today, indicated a man's status. A woman who wore a kerchief folded, shawl-like, to cover her neck, shoulders, and chest likewise signaled her modesty and social repute. Most white people did not think enslaved Black people were members of respectable society. A silk kerchief argued otherwise. Thus, enslaved people used consumer goods to announce personal identity, dignity, and aspiration.[81]

The ironworks also document the personal care that Black people gave to others—the proprietors subcontracted this work, too, outsourcing any anxiety connected to it. Ishmael, the Black man hired to do work for John Potts, was married and it was his wife (unnamed) who in 1740 was paid for nursing "Violett." Whether this person was ill or injured, or an infant needing breast milk, is not specified, nor is her relation to anyone listed in the accounts. Jenny received multiple gallons of molasses in 1742, hinting she may have been cooking for the workers, though maybe for a family. Likewise, in 1736 Cesar was paid for "digging Bettys Grave." These entries reveal the presence of women on ironworks, and, by implication, children, with families and networks of care that rarely interested enslavers.[82]

The indifference is proof that, whatever precarious autonomy Black workers were permitted, it was opportunistic, exploiting a colonial economy's historically unusual rise in economic growth but not designed to lead away from enslavement. Franklin's fireplaces were forged in decades marked by very low rates of emancipation. Enslaved people had to free themselves, surmounting the labor system's ambiguities and categorizing themselves as autonomous—the colony's history shows both the strict dimensions of an organic energy system and the human will to transcend those limits. Thus, on February 18, 1746, five years after the Pennsylvanian fireplaces had gone into production, Streaphon had accumulated £58 and 20 shillings on his account with Robert Grace and, with this, freed himself. As was not uncommon, he continued to work for Warwick. The forge's accounts for 1747 list him alongside the still-enslaved Cesar but specify that he received "Wages"—a racist ambiguity in which Grace regarded Black workers as categorically distinct from white ones (who were listed separately), even when one of them had just rendered him a payment greater than Benjamin Franklin's annual rent.[83]

Some contradictions in American history are ironic, unless they're tragic. It was wrong for white colonists to enslave people, but especially when they were so busy making claims to their own liberty. Iron may have been forging prosperity for many Pennsylvanians, putting the colony on the map of global economic consequence and laying foundations for US industrial might, but those stirring historical circumstances didn't improve the lives of many of the people involved. On average, white colonists were better off than their European counterparts, but this was, in part, because the wealth they were creating was shared unequally, if at all, with Native and Black people.[84]

Not far from the Chester County ironworks that made Franklin's stoves, John Patton's establishment in Berks County made a highly decorated fireback that celebrated freedom. The design shows Fame blowing a trumpet and bearing a cap on a pole, an ancient Greek symbol of emancipation. She flies away from a personification of Britain, bearing shield and spear, and toward a Native man, with bow and arrows. And, lest anyone gazing absently into the fire miss the point, a banner above Fame proclaims: *BE LIBERTY THINE*. It's doubtful any Natives displaced from Berks County appreciated the sentiment, nor the people we know Patton enslaved, whose work produced the iron used in the firebacks, if not the plates themselves. (The fireback is now at the Winterthur Museum in Delaware, set inside a mantelpiece bearing Benjamin Franklin's profile.)[85]

Although Pennsylvania's iron industry hovered between colonial enslavement and modern wage labor, it wasn't inevitably moving away from one and toward the other. Nor is it correct, however, to con-

IRON FIREBACK, FEATURING LIBERTY, MADE BY
ENSLAVED PEOPLE IN PENNSYLVANIA.
*(John Patton, Berks County, 1762–1764. Bequest of Henry Francis du Pont,
Courtesy of Winterthur Museum, Garden and Library)*

sider slavery a problem of its time, whose faults were invisible at the time—meaning to white people, obviously not to Streaphon (or Cesar). Rather, many white people preferred not to make slavery and its faults visible. "The Writer of this," Franklin said of his stove pamphlet, had "been present at a Furnace, when the Workmen were pouring out the flowing Metal to cast large Plates." Enslaved people may have been some of those workers, or busy elsewhere at the furnace, yet they go unmentioned. Most colonial American iron artifacts now displayed in museums should disclose they were made with enslaved labor, including Patton's liberty-bearing fireback—and the sole surviving Franklin stove that opened this book.[86]

Always, when a technical device promises to make life better, it's useful to ask: Better for whom? Long celebrated for his late-life contributions to white abolitionism, Franklin is now also known for personally enslaving several people and for profiting from printing "runaway" advertisements for Black fugitives in his newspaper. To these corrections, add another. Benjamin Franklin, the genius of heat, used the work of enslaved Black people to make himself famous, but never spoke of their work in forging his stove project or expressed regret for exploiting their labor, which was likewise essential to Pennsylvania's industrialization. Franklin's acceptance of slavery while he promoted the region's iron industry is yet another twist in his efforts to adapt an organic economy to climate change: he modernized consumers' energy efficiency while using an ancient form of labor, an adaptation he would eventually criticize, though without ever divulging the advantages he had gained from it, from keeping warm at home to becoming a scientific expert.[87]

4

ATMOSPHERES

Franklin's stove pamphlet was his first bestseller, the first of his writings to be translated and printed abroad. This was personally gratifying for him, and the perfect introduction to his more ambitious scientific writings, including his celebrated work on electricity. His pamphlet registered his interest in atmospheres even before he realized it himself. The word that appears most frequently in the pamphlet is "air" (ninety-eight times), with "room" in second place (seventy-five) and "fire" turning up third (sixty). The heat mattered, of course, but the tirelessly pirouetting air in the room turned out to be the star of the show. Having turned the outside in, using natural resources to warm his parlor, Franklin redeployed the insights he'd gained from that set of investigations to turn the inside out and visualize, for himself and others, how air circulated in his hemisphere's atmosphere.

His practical efforts would be all the more convincing because he connected them to universal principles of nature, and vice versa. It turned out to be compelling, sometimes even pleasant, for people to think of themselves as living in nested sets of atmospheres, natural and artificial, with an ability, therefore, to make themselves comfortably at home in the world wherever they were. Though, in the end, this raised the question of how human-generated climates might be affecting the natural one, with implications that weren't quite so pleasant.

The first record of Robert Grace's producing the Pennsylvanian fireplaces at his Warwick Furnace is from June 12, 1745, which had given Franklin's 1744 pamphlet a good interval in which to circulate. Grace produced one ton of fireplaces in 1747, then seven tons in 1751. This was a substantial increase in itself, and a leap up from the Potts and Potts 1742 production of seven stoves, weighing a total of 2,106 pounds, slightly more than one ton. Grace may have used that initial total to test the market, then raise production, the results measuring the difference between newspaper advertisement (for Potts and Potts) versus a promotional pamphlet (for Grace). The sevenfold increase in sales by 1751 was also interesting for its timing, itemizing fireplaces sold between July and October; clearly, people were planning ahead for winter. That was true of Franklin himself. On October 24, he paid for "4 Plates of ye New Fashion Stove," maybe two sets of airbox components to update two older firebox models (though possibly for another kind of stove entirely).[1]

From his shop and post office, Franklin helped expand the sales network for his fireplaces. Between 1744 and 1748, he sold several of them, plus items for their assembly. Grace's distributors, moreover, included several men with ties to Franklin, including two of his brothers, Peter Franklin (Newport) and John Franklin (Boston), and at least two fellow printers, William Parks in Williamsburg, Virginia, and James Parker in New York City. Franklin sold stoves directly to men with whom he did business, including the Pennsylvania papermaker William Dewees Jr. The New England sales were unusual, given the cost of shipping big iron objects. In Newport, Rhode Island, Peter Franklin bought eleven of the largest stoves in 1744 at £15 per unit, but his selling price of £27 and 10 shillings reflects a notable markup to cover transportation. In twenty years, he managed to sell only two stoves. Meanwhile, in Philadelphia, Grace's smallest stoves were, at £5, priced to move.[2]

It's harder to determine how Franklin's fireplaces competed with other heating technologies in Pennsylvania. Branded products, linked to specific makers, were not yet common. In eighteenth-century London, for instance, the only really successful consumer items with brand identities were medications. Only toward the end of the century would brand identity sell household goods, as with Josiah Wedgwood's famous pottery, and only in the nineteenth century would "patent" stoves be commercially successful.[3]

In Philadelphia, Franklin was competing, moreover, with many other kinds of stoves, plus firebacks. That multiple museums in the United States display numerous early American firebacks makes obvious their greater popularity. They were much cheaper than a full stove, generating less income for an ironmaker, but more affordable for consumers and therefore more commonly sold, at least in the first half of the century. In 1734, a resident in Germantown asked his brother, who was preparing to leave Silesia for Pennsylvania, to bring an entire five-plate stove: "they are dear here." The fact that full stoves were more quickly adopted for businesses and public buildings indicates that ordinary householders were reluctant to take on their initial expense, whatever their power to offset the cost through fuel savings.[4]

Franklin left a couple of hints that he didn't intend to compete in the local market on its own terms anyway. The ornamentation on his fireplace's front plate was evidence of this. The plate had foliage, somewhat like oak leaves (from the source of the best fuel wood), surrounding a fully risen and smiling sun. The design was different, and probably deliberately so, from traditional firebacks and even many stoves, with their scriptural and classical vignettes. The Pennsylvanian fireplace, with its symmetrical and ornamental design, instead resembled stylish English furniture. The foliage adapted rococo elements inherited from the seventeenth century; the sun resembled the many French ornaments honoring Louis XIV, the Sun King, which made their way into English designs and survived into the eighteenth century. But such a design for fireplaces was part of a new trend. Franklin's sun was copied on New England firebacks by the 1760s. Gradually, even German American firebacks adopted designs similar to Franklin's (and English style more generally), using rococo and secular images instead of themes from scripture and history.[5]

Even more ambitiously, the iron sun on the Pennsylvanian fireplace announced a kinship to the fire burning below it on the hearth. And for anyone who somehow missed the point but knew Latin, there was an accompanying motto, in effect a caption: ALTER IDEM, "Another, the same." In his pamphlet, Franklin printed some verse provided by a "Friend":

> ANOTHER Sun!—'tis true; but not THE SAME.
> Alike, I own, in Warmth and genial Flame:
> But, more obliging than his elder Brother,
> *This* will not scorch in Summer, like *the other*;
> Nor, when sharp Boreas chills our shiv'ring Limbs,
> Will *this Sun* leave us for more Southern Climes;
> Or, in long Winter Nights, forsake us here,
> To chear new Friends in t'other Hemisphere:
> But, faithful still to us, this *new Sun*'s Fire,
> Warms when we please, and just as we desire.

Given the extreme winters people had just endured, plus lingering suspicion of bizarre American climates, the verse offered powerful assurance that humans could correct for nature's flaws, either hemispheric or historic. Four years later, Franklin would deliver the solar analogy as a joke in his *Poor Richard* for 1748, mocking the exploded notion that the sun must orbit Earth. Anyone who believed that was like "a whimsical Cook, who, instead of Turning his Meat in Roasting, should fix That, and contrive to have his whole Fire, Kitchen and all, whirling continually round it."[6]

But if the sun beams over an entire whirling planet, the fireplace it adorned was specifically named *Pennsylvanian*. At the very least, and logically, this warned that adaptation to harsh weather needed to be tailored to particular places, including their climates. That certainly expressed a valuable insight. But if the fireplace was Pennsylvanian in its material impact, saving "Penn's trees" for posterity, whose trees were they really, and whose posterity was protected? Franklin accurately identified climate change and resource depletion as problems. But this carried an assumption that scarcity's cause—white settlement—didn't matter if there was a technical fix for it, making an emergency into an opportunity for some people at the expense of others, in this case, especially, the Lenapes.

Franklin's strategic disclosure about using a room in his own house to demonstrate his fireplace's powers made his device seem plausible and inviting, but not universally. His intended audience was people who also had that kind of indoor space and could make discretionary decisions about making it more comfortable. Those people were growing in number. Franklin was right about the accumulation of better-off consumers and therefore their impact on the fuel supply. (Affordability is still a factor in making changes in energy use, and the wealthiest consumers still have the greatest impact.) But was he right to suggest that technology could fix that?

Was he "right" in either of two senses, factually or ethically? In terms of the latter, Franklin's pamphlet defined the benefits of his stove only for Pennsylvania's free white population. And yet, in his newspaper, Franklin had quoted Native testimony about Pennsylvania winters as valid data about the past. And in the treaties he printed, he cited Indigenous insistence that the council fire bound them to colonists—if colonists agreed. All the more striking, therefore, that he ignores North America's original people in his stove pamphlet. When he assesses soldiers' ability to withstand winter conditions, he names Russians and Scandinavians, not snowshoe-wearing Wabanakis. His scientific work implied that *suffering* from climate change could become exceptional, something that happened to other people, including Indigenous people, almost always excepted from any definition of modernity. By citing their testimony about extreme cold without explaining what they did about it, he had already made them seem unable to cope. Then, he endorsed the value of being warm at home, in colonial dwellings that, fitted with improvements like his fireplace, would permit many culturally specific things: reading printed material beside a glass-paned window, kept warm by a fire that might be over on the far side of the room.[7]

Printed materials continued to circulate that precise message. An anonymous letter in the *Boston Evening-Post* for September 8, 1746, possibly written by the paper's printer, Thomas Fleet, praised Franklin's new device. (Fleet and Franklin had known each other since Franklin's apprenticeship; it's possible they wrote the letter together.) The "long tedious Winters of this cold Climate" had made Boston's firewood "scarce and dear." Adoption of Franklin's invention would dramatically shrink the need for fuel while maintaining *"Health* and *Comfort."* Misremembering the devices as *"Philadelphian Fire Places,"* the author insisted that, anyway, they deserved to be called "Mr. FRANKLIN's *Stoves,"* the first known appearance in print of the name that would eventually prevail. Franklin probably enjoyed that, plus the letter's proposal that a statue be erected to him, not unlike those to ancient humans whose inventions—wine and bread—had deified them as the gods of abundance, Bacchus and Ceres. Even before Immanuel Kant's astonishing comparison of Franklin to Prometheus, the colonial inventor was described as having divine powers.[8]

How serious was Franklin's commitment to conserving Pennsylvania's trees? He knew what such a project might entail because he printed a plan for it in his almanac for 1749. This was the new series, *Poor Richard Improved*, the improvement being an additional twelve pages, generating a third again of space for content beyond the almanac's core calendar. For 1749, the almanac began "by way of preface" with "an *essay* wrote by a celebrated *naturalist* of our country." This must have been Quaker botanist John Bartram, the colonies' most celebrated naturalist at the time; Bartram was an important correspondent of Linnaeus (who defined an influential system of taxonomical classification), and he established a significant botanic garden just outside Philadelphia. He would eventually be named botanist to the king. The essay he gave Franklin for the almanac was *"An Essay for the improvement of estates, by raising a durable timber for fencing, and other uses."*[9]

The essay's goal—saving trees—was what Franklin had claimed for his stove. But more clearly than Franklin, Bartram identified settler farming as a major reason for deforestation: "Our farmers have to clear the greatest part of their land for tillage and pasture, and partly for fuel and fencing." Once settlers axed the trees, they lacked wood to maintain infrastructure, including fences, critical for marking property lines and for controlling livestock. Farmers had tended to make their fences from oak, superb wood for building, though from a very slow-growing tree—hardest to replace. Rather than replant oak or try to import it, settlers should cultivate red cedar. Much of the essay is a guide to growing this cedar from seeds found near the sea, where the trees grew wild. They were in truth "a species of juniper," and "the most profitable tree for fencing," being fast-growing, hardy, and tolerant of many kinds of soil.[10]

By cultivation, Bartram was recommending European forestry techniques to "assist nature by art." Cedar seed had to be planted in prepared ground, in "upright and regular rows," not a natural pattern. After five years, they "will put forth their fragrant male blossoms; and the females will begin to produce their aromatick berries." The planted lines could be thinned as the trees grew, producing some useful wood while maintaining the overall stock. Their fragrant, flowering, geometrical groves "will be very ornamental." The effect was all the more striking precisely because farmers had removed most of the original trees. The plan imitated efforts in Europe to create artificial "wildernesses" on long-cleared land.[11]

Aromatic blossoms were nice, but utility was the payoff. Bartram sped through a verbal time-lapse of the red cedars' growth, from eight years (the right size for barrel hoops) to twenty years, when the trees were "big enough to make three [fence] posts, besides a good stake at top." This required "carefully trimming them every three or four years" to keep the trunks as straight as possible. The effort was worth it, Bartram insisted, because cedar fences "will last fifty years, or longer," with comparable results for other constructions, including boats; like "the Bermudians," Pennsylvanians might, after forty to fifty years of cedar cultivation, "build fine durable vessels thereof." Thus, cedar groves could become an "easy, ornamental and profitable" stock on any "old worn-out field."[12]

It was a fine idea. But it's not clear that even Bartram adopted it. Orchards of trees for wood, unlike for fruit or nuts, simply did not make sense to colonial farmers at a make-or-break stage, with limited interest in any nonedible or non-vendible crop. And the cedar project was a hard sell to landowners whose strategy was to maximize grain harvests for sale, generating income with which to buy supplies, including fencing material. Nor did colonial officials legislate or subsidize tree planting. Rather, the colony's main policy in relation to the natural world was to keep transferring land titles to white colonists, turning nature into units of material resources, for the most part under private control.

Franklin and Bartram both defined conservation of American trees as an ideal, proposing two ways to achieve it while insisting on their projects' practicality, pitched to individual property holders. Revealingly, Franklin's project would enjoy greater circulation, especially in Europe. Where, why, and how that happened reconstructs a revealing map of places that needed at least to think about conserving fuel—the colonies now included, despite the centuries of fantasies about new-world abundance. Europeans found a colonial American conservation scheme all the more convincing because it was *preventative*, designed to keep natural resources and human demand in some kind of balance, an opportunity missed earlier in parts of Europe.

On paper, Franklin's stove pamphlet and his later writings about heat would travel even into places he never visited, spreading confidence that life could be improved, and the future comfortably managed, whatever the variations in climate. But as elsewhere, benefits from any project of resource management were never all-inclusive.

Conservationists inspired by Franklin's stove assumed that American resources belonged to colonists, not Natives, interpreting new-world places according to the goals and desires of their white populations: their material needs, their material possessions, their very bodies as material entities. This meant that conservation intersected, to a perhaps unexpected extent, with the project of imperialism.

To further disseminate his fireplace pamphlet and heating technology, Franklin built on his initial list of distributors—Grace in Philadelphia, John Franklin in Boston, Peter Franklin in Newport, James

Parker in New York—with circulation of his project on paper and a new web of sellers. He acquired secondary agents in Boston, New York, and Philadelphia (using his illustrator, Lewis Evans, for that last city); he eventually had a seller in Yorktown, Virginia, as well as the one in Williamsburg. Franklin also circulated his pamphlet by itself, as a declaration of how science and technology could defeat physical discomfort. In February 1745, the *Boston News-Letter* advertised both the fireplaces and the pamphlet. Parker said in his *New-York Weekly Post-Boy* that he would throw in a free copy of Franklin's pamphlet with every fireplace sold. And Franklin sent a copy of the essay to his main scientific correspondent, Cadwallader Colden in New York.[13]

That was a shrewd move. Colden forwarded Franklin's pamphlet on to Johann Gronovius, professor and librarian at the University of Leiden, who had it translated into Dutch and printed in a 1746 edition, including Franklin's citations and reproducing the pamphlet's two plates on one fold-out sheet. "That invention," Gronovius declared of the fireplace, "hath found great applause in this part of the world, which is the reason that I could not hinder to let it be translated into Dutch, and no doubt soon into French," maybe meaning for the Francophone population of the Low Countries, or for the *savants* in France itself. Franklin had self-published his essay, and having it independently translated and reprinted with the endorsement of a European man of science was important validation of his fledgling reputation as a scientific experimenter, comparable to a peer-reviewed science article today. With a nice circularity, John Bartram, who would have read the original pamphlet, received two copies of the Dutch translation from Gronovius in 1746.[14]

The Colden-Gronovius connection was a gift to Franklin, but it was the perceived need to conserve fuel in the Netherlands that had prompted the Dutch translation. Given that many Dutch people already used stoves for heating, it's not obvious why a new fireplace design would have interested them. Stoves in particular, often ornamented with blue-and-white Delft tiles, were common sources of domestic heat. Moreover, the main fuel in the Netherlands was by this point no longer wood but peat. This industrious-revolution shift represented an interim point in a longer transition from wood to coal (and then onward to other fossil fuels). As people in the Netherlands reclaimed land by constructing their famous dams and dikes, they exposed peat fields from which they could extract fuel, fortuitous in an environment that had few remaining woodlands. Franklin's translator for his stove pamphlet thus added a small footnote to his description of fire: "Common fire of peat, wood, or coal."[15]

For a time, a wealth of peat made the Netherlands, compared to its neighbors, unusually energy rich. The slow- and hot-burning fuel was perfectly suited to multiple and lucrative industries: boiling dye for cloth, firing brick, brewing beer, making salt, refining sugar. These were urban or semi-urban activities, often conducted in port cities; the famous Dutch windmills were more commonly used in the countryside and for agricultural tasks. Peat was also good at keeping Dutch people warm and supplied with cooked meals, both crucial tasks as, fueled by industrial and commercial growth, the population grew in turn.[16]

But this energy advantage was temporary. It had carried the Netherlands through the seventeenth century and into the first decades of the eighteenth, at which point peat supplies were dwindling and the Dutch, following the English example, began to use coal, a more expensive choice, given it had to be imported. Gronovius was right: in 1746, Franklin's pamphlet arrived at the perfect moment to take advantage of Dutch curiosity about conservationist energy use. Perhaps for that reason, the translator made several small, practical changes. They omitted the original Latin of one of Franklin's quotations, simply putting it into Dutch, and deleted the sun poem at the end. Also, they gave more extensive descriptions of the engraved illustrations and for installing the false back inside the chimney. Each of these adjustments indicated the likelihood of real use, including possible Dutch manufacture of the fireplaces, in which case the Pennsylvania ornamentation could be replaced with whatever local makers or customers might prefer. In the Netherlands, Franklin's stove pamphlet found an engaged audience.[17]

The opposite was true in Sweden, where an account of Franklin's fireplace failed to circulate, but came close to doing so in a work by Pehr Kalm, not quite a decade after Gronovius's translation. Kalm, a Finnish-Swedish naturalist, had traveled in the British colonies from 1747 to 1751. He came as one of Linnaeus's "apostles," a research assistant who gathered specimens and information for the celebrated taxonomist. Though Franklin would not publish his famous electrical experiments until 1751, he was already, when Kalm arrived, a leading American figure in science, precisely because of his work on heat.[18]

It's clear from Kalm's diary that he sought a strategic connection with Franklin, and quizzed him about natural history, including North America's climate. He complained that the cold in the northern colonies surpassed the winter weather in the same latitude in Europe, worse even than in Sweden, which was considerably farther north. And yet despite burning prodigious amounts of wood, most colonists lived in houses "no warmer than barns." For these reasons, Kalm praised Franklin's work on heat. The Pennsylvanian fireplace was "a new type of stove, which not only provides plenty of heat, saves fuel and brings fresh air into the room, but is so constructed that the flame may be seen." As he recounted in his diary, Kalm attended a dinner at Franklin's house on December 8, 1748, where the fireplace was in use (another, unseen, was upstairs in Deborah and Benjamin's bedchamber). Lewis Evans, who had made the illustrations for the pamphlet, was also at the dinner and presented Kalm with a copy of the stove pamphlet annotated with his further thoughts. Franklin would loan Kalm a fireplace in the winter of 1749–1750, maybe the more easily installed original version, firebox without airbox. Kalm gives the first account of cooking with the stove, from making hot chocolate to roasting meat suspended above the firebox, where it cooked faster than in a regular fireplace.[19]

If Kalm had a doubt about Franklin's fireplace, it was its cost. Given the expense of acquiring one, colonists had been making cheaper mock-ups, using brick or tile to imitate the interior baffles, with only an exterior firebox of iron—less effective, but better than a regular fireplace. Franklin's design had spread in Pennsylvania and New York, Kalm related, but not so much into the countryside, where wood remained abundant. Also, the small boxlike iron German stoves, similar to those used in Norway and northern Sweden (Kalm explained), were a cheaper option for anyone who wanted fuel economy but also minimal household outlay for any new heating device.[20]

The original Swedish account of Kalm's travels was published in Stockholm, from 1753 to 1761, in three volumes. It would later be translated into German, English, and Dutch. But Kalm omitted his diary entries on the Pennsylvanian fireplace—the day-to-day notations for 1748 skip the dates when he had visited Franklin and made those observations. What this means is that Kalm, like Gronovius, had assessed his homeland's audience shrewdly. Swedish iron producers resented their Pennsylvania competitors for the British market; the less said about them the better. As importantly, Kalm might have concluded that the Swedes, unlike the Dutch, were not interested in innovations that saved fuel, because they were still rich in trees.[21]

Swedes retained much of their original wooded land and were pushing north into Sami forest regions in Lapland, an internal colonization of Indigenous territory akin to Pennsylvania's removal of Lenapes. Aside from some hydropower, the nation's main energy source was wood, burned directly or turned into charcoal. Sweden's energy wealth gave it unusual economic might, particularly in forging metals with charcoal—it dominated Europe's steel production. Other Swedish exports, including tar and potash, were directly harvested from trees, with the exports also requiring firewood for their production, an energy expense few other parts of Western Europe could afford. Not until the late nineteenth century, especially after 1890, would coal replace wood as Sweden's main source of industrial energy and for some domestic heating. There were earlier worries about availability of wood. But, as elsewhere, this focused on large trees for shipbuilding. Royal authorities wanted oak and beech protected for that use; peasants resented how this reduced land for pasture or farming.[22]

Blessed with extensive forests, Swedes' domestic heating came mostly from wood burned in fireplaces—the exact scenario Franklin associated with an older England and with earlier phases of colonial American settlement. Although Scandinavia also had peat bogs, the Swedes, unlike the Dutch, didn't harvest them intensively. Nor were they very interested in new heating technologies. Iron and tile stoves arrived later, with the late-nineteenth-century shift to coal. In short, the Netherlands and Sweden represented opposite trends in premodern fuel economies, with contrasting levels of interest, therefore, in the Pennsylvanian fireplace, from an American colony that itself existed midway between abundance and scarcity. In each case, there was an embedded comparison between natural resources and indoor conditions, setting expectations for what could be burned and in what quantities.[23]

Franklin had always considered his invention not only as a solution to the specific problem of a colder climate but also as a way to explore climate's general characteristics. Hence the small sun on his fireplace's front plate, declaring the device to be a biddable star that "Warms when we please." Human control over an indoor atmosphere and artificial climate required some knowledge of how the bigger and less obedient climate and atmosphere of Earth behaved, and how their properties might reveal yet more of nature's powers.

"Atmosphere" was a relatively new word, just beginning to be defined as a space composed of multiple and mobile elements and, by the end of the eighteenth century, as a set of chemical states, specifically, a mixture of gases. Drawing on Aristotle's definition of the celestial realm as composed of nested geocentric spheres, and on Ptolemy's hypothesis that Earth emitted heat and damp, medieval theorists had postulated three aerial layers above their heads: a hot and dry zone farthest away; a colder and wetter one closer in (which generated precipitation); and a wet and warm one hugging the Earth, the most variable of the three. Then, in 1608, the Dutch savant Willebrord Snellius merged the Greek words for "vapor" and "globe" to coin *atmosphæra*, bequeathing to all atmospheric studies a concept of flowing motion that could occur invisibly and yet on a planetary scale. That concept gave early modern studies of the weather a sci-fi dimension, as investigators reconsidered the forbidding empyrean, rendering its nearest layer into a friendly life-support system, one that might exist around other planets.[24]

The first time the word "atmosphere" appears in his surviving writings, in 1745, Franklin was describing pressure within another semi-enclosed system. In a much-delayed letter to Cadwallader Colden—"you should conclude me the very worst Correspondent in the World"—Franklin described his experiments with a device that, if made of glass, visually demonstrated how tubes of equivalent length were subject to equivalent "Pressure of the Atmosphere," as proved by the uniform motion of water within each length. He called this device a "syphon," though it was incapable (unlike a true siphon) of conveying liquid upward; it instead resembled the J-shaped glass tube Robert Boyle had used to calculate atmospheric pressure: by stopping up the tube's shorter leg and pouring mercury into the longer, Boyle measured the slow compression of air in the short leg. An inverse relationship between pressure and volume (basis of "Boyle's Law") was observed when the temperature was constant. The principle was not immediately relevant to a device that created heat; Franklin instead emphasized the pressure of accumulated warm air that could banish smoke up a chimney, without explaining (yet) the phenomenon in terms of atmospheric pressure.[25]

Franklin did just that, however, in 1753, to describe the motion of waterspouts at sea, which operated "exactly as Water would do when forc'd up by the Pressure of the Atmosphere" in a tube. Next, he conceptualized the movement of air (which raised the water within such a spout) as generated by heat, with layers of warmer air rising, becoming "specifically lighter than the superincumbent higher Region of the Atmosphere, in which the Clouds commonly float"—just as hot air from a (good) fireplace would fill a room. Franklin connected the room to the larger atmosphere, rather ominously, by observing that, should the air pressure necessary to generate a waterspout or whirlwind pass over a house, it could shatter the dwelling's windows, pluck away its chimney, burst out its walls, and suck up its floor "by the Sudden Rarefaction of the Air contain'd within such buildings, the outward Pressure of the Atmosphere being suddenly taken off."[26]

As Franklin was writing up his waterspout observations, connecting anthropogenic spaces and the larger atmosphere to consolidate his status in science, he deployed his stove for the same reason. In July 1749, he arranged for Robert Grace to send two Pennsylvanian fireplaces to John Mitchell in England. Mitchell was a medical doctor, naturalist, and Fellow of the Royal Society of London. At least one of the fireplaces was intended for his patron, Archibald Campbell, third Duke of Argyll; the transaction hints that Mitchell and the well-read duke had some familiarity with Franklin's stove pamphlet. Franklin knew he needed to make an impression on the Royal Society in order to be welcomed as a scientific investigator, and his stove pamphlet had not by itself opened that door. So, before shipping off the stoves, he'd sent Mitchell a letter, hoping it would be officially read at the Royal Society, if not published in their *Philosophical Transactions*.[27]

The letter of April 1749 was actually an essay with a long, formal title, "Observations and Suppositions towards forming a new Hypothesis for explaining the several Phaenomena of Thunder Gusts"; Franklin obviously intended this for an audience somewhat beyond folks in the market for a new stove. The inclusion of "F.R.S." (Fellow of the Royal Society) after Mitchell's name in the address was another nudge to have the piece read before the society—and it worked. This was a particular victory because the essay was designed to connect Franklin's expertise with heat to his emerging command of electricity. He explained thunderstorms as collisions between electrified and nonelectrified materials, including those in clouds. He redeployed his description of hotter air rising above cooler, which accounted for the transit of clouds at varying heights. In the Northern Hemisphere, air was heated in the tropics and, rising far above the Earth's surface, trended to the north, "the denser Northern and Southern Air pressing into it's Place," eventually descending, "that the Circulation may be carried on."[28]

Franklin's subsequent work would continue to compare artificial heat to the natural kind, and the motion of air within a small space to circulations over entire hemispheres. He also continued to use human beings' tangible experiences as a primary point of reference. For example, when he investigated evaporation, he explained that it made the body feel colder than the ambient temperature a thermometer might register. And whenever he returned to the question of heat, it's obvious that his initial experiments in a room equipped with a fireplace had primed him to explain thermal variation as the product of convection—heat-generated circulation within definable atmospheres—though at this point he began to conjecture larger spaces, connecting entire continents via whole oceans, and taking in the nature of the atmosphere far overhead for "both Hemispheres."[29]

So, too, Franklin's electrical experiments progressed logically from his investigation of heat. He accepted the contemporary definition of electricity as a kind of flowing fire; he was already an expert on that, having theorized similarly about heat, another subtle or invisible "fluid." His major contribution to the physics of electricity built on that logic, conceiving of an electrical fluid as occurring in two states, positive and negative, set into action as it transited balances between these two kinds of charge. Moreover, and starting with the first report (of 1747) on his and three collaborators' experiments with electricity, Franklin noted that it could form "Atmospheres" around objects, "making them look beautifully; somewhat like some of the Figures in Burnets or Whiston's Theory of the Earth," two reigning interpretations of the planet and its atmosphere (recall Burnet's argument about a planet ravaged by the scriptural Deluge). In his second report on his investigations, Franklin was confidently using the term "an Electrical Atmosphere." Finally, he described electricity as starting fires in the conventional sense. At a spring picnic to celebrate a successful round of experiments, he and his three colleagues lit a fire with an electrical spark to roast an unfortunate turkey they'd killed with an electrical shock.[30]

Amid their wealth of discoveries, Franklin and his colleagues found that electrical fire was a source of energy capable of creating motion. Franklin called the device they rigged up to prove this "an electrical Wheel," one that "turns with considerable Strength." They made it from a selection of consumer items that would have been out of the question for people of their buying power even fifty years earlier. A circle of thin, light wood, twelve inches in diameter, was positioned horizontally, with thirty narrow vanes of window glass slanting from its edge, four inches apart, each capped with "a Brass Thimble." Now for the power source: two Leiden jars, early capacitors, made of glass bottles filled or lined with metal. These were charged with electricity from an electrostatic generator and placed opposite each other at the rim of the wheel. Each jar had an extending wire that stroked the thimbles, sparking the wheel into motion. Franklin marveled that it could spin twelve to fifteen times a minute "and with such Strength, as that the Weight of 100 Spanish Dollars, with which we once loaded it, did not seem in the least to retard it's Motion." (Franklin was, at the time, shipping dollars into Pennsylvania as prizes for a lottery; not for the last time, science needed money.) A Spanish dollar weighed just over 27 grams if unworn by use or untrimmed by coin clippers. Assuming each of the experimenters' dollars was close to full weight, the electric engine could twirl a burden of about 2,700 grams or 5.95 pounds.[31]

It was such an unexpected discovery that the Philadelphia experimenters didn't quite know what to do with it. The electrical wheel had enough power, Franklin estimated, to turn a "large Fowl" on a spit to cook it. He therefore called the invention an *"Electrical Jack"* after the mechanized turnspits that operated in fireplaces. Sometimes, these were turned by dogs. Behold: the world's first electrical engine operated at one dogpower (dp). But Franklin recognized that it wasn't itself a power source. As with a mill turned by wind or water, it had "a foreign Force, viz. that of the Bottles." So, the Philadelphia crew tried next to build a *"Selfmoving Wheel."* This circle of glass, thinly gilded on both sides, except for a bare two-inch rim, was also mounted to rotate horizontally, surrounded by twelve "Pillars of Glass" topped with thimbles. Two lead bullets fixed to the top and bottom surfaces of the disk at its ungilded edge had wires communicating with the gilding and able to swipe the thimbles. When the disk was "well chargd it begins to move," revolving up to twenty times a minute for half an hour. Perhaps this could perform some light task—one of the experimenters used it to ring chimes.[32]

In essence, this invention was half battery, half engine, its glass-and-metal sandwich able to store electricity (like a Leiden jar) and release it to spark motion in a small device for a short interval. Franklin believed the glass held the charge, not the gilding, based on the fact that the sparks generated while the device revolved tore tiny holes in the gilt, "which shews that the Fire is not accumulated on the Gilding, but is in the Glass itself." But maybe because electricity's mechanical power was so unexpected, and slight, few subsequent investigators looked into it. Not until Michael Faraday developed a rotating electrical motor in 1821 would the concept begin to seem promising. Even then, however, electric batteries cost twenty times more than steam engines. For a very long time, the Philadelphian electrical wheel went nowhere.[33]

In the meantime, Franklin's most successful electrical invention was the one that saved lives: the lightning rod. Alongside the household and shop fires that Franklin noted in *The Pennsylvania Gazette*, he related similar accidents with lightning. In Virginia, for instance, "a Flash of Lightning" killed a man standing at the door of his house. Two workmen who sheltered from rain in a shed on Society Hill in Philadelphia were both "struck down by a Flash of Lightning" and one man killed. A Pennsylvania thunderstorm in August 1743 "burnt several Barns."[34]

Lightning was all the more unnerving because it defied categorization. It was hot, like fire, and bright, like other forms of light. But its co-occurrence with water (meaning rain) and with loud blasts in the air were puzzling. Down on Earth, meanwhile, electricity also had those properties—was lightning a kind of electrical fire? Only when he confirmed that lightning was in fact *atmospheric* electricity did Franklin manage to impress European men of science. Despite his sending reports of his indoor electrical experiments to the Royal Society, the FRS worthies hadn't published them in their *Philosophical Transactions*. But when Franklin discussed lightning, in a letter in spring 1751, he won his debut in the journal. He was both propelling and benefiting from interest in the atmosphere: that topic appears eight times in volume forty-six of the *Philosophical Transactions* for 1749–1750, once referring to the reading of Franklin's paper on thunder gusts, then sixteen times in the next volume (1751–1752), eleven times in relation to what he had said or was said about him. A colonist's testimony about the great American outdoors was more gripping than his controlled experiments indoors. Despite his likely exasperation at being treated more like one of Linnaeus's docile informants than a theorist in his own right, Franklin—the American-born Briton—took the hint.[35]

His resulting kite experiment is often dismissed as a clever, semi-scientific exercise. Clever it was: it established that electricity was not just an interesting anomaly emitted by ruffled kittens or rubbed amber, but a gigantic force streaming through the atmosphere. By devising an experimental protocol in 1750, using a metal point raised to a great height (hill, mountain, church steeple) to conduct an electrical charge down to an observer, Franklin had defined his ambition in relation to atmospheric physics. Although the version of the experiment that uses a high place had appeared in his *Experiments and Observations on Electricity*, Franklin printed a variation using a kite in his *Pennsylvania Gazette* in fall 1752, bringing lightning down to Earth, diminishing the heavens' terrors. As a thunderstorm materialized, someone could verify that electricity occurred in the atmosphere by launching a sensor assembled from available consumer items: string, ribbon, strips of cedar or other light wood, a small metal object to collect an electrical charge (a key, say), and a silk handkerchief—yes, the very item an enslaved ironworker might save a year to buy, a hint that while DIY investigation of the atmosphere was possible in this era, it wasn't equally possible for everyone.[36]

Franklin thought that electricity might be only one atmospheric element. He noted that the atmosphere had circulation, including of ordinary heat. In an essay written in parallel to his kite experiment, he described how heat and moisture gave different parts of the "Atmosphere" varying weights, able to rise or descend relative to each other in a constantly shifting, complex pattern. This piece also interested the Royal Society and appeared in its *Philosophical Transactions*. Finally,

Franklin connected his observations of atmospheric electricity back to the ancient and medieval theory of a celestial zone of fire:

> An Electric Atmosphere cannot be communicated at so great a Distance thro' intervening Air, by far, as thro' a Vacuum. Who knows then, but there may be, as the Antients thought, a Region of this Fire, above our Atmosphere, prevented by our Air and its own too great Distance for Attraction, from joining our Earth? Perhaps where the Atmosphere is rarest, this Fluid may be densest; and nearer the Earth, where the Atmosphere grows denser, this Fluid may be rarer, yet some of it be low enough to attach itself to our highest Clouds, and thence they becoming electrified may be attracted by and descend towards the Earth, and discharge their Watry Contents together with that Etherial Fire. Perhaps the Aurorae Boreales are Currents of this Fluid in its own Region above our Atmosphere, becoming from their Motion visible.

"There is no End to Conjecture," Franklin concluded. "We are but Novices in the Branch of Natural Knowledge," or, as he comically put it, he was still "much in the *Dark* about *Light*."[37]

That was no excuse for not putting the science to use, however, while slowly enlightening the gloom. It's notable that in 1752, while he was investigating lightning, Franklin helped consolidate the older plan of fire prevention, the Union Fire Company, with a broader insurance scheme. (He did this in the wake, and knowledge, of New York's hysterical reaction in 1741 to fires that white residents assumed were part of a slave conspiracy against them.) The Philadelphia Contributionship for Insuring Houses from Loss by Fire did not require active firefighting, just prepayment for emergency services should the worst occur. This was better suited to the wealthier men of the city, including Franklin, who could hire others to roll up their sleeves and get out the axes and water buckets.[38]

But why not prevent fires in the first place? Cue another Franklin invention: the lightning rod. That rooftop device was often supported by a chimney, a nice convergence of fire-taming technologies. Projecting above the building, the rod could take a lightning strike and convey the charge down into the ground, preventing damage to masonry or ignition of flammable materials. Franklin described the rod for the first time in his almanac for 1753, telling as large an audience he could marshal about controlling the natural world with technical devices. "Poor Richard" assured his readers that God did not send lightning to punish them and, therefore, it was perfectly appropriate to ward it off: "It has pleased God in his Goodness to Mankind, at length to discover to them the Means of securing their Habitations and other Buildings from Mischief by Thunder and Lightning." Most cleverly, Franklin described how to make a lightning rod from two kinds of prefabricated, inexpensive hardware: a length of the iron that builders cut down into nails, topped with thick, commercially available brass wire filed to a point at the top. Another economical design made of rod and chain could protect a ship, with the chain conveying any electrical jolt down into the sea.[39]

Franklin's electrical work consolidated his scientific reputation. Between heat and electricity, he'd mastered experimentation, and science made him a celebrity. This was when (and why) Immanuel Kant declared Franklin the Prometheus of the modern age, taking "fire" from the heavens to protect humans. But look what had happened the last time, Kant warned. Electricity would not, in this era, transform everyday living, and coal and steam were only beginning to do that. But control of atmospheric phenomena symbolized a new power over nature, as embodied in the Modern Prometheus who'd somehow sprung up in Penn's Woods. In a 1763 portrait, Franklin connects two atmospheres, one outside, where a thunderstorm rages, and one in the room he occupies, where a lightning rod and wire have conducted the electricity indoors for the eager current to chime a waiting bell.

MASTER OF HIS
ATMOSPHERES,
INDOORS AND OUT.
*(Mason Chamberlin [artist] and
Edward Fisher [engraver],
B. Franklin of Philadelphia
[London, between 1763 and 1785],
Library of Congress)*

It looks serene, it seems quite cozy, but don't be fooled. At the time, Franklin's mastery of atmospheric electricity represented power over death and therefore—maybe?—over life, and that is how a colonial printer, an American founder, became the gothic namesake and dark spiritual godfather of Dr. Victor Frankenstein.[40]

The fame Franklin gained from his work on electricity enriched his reputation as an investigator of heat, though it also began a pattern in which his writings and inventions circulated without his approval or supervision. With scant acknowledgment, British entrepreneur James Durno coolly appropriated his stove design—and lifted language from his pamphlet to promote the results. Durno published his *Description of a New-Invented Stove-Grate* in London in 1753, more than a decade after Franklin's pamphlet had appeared in Philadelphia, though, strategically, two years after publication of the Philadelphian's *Experiments and Observations on Electricity*. Durno, a London artisan, had chosen his source wisely and, maybe to Franklin's consternation, scored a very good point about the Pennsylvanian fireplace's key feature—and flaw.

Durno opens his pamphlet with Franklin's emphasis on climate and place, stealing the exact phrasing, substituting only one word. "In this *Island*," he explains, the British "are obliged to keep Fires to sit by near Eight Months in the Year." With wood prohibitively expensive and coal subject to a special duty, heating was costly, especially in London. Anything to keep rooms warmer at less expense "and equally chearful as with the common Fires; will meet with the Favour of the Publick." The framing of his argument is identical to Franklin's description of a fireplace with visible flame but requiring less fuel. In greater contrast to Franklin, however, Durno offers his device's benefits not "as uncertain Theory" (Franklin had done all that work already), so his discussion of heat and circulation is accordingly briefer. He emphasizes that his fireplace was used "in a common Room at the Inventor's House, where several curious and ingenious Persons have been to see and observe the Effects of it." Unlike Franklin, he gives the dimensions of this room, plus the locations of its doors. He thus more precisely establishes his stove's reduced fuel use: only a fifth of the coal he used to burn, yet with greater diffusion of heat in a space 26.5 feet long, 13 feet wide, and 10.5 feet high.[41]

To further specify his invention's advantages, Durno, like Franklin, assesses other kinds of heating, though emphasizing the type of fuel each could use, which Franklin hadn't considered. He agrees with Franklin's objections to German and Dutch stoves, which give no sight of the fire, don't circulate air, and tend to overheat rooms. French stoves, shaped like low chests of drawers, have the same drawbacks. Heating that does not provide ventilation gives people headaches and, if burning coal or peat, fills a room with a sulfurous funk. He notes that Gauger's interior "Turnings and Windings" are adapted to burning wood, "the Heat whereof is more diffused than that of Turf or Peat." But Gauger's baffles are expensive and hard to install, particularly in older chimneys.[42]

"The *Pensilvanian* Stove-Grate comes lastly to be considered," Durno concludes. This had been "contrived about Twelve or Fourteen Years ago, and particularly described by Mr. *Franklin* of *Philadelphia*" in 1744, leaving out Franklin's role as the actual contriver. Durno grants that the American invention "must infallibly cure most of the Inconveniences" of other models, if its interior baffles are kept clean, but it was made to burn wood, the common fuel "of that Country." This wasn't feasible in much of England, certainly not in cities. Following Desaguliers, Durno had modified his grate to burn sea coal. He had also adjusted it for easier cleaning, a concern with coal, which burned hotter but left more soot; Durno's design therefore had a brickwork rather than metal firebox. Like Franklin, Durno says his system prevents people from being "burnt before" while "ready to freeze behind"; unlike him, he advises using a thermometer to gauge how evenly his grate would heat a room with less coal. Finally, and again like the Pennsylvanian fireplaces, Durno's devices came in three sizes for different rooms and budgets.[43]

At the very least, Durno's imitation of Franklin (a sincere form of flattery?) amplified the chant of praise for new heating devices—if adapted to coal. Franklin's invention was thus put into more general circulation within Europe, though no longer with his original intent of preserving trees. That trend was also obvious, one year after Durno's pamphlet appeared, in a massive, illustrated, and influential design compendium, Thomas Chippendale's *The Gentleman and Cabinet-maker's Director: Being a Large Collection of . . . Designs of Household Furniture in the Gothic, Chinese and Modern Taste* (1754). Chippendale offered multiple designs, with grates built for coal, signaling that trees as energy sources were becoming a distant memory in England.

In Pennsylvania, the trees' original owners were also being effaced. Franklin had left them out of his analysis, noting only that Penn's woods were not, in fact, inexhaustible. The Europeans who would read and cite his stove pamphlet would take that point, but with diminished awareness of the Pennsylvanian fireplace's imperial heritage, especially its Walking Purchase pedigree. In the end, the American phenomenon that Benjamin Franklin had successfully inserted into a scientific conversation about resource conservation was himself, the clever American inventor. As his subsequent work on resource consumption would make clear, that suited him just fine.

5

GROW OR DIE

"Was the Face of the Earth vacant of other . . . Inhabitants," Franklin speculated in 1751, "it might in a few Ages be replenish'd from one Nation only; as, for Instance, with Englishmen." Interesting hypothesis—not a disinterested one. It comes from what is probably Franklin's most influential work, *Observations Concerning the Increase of Mankind*, in which he'd become one of the "political arithmeticians" he mentioned in his stove pamphlet. Franklin wrote his essay, quite openly, as propaganda. He wasn't describing the increase of "mankind" generally, but of white British colonists specifically, as economic producers and consumers, and the more of them, the better. And he waited to publish his essay until 1755, when the local conflicts of the French and Indian War were blowing up into the global Seven Years' War, fought to define British, Indigenous, and French rights to American land and other resources.[1]

Franklin's population essay is an origin text for growth economics, the theory that societal improvement requires an expanding material plenty, and it's notable that a colonist's comment on new-world abundance helped generate that proposition. But it was as if Franklin tried to write a history of his colony in reverse: he'd recommended Pennsylvanian fireplaces to a settler population running short on natural resources and needing to conserve them—*then* he looked back at the natural profusion that had encouraged settlement in the first place. He didn't join the two stages, not even to speculate about when one scenario might give way to the other, nor estimate how many people might be too many. But his publications, put together, clearly warned that a rapidly expanding colonial population's material demands could, in their power to shape American economies, be even more influential than the changing weather.[2]

The colonists and their growing consumer economy were most obviously damaging by crowding out Native populations, which Franklin justified by describing Indigenous people as less economically important and probably destined to vanish away. He wasn't subtle in making this point, stressing the tendency of colonists to thrive while Native populations declined, raising the possibility that they, like ancient plants and animals long reduced to fossils, might go extinct. He welcomed this prospect—colonial growth paralleling Indigenous death—even as he continued to print treaty negotiations with Natives, documenting their determination to survive despite the mounting warfare and violent removal that were the real reasons for their population struggles: human-created crises, not natural ones.

Franklin never reconciled the contradiction between conservation of natural resources and economic growth—no one really has. The tension lingers; the ghost of a homespun Poor Richard haunts the shiny machinery of the modern global economy, questioning any assumption that economic progress must be linear. Franklin's working backward, from scarcity to plenty, may simply be the result of his addressing different situations at different times, recommending conservation during a bitter winter (and rocketing fuel prices), then extolling the new world's bounty when he wanted to insist on Britain's destiny to command it. At the very least, he was astute in intuiting a functional or historical relationship between profusion and scarcity, with new-world abundance the tasty bait in an inevitable trap.

Maybe it took an American colonist to see this tension between growth and dearth—quite possibly the most powerful force in American history. Precisely because of colonial population growth, the industrious revolution, leading into the Industrial Revolution, would have enormous historical impact, eclipsing the subsequent American Revolution. All the attention paid to white colonists' rebellion against British taxation of their consumer choices, for example, is the tail wagging the dog: consumerism was the real beast, and energy and other resources its meat and drink.

Franklin's first major survey of population growth was prompted by a new piece of imperial legislation, the Iron Act of 1750, that regulated colonists' economic activities. British ministers wanted the law to protect a captive colonial market for British manufactured goods, including anything made from iron. Colonists were still permitted to export—to Britain—crude iron; in fact, the act dropped duties on iron in pigs or bars to encourage that trade. But colonists could no longer establish ironworks capable of processing the raw material—rolling it out, slitting it into workable pieces, plating it, molding it, or turning it into steel. Governors were required to submit lists of any existing American ironworks that, legally, could continue that kind of work. Obviously, Robert Grace's operations would be on Pennsylvania's official list. He could even profit from the smothering of competition; his business records show that he would produce Franklin's fireplaces into the 1750s. Franklin himself, however, worried that the new act was the first of more to come, making colonists into political and economic subordinates.[3]

The gist of Franklin's essay was his estimate (later substantiated) that the white population in Britain's North American colonies was growing much faster than its British counterpart, doubling every twenty to twenty-five years, a rate unheard-of in Europe. So "in Proportion to the Increase of the Colonies, a vast Demand is growing for British Manufactures, a glorious Market wholly in the Power of Britain," the fruits of a continuing industrious revolution. By rapidly expanding a colonial consumer base, the trend would enrich the whole empire, no government regulation required. It was counterproductive to "restrain [colonial] manufactures . . . a wise and good Mother [country] will not do it." But why the growth differential in the first place? Compared to white colonists, Franklin explained, Europeans were generally older when they married, due to lack of land and the autonomy it offered within what was still a largely agricultural society. Blessed with an abundance of land, colonial couples married younger and produced more children.[4]

Although the essay initially focuses on phenomena that now belong to the human sciences, particularly economics and demography, Franklin's more challenging assertion was that the rate of population growth was mostly determined by natural resources, especially food, with little human awareness of what was happening to them (let alone ability to plan for it). People cannot increase "beyond the Means provided for their Subsistence"—in ancient Wales, he wrote, "the People bore a Proportion to the Produce." In this, he compared humans to anything else living: "There is in short, no Bound to the prolific Nature of Plants or Animals, but what is made by their crowding and interfering with each others Means of Subsistence." Either "Fennel" (for instance) or "Englishmen" could replenish a planet somehow stripped of competitors.[5]

In defining human increase as dependent on the availability of natural resources, Franklin was analyzing the upper bound of an organic economy, in which human labor was the main motive force and source of value. He assumed that North American resources would be economically developed at a level comparable (eventually) to Europe.

And by comparing colonial and Native populations, Franklin assumed that political decisions, including those regarding the distribution of resources, should be based on calculations of "need" at the population level, with smaller social groups counting for less, particularly if they did not develop natural resources, thus compromising their ability to increase. In his dismissal of Indigenous people as migratory hunters who merely skimmed the land, for instance, he claimed they were "easily prevail'd on to part with Portions of Territory to the new Comers" who "furnish'd them with many Things they wanted." Here, Franklin was pretending not to know anything about the Anglo-Wabanaki Wars or the Walking Purchase, in which Natives had resisted giving up land. In the end, he implied they would at best be a tiny minority.[6]

But as Franklin knew very well, resistance to colonialism was continuing. Accounts from a new Haudenosaunee settlement called Shamokin, on the site of a Saponi village in the upper Susquehanna Valley, describe a reality of displacement and deprivation. Forced to abandon much of Pennsylvania's coastal area after the Walking Purchase, many Lenapes tried to demonstrate sovereignty over interior lands they thought the younger Penns had reserved for them. That was why they fled to Shamokin, seeking a political alliance with the Haudenosaunee. A settler account of 1745 describes a town of fifty houses and about three hundred people speaking multiple languages, especially as Shawnees and other Natives continued to arrive, later joined by German-speaking Moravian missionaries and English-speaking soldiers at nearby Fort Augusta. Early conditions were harsh. An outbreak of fever in 1747 struck both Natives and Moravians; late frosts and high winds diminished corn harvests; deep snow impeded hunters. Refugees and visitors depleted stocks of food. One white newcomer was apprehensive in 1745 "that the Indians here are still so cold and dead," stunned into a state of bare survival, at best.[7]

Although white observers didn't understand what bound people together in Shamokin, the Natives themselves insisted that they had a *collective* right to be there. Their proof was the council fire the Haudenosaunee had kindled at Shamokin, around which all could gather: "We will never part with the Land," one Oneida leader told Pennsylvania officials, "our Bones are scattered there, and on this Land there has always been a great Council Fire." Equally, it made little sense to Natives that white people lived in separate houses with private fires. When the Moravians built a house for their missionaries in 1747, Natives gathered to watch the strange thing go up, chimney and all. Later, in 1775, a visiting Presbyterian minister was astonished when two Indigenous boys burst into the settler house where he was staying with fresh fish they wanted to cook and eat. After all, Native adults would have let them do that, wood and fire being absurd to regard as private property. That ethos of sharing resources contrasts to Franklin's (and others') insistence that economic growth required nature to be divided, privatized, and developed through labor, resulting in population increase.[8]

Even as Franklin criticized Natives for their unwillingness to develop natural resources, his population essay did denounce enslavement as economically inferior to free labor, a bolder declaration about one of the era's major energy options: using organic resources (like food) to power human beings. Enslaved people, by definition coerced into working, could not be as productive as free workers. Their only advantage was their continuity: "hired Men are continually leaving," while "Slaves may be kept as long as a Man pleases." That came at the cost, however, of sapping energy from white people, because work itself, Franklin argued, gave energy. "Whites who have Slaves," he said, "not labouring, are enfeebled" and have fewer children. And in families with enslaved Blacks, "the white Children become proud, disgusted with Labour . . . rendered unfit to get a Living by Industry." This was not a moral statement defending enslaved people themselves, but it was the beginning of Franklin's reassessment of an economic practice he hadn't questioned earlier.[9]

He made that point even more strongly in his analysis of the Caribbean colonies, arguing that, dependent on captive workers, they represented an utterly backward system. Enslaved Black people, constrained from making free consumer choices, constituted the smallest segment of a colonial market for British manufactures, whatever the profits from the sugar and other commodities they produced—their work was a dead end in the consumer revolution. And Franklin openly stated his preference for people he deemed white (Northern Europeans and their descendants), worrying that globally they were vastly outnumbered by those who were "black" (Africans), "tawny" (Asians and Native Americans), or "swarthy" (Southern or Eastern Europeans). He complained that enslaved people from Africa had "blacken'd half America" even as some "tawny" Natives remained. But he hoped that "White People" would increase and continue "*Scouring* our Planet, by clearing America of Woods, and so making this Side of our Globe reflect a brighter Light" to any extraterrestrials who might be out there observing Earth from space, a possibility Franklin had speculated on in his youth.[10]

The zoom into outer space is a big clue about Franklin's reason for reconsidering conservation. In his stove pamphlet, he'd made calculations based on the extent of the Walking Purchase. But in his *Increase of Mankind*, he sprawled over a different geographic scale: "North-America." With a continent at stake, the British should be "careful" to "secure Room enough, since on the Room depends so much the Increase of her People." Hemispheric domination was the goal, and thus the similarity (and historical connection) between colonial history and science fiction, sharing, as they do, a fantasy of some vast tabula rasa for someone's huge project, with an embedded fantasy that there are no other beings able, or entitled, to interfere with the scheme.[11]

Franklin's endorsement of continental terraforming, destroying North America's woods so completely that the handiwork could be admired from outer space, contradicts his stated concern elsewhere to preserve those very trees, because with so many more trees on an entire continent, any environmental reckoning could be postponed. Altogether, the essay, from which Franklin would later excise the worst of the racism, slowly realizing its offensiveness, sits uneasily in relation to his pamphlet on the Pennsylvanian fireplace, which, by leaving out the ugly details of enslavement and Native removal, gave the stove project a benevolent appearance, matching its ethical prompt to conserve resources.

The Iron Act may have been Franklin's specific inspiration for this other essay, but his response is indiscriminate in its threat that white colonists might be able to simply outbreed everyone else, British, Black, Indigenous, or French. In this essay, at least, settlers' proto-national power as consumers, steadily reproducing in a semiautonomous part of the British Empire, was an economically positive trend that outweighed their negative impact in gobbling up American natural resources (which Franklin had just described in his stove pamphlet). This was when Franklin used the chilling concept of infinity to describe their consumer appetites. In a 1753 letter to a friend who commented on his population essay, he said that Natives had "few but natural wants," in contrast "with us," who had "infinite Artificial wants." Through those wants, the industrious revolution would enrich and unite Britons and colonists: "What an Accession of Power to the British Empire by Sea as well as Land!"[12]

Franklin's *Increase of Mankind* would rival his *Experiments and Observations on Electricity* and his *New Invented Pennsylvanian Fire-Places* in its impact. It's significant that his unresolved tension between growth and scarcity invited completely different emphases among his interpreters—Adam Smith leaned into the growth and Thomas Robert Malthus the scarcity. (Through Malthus, Victorian evolutionists Charles Darwin and Alfred Russel Wallace would likewise stress nature's cruel finitude.) Franklin had maximized his essay's effect by delaying its publication until 1755, when it appeared, anonymously, alongside another work, William Clarke's *Observations on the Late and Present Conduct of the French: With Regard to Their Encroachment upon the British Colonies in North America*. Packaged together in London, the two pieces did some very loud saber-rattling, taunting the French about colonial and British might. (The British would seize parts of Acadia, in the French maritime provinces north of New England, in June 1755.) In quick succession, as well, Franklin's essay reappeared in two widely read British periodicals, the *Gentleman's Magazine* (London) in 1755 and *The Scots Magazine* (Edinburgh) in 1756. Thereafter it was reprinted under Franklin's name, often with its openly racist sections excised.[13]

But the essay's guiding assumption, that free white people were the British colonies' greatest weapon, was unredactable. Its publication in 1755 aimed this grenade, patriotically, at the French. This was an ex post facto repurposing of an analysis that had described Natives and enslaved Black people as material resources, and possibly only temporary ones. In fact, the repackaging exactly paralleled the unfolding conflict. The Seven Years' War, lasting from 1756 to 1763, grew out of the so-called French and Indian War, a proxy war within North America that began in 1754 and then merged into the declared conflict between Britain and France. The colonial antecedent, as with the British seizure of French Acadia, was quite openly a war over America's natural resources—the land and anything on it. And the international rivalry rehearsed every possible way of imagining the Americas, from the fantasy of abundance to the fears over bizarre climates and looming resource scarcity.

Within both social science and natural science, Franklin's work enshrined the hypothesis that humans are just another species whose numbers rise and fall in relation to material conditions. Subsequent work would outline what's been called the "Malthusian trap," in which populations lured into growth cannot sustain it. Franklin had admitted this exact hazard about firewood in his stove pamphlet. But he ignored it in his work on population. The later essay instead questioned any legislation that tried to guide economic decisions, inspiring liberal preference for unregulated consumer behavior. In this regard, his stove pamphlet had prefigured his later work in economics, given its premise that conservation of natural resources was something consumers might be persuaded to do, but should never be required to do.[14]

Demography is seldom neutral. Biopolitics, as it's been called, ranks populations by their perceived value to a political entity, whether a nation or an empire with assembled colonial territories. Reading Franklin's stove pamphlet and his population essay together, it becomes even more obvious that his efforts to conserve wood prioritized his fellow white settlers. The Natives who'd testified about past winters? Franklin assumed they'd be supplanted by the people who installed his stoves. The enslaved Black people who made iron, including his stoves? Franklin declared their labor substandard, while insisting that its drawbacks mainly affected the Caribbean colonies. And yet when he first printed his stove pamphlet, Pennsylvania was approaching a time of maximum investment in enslaved labor, the nightmare that the Lenapes had feared and tried to prevent.

By 1767, one in five of Philadelphia's households would include at least one enslaved person, who collectively comprised an estimated 8.8 percent of the urban population. It's worth noting that Franklin was in the unfortunate vanguard of this trend; he and his wife held an enslaved man in their household as early as 1735. The regional total was much less than in the southern and Caribbean colonies; a slaveholding household in Philadelphia tended to have only one or two such workers. Still, this was greater than in other parts of the colony. Only 4.2 percent of taxpayers in outlying Chester County owned enslaved workers, and 1.2 percent in Lancaster County, as opposed to the estimated 20 percent for Philadelphia in 1767. On the other hand, the first federal census of 1790 would calculate that slaves were 8.7 percent of the state's population, indicating that some rural estates—including ironworks—might eventually have had more enslaved Black people than typical city households.[15]

The colonial iron trade in the region stretching from New Jersey to northern Virginia was expanding precisely because of its growing exploitation of enslaved workers, something that evidently didn't trouble the Iron Act's architects (or critics). From 1725 to 1750, up to half of the labor force at some Pennsylvania ironworks was enslaved. Even Quaker ironmasters held slaves, including those Franklin worked with. Franklin left this not-so-small detail out of both his stove pamphlet and his analysis of populations and economics in North America.[16]

The white people who bought Pennsylvanian fireplaces represented, as well, a specific and better-off clientele. The total number of fireplaces sold seems to have ranged in the low hundreds (not including all the likely knockoffs)—a fair number, given the price, but hardly mass adoption. After all, for anyone struggling to pay for firewood, buying a fireplace insert priced at a minimum of £5 was impossible. Estimates of average household income in colonial Philadelphia put it at £20 and 1 shilling per year in 1725, rising to £23 and 6 shillings in 1750. So, a Pennsylvanian fireplace represented 20–25 percent of a city-dweller's likeliest annual income. (In Philadelphia today, installing a heat pump to conserve energy consumption requires only about 10 percent of median household income.) Moreover, anyone in colonial Philadelphia willing to buy a fireplace insert would be among the minority who owned or had a long-term lease for their home or workshop—otherwise, why pay to modify its chimney? The poor who went "shud'ring" forth in dead of night to pry firewood from someone's fence were not going to avoid that fate because some wealthier people were burning less wood. Premodern, climate-induced scarcities were neither suffered nor solved equally, least of all in colonized places.[17]

And despite Franklin's claim that his fireplace inserts conserved wood (and therefore trees), their potential benefit could not be measured in per capita consumption alone. Each household had limited economic impact. The problem was not just reducing the units of anything consumed but also the number of consumers. Could fuel efficiency based on stove improvement really keep pace with population growth?

If they performed as Franklin said they did, the fireplaces did reduce use of fuel, enough to offset the wood required to produce them. Recall that the average household in the northern colonies consumed between 20 and 40 cords of wood (a cord being 128 cubic feet) per year, so maybe 30 cords a year in Pennsylvania, for an annual household total of 3,840 cubic feet. If, as Franklin promised, adopters of his fireplace could look forward to buying only a quarter of the fuel they once did, they were saving 2,880 cubic feet of wood per year. Of course, they probably did their cooking in the kitchen, not the parlor, and might have had other fireplaces or stoves in other rooms anyway, as the Franklins did in their bedroom.

Even so, the amount of wood used to forge an individual Pennsylvanian fireplace is dwarfed by its annual fuel savings. It took about 1,800 cubic feet of wood burned down into charcoal to produce sufficient crude iron to be processed into a ton of wrought iron; casting iron, as for stove plates, used less fuel. It's not clear what an average Pennsylvanian fireplace weighed—again, they came in several sizes and the airbox model obviously weighed more than the first version with just the firebox, which had been around 300 pounds, though the surviving firebox at the Mercer Museum, possibly a larger model, weighs 500 pounds. Let's work with that weight, another high estimate, but the only real data that survives. These 500 pounds of iron would have required 450 cubic feet of wood to produce. A single year's use of the stove would offset that quantity, given the 2,880 cubic feet of wood saved each year. A cost analysis of fuel consumption also bears this out: with firewood priced (in the 1740s) at about 12 shillings a cord, a three-quarters reduction in fuel represented an annual saving of £13 and 10 shillings, enough to justify a £5 fireplace in its first year of use. Even if Franklin's own estimate was on the high side, the gap between it and the fuel costs for making his fireplace is enough to assume its efficiency.[18]

This pleasing net gain was eclipsed, however, by the colony's rapid population growth, the accumulation of white consumers who wanted to live in comfortably warmed houses. One such new household, burning an average of 3,840 cubic feet of wood each year, could obliterate one Pennsylvanian fireplace adopter's conserved 2,880 cubic feet. And there were many such households. In 1730 there were 51,700 settlers; in 1740, 85,600; and then in 1750, 119,700. Hundreds of new households were demanding firewood for every dozen that invested in conservation. And if anything, Pennsylvania's iron plantations, lying in what would become Philadelphia's extended suburbs, were in large part the cause of the city's firewood shortage, compromising the local fuel supply for the colony's largest population center. In 1749 and again in 1754, rioters attacked woodcutters from the Union Iron Works for taking down so many trees for charcoal, depriving others of any access to the wood.[19]

Franklin had acknowledged the problem of population growth in his stove pamphlet—wood was vanishing "as the Country is more clear'd and settled"—only to rebrand the trend, in his work on "political arithmetic," as a distinctive colonial strength. He was closer to the mark the first time. Three-quarters of all fuel wood ever consumed in the British North American colonies and in the later United States, for instance, would be burned for domestic consumption. Even after a shift toward coal, wood remained an energy mainstay, forging half of US iron as late as 1850, powering steamboats and railroads until the 1870s.[20]

Often regarded as an economic laggard and a copycat, the North American colonies, due to their population dynamics, were actually on the cutting edge of industrialization—with dire consequences. The tens of thousands of wooded American acres being claimed for settlement, agriculture, and industries like iron production were proportioned, not to the more compact and urbanized factory operations of the nineteenth century, especially in Europe, but instead to the lingering fantasy of endless new-world plenty. This imperial context was essential to conceptions of modern economic growth; so, too, was the piety that white people were a self-generating engine of labor and consumer power. For the moment, this scenario still existed within the constraints of the old organic economy. The small sun ornamenting each Pennsylvanian fireplace was a confession, of sorts, about the energy source that had brought it into being, the solar cycle that grew fuel-producing trees, though never as fast as colonists needed them. This was when a very serious question should have been asked and not dodged: What can be done *while there still is time to do it*?

One solution would have been resource conservation, which Franklin had invited white Pennsylvanians to consider as a possibility while they were shivering through a series of bad winters. Some of them had already contemplated that strategy, including at least some ironworks' proprietors. James Logan, the acting governor who'd defrauded the Lenape in the Walking Purchase and who invested in iron production, was one example—his big private library (from which Franklin borrowed books) held a fifth edition of John Evelyn's conservationist *Sylva*, with annotations in Logan's hand. He read it, but he never acted on it. Another solution was importing coal, a firmer shift away from an organic economy, if for the moment blended into it. A 1748 *Pennsylvania Gazette* advertisement for goods on a ship from Dublin, for instance, had listed "servants" alongside mineral coal. Similar ads in the 1750s continued to offer white indentured servants and coal, or else food and coal—in each case, an organic energy source supplemented with an inorganic one.[21]

Colonial expansion into areas with fresh resources continued to be the most tempting solution, whether done with care or in haste, and even if the incursion risked armed conflict. Franklin's own scientific investigations, and those he sponsored, mapped out the expanding zone where colonists assumed Natives mattered only for what they might briefly offer to settlers before ceding the territory in question. In the case of Pennsylvania, this land extended westward to what would become militarily strategic Fort Pitt, later industrially powerful Pittsburgh—the road to economic development ran through an imperial landscape. Making Pennsylvania's interior fit for a rapidly growing settler population was the first step in building that road, also extending Franklin's logic of planning for American economic growth on the scale of entire continents and oceans.

Franklin's next scientific work, like his population analysis, represented the natural world as having universal characteristics, above and beyond everything on Earth, because he was looking up at the air. But in doing this, he repurposed his studies of heating and convection. He described the atmosphere analogically, both as kin to a domestic interior and as mappable in relation to the world beneath it, which happened to be colonized territory.

He began this phase of his scientific investigation, in the 1750s, by considering storms that gusted up the coast of North America. They moved northward, from southwest to northeast, even though the prevailing wind was blowing southward. That was because the wind, stirring things near the ground, was only one relevant force. The other was heat radiating from above, making the system into a three-dimensional force field. In the Gulf of Mexico, the air was "heated by the Sun" and, being thus "rarified," drifted northward, where it met the "chill'd and condens'd" air over the British colonies. The collision sent the warm air up while the cool draft sank, moving counter to each other, warm to the north and cold to the south. Convection again: the two thermal forces generated a dynamic system that corkscrewed north, even as someone who wet a finger and stuck it in the air could detect a definite southward trend.[22]

Franklin's description cleverly expanded on existing theories about prevailing winds, while comparing nature's patterns to those humans could generate themselves. Naturalists had used thermal variation to explain trade winds, for example. Franklin knew about William Dampier's work on trade winds from the previous century, and he might have read George Hadley's elegant 1735 summary, with its hypothesis that Earth's rotation was also a factor, generating wind patterns that moved in opposite directions on either side of the equator. And yet this gigantic atmospheric phenomenon, Franklin said, resembled what anyone could feel in a cozy room with a fireplace. "A Fire in my Chimney," he declared, also created an upper current of heat that eventually met a cooler "Current of Air constantly flowing from the Door to the Chimney," as if the entire coastline of North America were a huge, heated parlor. The vision was one of intrinsic harmony between natural and human-generated thermal systems, another macrocosm-microcosm analogy that solicited confidence in natural science.[23]

Might colonization itself disrupt this harmony? Two of Franklin's friends reached different conclusions. Each traveled into Pennsylvania's interior, to places that had no parlors but still plenty of trees—because they still had Native populations. The two men's forecasts reflected differing respect for Indigenous rights to the land. But both showed a dawning awareness of deep time, the geological chronology embedded in Indigenous histories, now becoming part of Western science, especially as fossilized evidence of a long-dead past was coming to light. Such assessments of North America's interior make clear that, as war loomed, the contested lines on maps weren't just abstractions, but represented real places with coveted natural assets. Further, both men were using natural science, but differently, one to criticize imperialism and the other to promote it.

John Bartram, the Quaker botanist with the project to plant cedar trees, prized the interior lands as they were. In 1743, Bartram joined a party from Philadelphia bound for a meeting at Shamokin to negotiate a peace treaty between Virginia's colonial officials and the Haudenosaunee. He then headed north to continue collecting plants and other specimens, relishing "this journey into the heart of a country, still in the possession of it's original inhabitants." He gives long, detailed accounts of the land and of the trees, many types of trees, which signal types of land beneath them: "middling oak land but stony," he says of one place. In a work of seventy-one pages, 59 percent of them contain a reference to living trees; 75 percent refer either to trees or things made from trees, from cooking fires to canoes.[24]

Far more revealing are Bartram's details about connections among physical places and their Native residents, plus the Natives' knowledge and cultivation of those places. While approaching the Onondaga seat of the Haudenosaunee (now in New York State), he describes "good land producing sugar-maples, many of which the *Indians* had tapped to make sugar of the sap, also oaks, hickery, white walnuts." To these, Onondagas had added Eurasian cultivars, "plums and some apple trees, full of fruit." Another "old town" had peaches, plums, and grapes. Other artifacts were not just old but ancient. Bartram notes where people make salt by boiling water from the bank of a creek, which "inclines me to think that there is a body of fossil salt here abouts." He finds stone fossils near one Native town, on a stream running east into the Susquehanna River.[25]

Altogether, Bartram describes a North America where Natives are autonomous and creative presences. It's a contrast to Franklin's prediction that those people would cease to exist, and their forests likewise.

Bartram's famous son, William, would describe this existential harmony so resonantly in his *Travels* (1791) that his book was one inspiration for the Edenic Xanadu in Samuel Taylor Coleridge's "Kubla Khan" (1816). But Xanadu's prototype was not in Pennsylvania, where the younger Bartram could no longer find autonomous Indigenous nations; he'd gone south, into Seminole, Cherokee, and Creek territories, in order to witness a surviving "terrestrial paradise."[26]

A quest to erase any such paradise appears in a work Franklin printed the year his population essay appeared in London. These were essays by another friend, surveyor and cartographer Lewis Evans, who had drawn the illustrations for Franklin's stove pamphlet—also, more ominously, the map for the Walking Purchase. Evans had traveled with Bartram on the 1743 diplomatic mission to the Haudenosaunee. Like Bartram, he exulted over the interior's ample resources; unlike the botanist (and more like Franklin), he looked forward to their removal from Indigenous stewardship. Evans marveled at the "Ocean of Woods" in North America, so vast it defied reconnaissance, with only "little [topographical] Inequalities, not to be distinguished, one Part from another, any more than the Waves of the real Ocean." The shoreline of a primeval, actual ocean bounded this forested expanse. "A very regular Curve" of upward-projecting land, like a reef, cut south from Manhattan through Virginia and North Carolina. "This was the antient maritime Boundary of America."[27]

In contrast to Bartram, Evans did not use Indigenous patterns of settlement and cultivation to map the region. The Haudenosaunee were the only population he thought worth describing at all. Because they had "subdued" the Lenapes, he said (falsely), they had won the right to their former country. Other Native groups were either "almost extinct," like the Hurons; "mostly destroyed by the English," as in New England; or else "exterminated" by the Haudenosaunee. Obviously, Evans was invested in how alliances with the Haudenosaunee might protect white Pennsylvanians' interests within this uncharted sea of trees.

Sending British settlers westward into the Ohio Valley—which was "naturally furnished with Salt, Coal, Limestone, Grindstone, Millstone, Clay for Glass-houses and Pottery"—would be a useful imperial bulwark against the French.[28]

Evans's vision of territory cleared of any serious Indigenous presence also featured in his two better-known works, maps produced in 1749 and 1755 (which Franklin would recommend to imperial authorities). In the 1749 map, Evans fills Native territory with text describing its nonhuman features, with only one mention of its people. He focuses on the region's storm systems and its natural history. Evans may have written some of this material with Franklin; it describes atmospheric electricity and thunderstorms, based on work Franklin wouldn't publish until 1751, and it describes Pennsylvania's allure for white settlers—"Opportunity & Materials are never wanting to furnish the Industrious w.th Profusion"—much as Franklin did in his work on population. The descriptions likewise touch on modern science, though Evans mostly credits the scriptural Deluge for leaving marine fossils at high elevations.[29]

In his next map, of 1755, Evans likewise represents the Haudenosaunee as the region's only real Indigenous presence. He calls them "the Confederates," implying that their power resulted from a population enlarged through the confederation of their houses. A caption above Lake Ontario marks his belief that Native nations were otherwise vanishing. The map codes the capital letters of their names, by color, to signal their presence or absence:

White ABC. Nations extinct,
Blue ABC. Antient Seats . . . or Nations nearly extinct,
Black ABC. Nations still considerable.

A glance revealed which color prevailed, signaling Natives' tendency, according to colonists, to leave little trace on the land. Meanwhile, obedient to colonial officials' instructions to identify mineral resources, Evans studded the map with the locations of what lay beneath the land, whether as scientific curiosities or usable commodities: *Coals... Freestone... White Clay... Coals & Whetstones... Salt... Coal... Coal... Petroleum... Elephants Bones.* This was a map showing the way to industrialization—when Evans published it, he emphasized that it notes the locations of coal. Extinction haunts the map, both in the form of fossils and in what Evans regarded as the welcome prospect of entire Indigenous nations vanishing, an early statement of the "extinction narrative" still used to describe Native Americans.[30]

Rather than acknowledge Indigenous knowledge of geological time, Evans claimed his own expertise about ancient fossils. In fact, Native men had identified the place where what Evans called "ele-

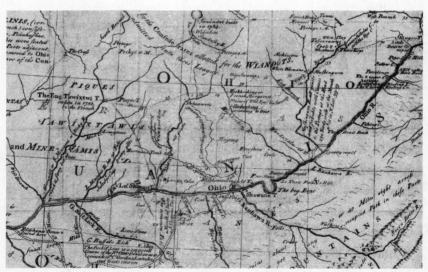

ON THIS MAP, THE OHIO RIVER IS MARKED ON ITS LOWER END (LEFT) WITH "ELEPHANTS BONES"—MASTODON FOSSILS—AND RUNS UP TO "PETROLEUM" ON THE RIGHT, PASSING DEPOSITS OF COAL ON THE WAY.

(Lewis Evans, A Map of the Middle British Colonies in North America. First... Published in 1755 *[London, 1776], detail, Library of Congress)*

phant" bones could be found—and knew them to be artifacts of large animals no longer living. Evans's tiny notations, from coal to fossilized bones, would have enormous significance for colonial and European interpretations of the natural world, by theorizing human and animal extinctions. But in stressing human extinction for Natives only, Evans supported the theory that America's Indigenous people were doomed, unable to thrive in a land of plenty, as Franklin had emphasized.[31]

Franklin did disagree, if selectively, with Evans's (and others') assumption that white settlement of North America's interior would improve its physical nature. Some colonists and naturalists continued to predict that the deforestation of North America would refine its climate—made warmer and better suited to commercial agriculture. But Franklin questioned whether warming the Earth by destroying forests was either possible or optimal. His skepticism underscored yet again his concern over demand for firewood, as the settler population increased—which he championed, though (in another of his contradictions) worried might spoil the nature of the continent, this time cocking an ear and hearing a distant fire alarm. He had encouraged the cultivation of trees, after all, when he published Bartram's project for planting red cedar. And even earlier, in his *Poor Richard's Almanack* for 1744, Franklin had congratulated any property holder "Whose Trees in Summer yield him Shade, / In Winter Fire." From principles that explained this pleasant scenario, Franklin felt able to describe atmospheres at the scale of an entire continent, moving from a small colonized space to a much larger imperial one.[32]

Building on that observation later in 1758, Franklin suggested that forests might cool if not actually create Earth's entire atmosphere. They did so, he said, not just by shading the Earth's surface, but through what would later be called transpiration, the movement of water into and (through evaporation) out of plants. Franklin knew the work of English experimenter Stephen Hales, who in the 1720s and 1730s had described transpiration in plants and respiration in animals. And Franklin had just been corresponding with Connecticut clergyman and Yale professor Ezra Stiles, who'd related to him a "new Theory of Dew" as a nightly exhalation over "the whole Globe thro' Vegetables." Franklin went further, hypothesizing that plants might engender the planet's entire atmosphere. Indeed, he thought animal bodies, including his own, generated atmospheres of warmer and wetter air. Might Earth's prodigious plant life collectively exhale vapor on a global scale?

> Would not the earth grow much hotter under the summer sun, if a constant evaporation from its surface, greater as the sun shines stronger, did not, by tending to cool it, balance, in some degree, the warmer effects of the sun's rays? Is it not owing to the constant evaporation from the surface of every leaf, that trees, though shone on by the sun, are always, even the leaves themselves, cool to our sense?[33]

Franklin conceded that the scale of the phenomenon defied easy measurement. "When a Country is clear'd of Wood," he believed, "the Sun acts more strongly on the Face of the Earth." "But whether enough of the Country is yet cleared to produce any sensible Effect," in North America, "may yet be a Question." Terraforming takes time. "Observations have not been made with sufficient Accuracy, to ascertain the Truth of the common Opinion, that the Winters in America are grown milder." What if they were, but overshooting the mark, with conditions becoming far too hot? Or else staying on the cold side, only now with less available firewood? Be careful what you wish for, Franklin was warning his fellow colonials. They were beginning to comprehend climate change for the first time in history, but lacked full knowledge of how and why it might occur, including for anthropogenic reasons (meaning because of them). Franklin had already, in his stove pamphlet, suggested that fighting frigid winters could be achieved while conserving trees. He wasn't saying colonizers should stop taking over North America's forested areas—but they might want to think carefully about how they did it.[34]

In positing that "Englishmen" might repopulate an entire planet, Franklin was imagining war by other means. Starting in 1755, when he first published his population essay, British colonists, with Britain's support, had been invading—and winning—territories the French regarded as theirs, even if still occupied by Indigenous people. In 1756, France and Britain formally declared themselves at war. By 1758, when Franklin began to ponder the vegetable origins of the atmosphere, the French had made major concessions along the border of what would become Canada. And the war was creeping into all continents and oceans. The Seven Years' War is for that reason now called the first world war (with the twentieth century's "World War I" being a slight misnomer).

Within that global history were many local stories, and a tragic tale for Pennsylvania. Both Lenapes and Quakers had shunned the violent means by which empires were formed elsewhere. But Franklin rejected pacifism, recommending the use of public revenue for military preparation at least since 1747, when he published a pamphlet, *Plain Truth*, warning of the danger of invasion. Next, he recommended pan-colonial military cooperation and a shared government when he attended the Albany Congress in 1754, even before war spread. (At that meeting, primarily with the non-pacifist Haudenosaunee, the British stressed the metaphoric chain that bound the two peoples together, until Native delegates pointedly referred to the council fire as another symbolic bond.) One year later, against Quaker protests, Franklin called for a Pennsylvania colonial militia. And with the onset of the French and Indian War, he helped construct Pennsylvania's first military fortifications, a battery of outposts to regulate the colony's interior. For this task, he commanded troops, and was colonel of Philadelphia's militia. He'd become William Penn's antithesis, the Lenapes' nemesis.[35]

Franklin's parallel musings about the atmosphere show an astonishing new dimension for science, the yawning *above* connected to the small humans *below*. In his *Poor Richard* for 1757, Franklin said that nothing mattered more than "Health," which "depends so much on the Air we every Moment breathe" that the choice of a wholesome place for a dwelling was paramount—but then the home was wholesome, inside and out. With less anxiety about American environments corroding their European constitutions, and greater confidence they would in the end replace Native populations, colonists could fling wide their windows and fill their lungs. But Franklin's thoughts on air and place also reveal his failure to see the full implications of his wartime hypotheses. The air surrounding the planet was of common benefit, but if the trees and plants that produced it were removed from nations that actively conserved them, the benefit would decrease.[36]

More immediately, the conflict brought economic hardships, including rising fuel costs. One can see why the possibility of hemispheric warming was so bewitching. The winter of 1759–1760 sent firewood prices up again, disastrously. Once again, the shivering poor sneaked out at night to steal fuel. As a January 1760 notice in *The Pennsylvania Gazette* put it: "HELP! HELP! HELP!" Firewood, the newspaper said, cost 10 shillings and 3 pence a cord: "a Price never heard of . . . The widow hears a noise in her Yard, rises from her Bed at Midnight, from her Window sees a Thief, and asks him what he is doing; he answers, *I must have Wood*." In the morning, the widow surveys her plundered woodpile, aghast at what it will cost to restock—"the Rich engross [fuel] . . . when perhaps two Hundred families have not a stick to burn."[37]

But even the poorest white settlers were better provided than Pennsylvania's original people. During a 1750 visit to Pennsylvania, Lutheran minister Gottlieb Mittelberger said that older Natives "complain and say that, since the Europeans came into their country, they were so frequently visited by heavy snow-falls, severe frosts, and torrents of rain, of which they had known nothing before the coming of the Europeans." Indeed, starting with English settlements in Virginia in the early 1600s, the invasion coincided with some of the worst intervals of the Little Ice Age. Mittelberger's Native informants could have added that their losing access to land, including its firewood and fur-bearing animals, had made wintertime decidedly harder.[38]

The rising cost of firewood in Philadelphia meanwhile hints that improved heating technologies were failing, at the population level, to offset the decline of fuel supplies. Firewood in Philadelphia had cost 8 to 10 shillings a cord between 1700 and 1735, when Franklin had first arrived. Prices rose to 10 to 13 shillings from 1736 to 1750, 12 to 18 from 1751 to 1756, then a bigger leap up to 18 to 30 shillings from 1757 to 1760 (within the formal years of conflict in the Seven Years' War), edging up to as much as 40 shillings (£2) by 1765, and then a subsidence, averaging from 18 to 30, from 1766 to 1776. (After 1776, new wartime disruptions—British invasion, loss of male labor to the armed forces, inability of the new state governments to keep order—would send the economy into chaos.) And if anything, fuel costs were worse in port cities to the north, where settlers had cut down the nearest woods even earlier.[39]

By the winter of 1760–1761, the situation declined again, not because of another temperature dip, but by postwar economic disruptions. To fill labor needs, Philadelphia merchants imported the largest number of Black slaves in the city's history, more than one thousand people from 1759 to 1766, even as free workers lacked employment and fell into poverty; Franklin's lament about the convenience of forced labor reflected a grim truth. Philadelphia authorities organized a Committee to Alleviate the Miseries of the Poor, whose members went door to door, soliciting contributions from some citizens and distributing firewood to others. When low temperatures returned in the winter of 1764–1765, people would plead for the restoration of the committee. If colonial society was supposed to demonstrate the material superiority of European economic norms, winter and war exposed their flaws, even for white people, let alone enslaved Black people who couldn't control their standard of living, or Natives removed from zones of ancestral plenty.[40]

At this point, Lenapes defined a new kind of resistance through cultural revival. In 1761, the prophet Neolin, a Lenape man living west of the Ohio River, began to preach about his visions, in which the Master of Life recommended a life independent of white settlers. This meant rejection of trade goods, especially alcohol, but also forged metal items (including agricultural tools) and woven cloth; true Lenapes should hunt animals and wear their skins. At its core, this was a rejection of colonial consumers and their material way of being.[41]

Neolin's prophesies inspired an international Indigenous alliance that eventually extended to the Anishinaabeg of the Great Lakes, reflecting widespread outrage over white colonizers' persistent betrayals. Questioning his nation's identification as peacekeeping women, an eastern Lenape leader, Teedyuscung, who'd witnessed the Walking Purchase, was by 1755 advocating military action against white settlements—though not Quakers or other white pietists. At this point, French imperial officials were losing allies in the Great Lakes region, principally among the Ottawas; the "Covenant Chain" between the British and the Haudenosaunee was also breaking apart. Lenapes were being pushed to Pennsylvania's margins and toward similarly disaffected Indigenous nations. Meanwhile, white Pennsylvanians hungering for western land were becoming impatient with the peaceful engagement with Indigenous people that was part of their colony's Quaker legacy.[42]

Franklin would attend and record diplomatic meetings between

Pennsylvania and the Lenapes that turned acrimonious, especially given that Quakers no longer controlled colonial policy and Lenapes were reconsidering their previously pacifist position. Tension had been obvious during the 1756 and 1757 meetings in Easton. Pennsylvania settlers' pressure on western land was exacerbated by rival claims from Virginia and Connecticut; the Connecticut investors had formed a Susquehanna Company to develop land on that river.[43]

Wary of competing claims that might turn violent (as was already happening toward the north) and exasperated by the younger Penns' openly anti-Native policies, Teedyuscung demanded a reckoning. During the 1756 meeting at Easton, he confronted the white delegates about the Walking Purchase. "This very ground that is under me," he said, "(striking it with his Foot) was my Land and Inheritance, and is taken from me by fraud." He would later insist that a white clerk at the 1757 negotiations give him a written transcript of the proceedings, which colonial officials initially protested as unprecedented, then agreed to. He also asked for "a certain Tract of Land fixed, which it shall not be lawful for us or our Children ever to sell . . . a certain Country fixed for our own Use, and the Use of our Children forever." At this place, Lenapes would adopt a way of life modeled on that of the colonists, with year-round residence; he requested help "in building Houses." Teedyuscung sketched out the desired territory, about 2 million acres centered on the Wyoming Valley, later the heart of Pennsylvania's anthracite coal region. Without making a firm promise about the land, colonial officials agreed to build the houses. White workmen constructed eleven of them with squared logs by early June 1758. Similarly, Lenapes at a 1758 conference in Crosswicks, New Jersey, won a reservation for themselves, though at the cost of ceding all other land.[44]

As Teedyuscung became aware, however, his proposal that Lenapes live like white people ran counter to Neolin's and others' invocations of a precolonized Indigenous past. And it would never convince most colonists of Natives' equality with them.[45]

The Lenapes' new settlement in the Wyoming Valley was, accordingly, under pressure both from white settlers and from the Haudenosaunee, particularly as the Seven Years' War strained every alliance. Teedyuscung was apprehensive at having the region's diplomatic center shift either to a British fort to the west or to the Onondaga house further north. He insisted that Pennsylvania was the original site of the southern Lenape council fire. The Haudenosaunee had for some time been questioning Lenapes' decision to take a female role in maintaining peace. Haudenosaunee men knew Europeans had little such respect for women, that Lenapes' status as honorary women could, within a context of imperial warfare, diminish them. As they would tell them at a diplomatic meeting in 1756:

> *Cousins, the* Delaware Indians, YOU will remember that you are our Women; our Fore-Fathers made you so, and put a Petticoat on you, and charged you to be true to us, and lie with no other Man; but of late you have suffered the String that tied your Petticoat to be cut loose by the *French*, and you lay with them, and so became a common Bawd, . . . notwithstanding this, we have still an Esteem for you, and as you have thrown off the Cover of your Modesty, and become stark naked, which is a Shame for a Woman, you must be made a Man; and we now give you a little Power, but it will be some Time till you shall be a complete Man.

The phrasing was crafted to mock Lenape men before a white audience, using Western standards of female behavior, but the Haudenosaunee were also issuing a plea: stand with us as *warriors* against the invaders. But if Teedyuscung was already encouraging Lenape attacks on some colonial settlements, he did not intend the new resistance to white imperialism to mean subordination to any Natives, including the Haudenosaunee, with a longer history of that defiance.[46]

As the Seven Years' War reached a crescendo, the language of treaty negotiations reveals a struggle for dominance and autonomy. In their 1736 treaty (the first Franklin had printed), the Haudenosaunee and Pennsylvania delegates had deployed three key terms: the council fire around which the men gathered, the chain that bound the Haudenosaunee to the British, and the road or path that had to be cleared of obstacles so the two peoples could reach each other. Three times, different speakers summoned all three, "the Fire, Road and Chain of Friendship," as a bundle of political idioms. Colonial negotiators did this twice (starting with Proprietor Thomas Penn himself), and a Haudenosaunee delegate once, this ratio possibly indicating different levels of enthusiasm for the alliance, though it was in fact a Seneca speaker who had introduced the three terms in a careful sequence.[47]

In subsequent treaties, Haudenosaunee and Lenape representatives tended to prefer different political terms, with an especially striking divergence over the chain and the council fire. After the 1736 treaty, Franklin printed twelve further sets of negotiations, three between colonists and the Haudenosaunee (in 1744, 1745, and 1748), three between colonists and the Lenape (with other non-Iroquoian allies, in 1747, 1753, and 1756), and six involving both the Haudenosaunee and Lenape (sometimes with other Native nations, in 1742, spring 1757, summer 1757, 1758, 1761, and 1762).

While these printed versions cannot be considered exact transcriptions, several factors argue for the accuracy of Franklin's accounts. Surviving notes from several other white informants indicate that Franklin avoided some of their biases. This was the case, for example, with one notetaker's tendency to substitute "conference" or "council" for "council fire," a term Franklin retained. Moreover, despite Franklin's preference for white settlers over Natives, he shrank from overt coercion to gain land, in contrast to his war-mongering against the French. In 1757, when he was a member of the colony's Provincial Assembly and a commissioner on Indian affairs, he and three colleagues responded to Teedyuscung's accusations about the fraud embedded in the Walking Purchase. The white men joined in denouncing the "Hearsay Accounts of the Walking Purchase," which were "universally given up as unfair, and not to be defended." Finally, Franklin had himself used the term "council fire" in the texts he'd helped draft for the governor's speeches at treaty meetings.[48]

And Franklin's printed versions of treaty negotiations frequently quote invocations of council fires. The term occurs 103 times in the speeches, as opposed to 68 times for a road or path leading to amity, and 67 times for a chain binding people together. For Indigenous peoples, speaking of a council fire emphasized their autonomy and sovereignty. Quaker members of Pennsylvania's Friendly Association for Regaining and Preserving Peace with the Indians by Pacific Measures recognized this in a medal struck to commemorate the 1757 treaty negotiated at Easton. The reverse of the device shows a Lenape man beneath a sun, and a Quaker representative under a tree, holding a peace pipe out over a treaty fire. (Franklin and members of the society gave copies of the medal to Natives as tokens of peace.) One year later, in 1758, the joint conference with both Lenape and Haudenosaunee at Easton had the highest incidence of "fire" recorded (of the total occurrence of the main three metaphors, council fires constituted 55 percent, as opposed to 12.5 percent for chain), possibly because colonial representatives agreed to meet separately with the Lenapes for part of the proceedings.[49]

By this treaty meeting, around the time Franklin was considering whether North America's deforestation might change the climate, the war was reaching its crescendo, with British victories mounting, ominously for Native people, representing a shift in the geopolitical balance to which they'd adapted their diplomacy. It was in 1758 that the

French had to relinquish their short-lived outpost where the British built Fort Pitt—a bright new pin thrust into an imperial map. At the Easton conference that year, Teedyuscung took great care to get in the first word, reminding the British and colonists that they already had a peace settlement between them and the Lenapes, lest there be any attempt to renegotiate the terms of that. Pennsylvania's governor then bid everyone "heartily Welcome" to "this Council Fire." It took seventeen days, but the Native delegates were finally satisfied that "the Limits of the Lands" belonging to their nations had been confirmed and the proceedings ended with "some Wine and Punch . . . with great Joy and mutual Satisfaction."[50]

But the war wasn't over. It would last another two years, and formal peace between France and Britain wasn't finalized until 1763—threatening earlier accords with Native peoples. To be on the safe side, in the last treaty between colonial and Indigenous representatives, negotiated in 1762, "fire," "road," and "path" began to refer to physical things that identified Indigenous material reality, directly imperiled by the still-unresolved Anglo-French contest. "Road" appeared multiple times because of debates over the literal routes connecting Indigenous and settler territories, especially new passages that could weaken Native sovereignty. A Seneca speaker was most vehement in telling the colonists, "we entirely deny your Request; you shall not have a Road this Way."[51]

Thomas King (Oneida) restated the significance of fire for Lenape sovereignty, though in a way that denied their diplomatic standing, their power to maintain a council fire. Responding to what he said Teedyuscung had claimed about lack of a fire, meaning a homeland, he protested: "Cousin, there is certainly some Fire, for I made one here for the *Shawanese, (Cacawasheca)* and I made another Fire at *Shamokin*, for *Alammapis*; another Fire I made at *Wighalousin*; another Fire I made at *Diahoga*. All those Fires are there yet." This was not a concession to those nations' self-determination, however, as King emphasized in presenting the Lenapes with a diplomatic gift of wampum, shell beads woven into a ceremonial belt:

> By this Belt I make a Fire for *Teedyuscung*, at *Wyoming* [valley]; I tell him to sit there by the Fire Side, and watch that Fire; but I dont give it to him, for our *English* Brethren cast an Eye upon that Land; therefore I say to *Teedyuscung*, watch that Fire, and if any White People come there, tell them to go away, for that Land belongs to your Uncles, the *Six Nations*. The *Six Nations* want to keep up that Fire.

This speech denied the Lenapes' ability to keep a *council* fire or be useful at one. Lenape speakers referred to that kind of fire eleven times during the proceedings; the Haudenosaunee only once. King was telling Lenapes to stay home at their family's hearth, not venture into politics.[52]

One year later, Teedyuscung was assassinated, at his fireside and by fire. On April 19, 1763, his new settler-style house, on the land he thought offered sanctuary for his people, was ignited from the outside and he burned to death. Twenty other buildings were also torched and the whole town burned down. Although colonists blamed the arson on the Six Nations, it's implausible that the Haudenosaunee, having told Teedyuscung to go home to his matrilineal fire, would then use fire to kill him. That would have discredited their own societal organization, which valued fire in the same way, as in the original structuring of Sky Woman's household, restored within her new home on Earth.[53]

It's much more likely that agents of the Connecticut-based Susquehanna Company, who wanted the land for white colonists, committed the atrocity. They had an obvious motive and a better comprehension of how buildings made of beams of wood might be vulnerable. The attack is a horrific reminder that technology, including housing and a built environment, may be dangerous without extended infrastructure and a social structure coevolved to support it. In Philadelphia, white people contracted with fire companies to protect their own and each other's houses, a system Franklin had helped to organize. This was precisely the protective societal system that the white workmen who built the Lenapes' houses did not install. Nor was a council fire protection against invasion by people—fake allies—who denied its significance for diplomatic relations and respect. Teedyuscung's assassination was the first of a series of violent attacks on Natives in Pennsylvania, vengeful punishment for the Lenapes' renunciation of pacifism. The time bomb set when William Penn started to buy Native land finally went off.

British victory in the Seven Years' War, and withdrawal of French officials and traders, increased Indigenous peoples' unease over being consigned to one imperial power—the one with a fast-growing population and its ravenous consumer appetites. Rather than trust and ally with any white people, why not join forces with each other? Inspired by Neolin, Ottawa leader Pontiac sought in 1763 to protect Natives from any further colonial incursion, this time through a war on the colonists, continuing the battle over natural resources. British leaders had been talking up the idea of a council fire relocated to Detroit to coerce Natives in the Ohio territory and Great Lakes into allying with them. Pontiac responded by laying Lenape war belts made of wampum before the Anishinaabeg Council of the Three Fires (Ottawa, Ojibwe, Potawatomi) in Detroit. His action identified their council fires as independent of either Haudenosaunee or British interests, and their military support as crucial to Indigenous autonomy. The result was the largest Indigenous confederacy yet seen, with coordinated attacks on British forts around the Great Lakes and in the Ohio Valley, including western Pennsylvania.[54]

For British officials, this fresh outbreak of war warned that the peace being brokered in Paris might be irrelevant. Colonizers had been too aggressive against Natives and against French settlers. Keeping the peace in an expanded territory that included unruly British subjects, as well as aggrieved Native nations and defeated French Catholics, would require special effort. In late 1763, George III issued a proclamation forbidding further colonial expansion (without royal permission) past a western line drawn through the Appalachian Mountains. This was intended precisely to protect Indigenous people; Canada's First Nations now consider it legal recognition of their sovereignty. For white settlers used to predictions of their inevitable dominance, the proclamation seemed to set their interests and rights below those of Natives, jeopardizing their ability to acquire more resources through expansion.[55]

The new policy, coming during Pontiac's War, incited colonial attacks on Natives, as if Teedyuscung's assassination had lifted the old Pennsylvania prohibition against direct violence toward Indigenous people. A land once so famous for its peace that (Voltaire had marveled) mere words without oaths kept harmony became the most blood-soaked in British North America. In 1763, a vigilante group called the Paxton Boys murdered nineteen Susquehannocks, most of them Christians, some of them women and children, all from a community whose land rights were well documented, permitting their coexistence with white settlers. Pennsylvania officials, including Franklin, denounced the action as murder. In his *Narrative of the Late Massacres*, Franklin diverged from his tendency to rejoice in Natives' disappearance and said the white men were the savages, destroyers of the colony's concord: "O Pennsylvania!" But none of the Paxton Boys ever faced charges. It's all the more insulting that they're nicknamed for a place the Susquehannocks themselves had named. *Peshtank* means a place of springs, where the water stands, a reminder that Indigenous lives flowed from living places in the natural world.[56]

An end to Pontiac's War was negotiated in 1765 over a council fire, if uneasily. Having repelled Pontiac's attempt to take Fort Pitt,

Colonel Henry Bouquet proceeded to the forks of the Muskingum River, now in Ohio, where he met Lenapes, Shawnees, Wyandots, and others allied with Pontiac to begin negotiations. With an assemblage of threats, promises, and acts of consolation, peace came. The artist Benjamin West commemorated the event with an image of men from the different sides gathered around a council fire. What West didn't represent was the fact that Bouquet was one of the British officers at Fort Pitt who, in trying to set off an epidemic among the Natives, had offered them "gift" textiles contaminated with smallpox. The strategy

INDIGENOUS ORATORY AT A COUNCIL FIRE.
(Benjamin West, The Indians Giving a Talk to Colonel Bouquet in a Conference at a Council Fire near His Camp on the Banks of Muskingum in America, in October 1764, 1765. *[Made between 1765 and 1766]. Yale Center for British Art, Paul Mellon Collection)*

was fresh proof of whites' expectation that, one way or another, Natives must surely die away as colonial populations grew and moved west.

Despite his outrage over the Paxton massacre, Franklin continued to believe that Native populations were less economically valuable than white settlements. In 1769, he would be one of the organizers of the Grand Ohio Company, carved from Haudenosaunee land concessions made in the Treaty of Fort Stanwix (New York) in 1768. The treaty tried to move the king's proclamation line westward, favoring the colonies of Pennsylvania, New Jersey, and Virginia, whose delegates signed the treaty. In return, the Haudenosaunee received the highest payment the crown would ever make to an Indigenous American nation, at the price of concessions that hurt Haudenosaunee allies, including the Lenapes. The Grand Ohio Company proposed developing 2,400,000 of these acres, "vacant and unsettled by any of Your Majesty's Subjects" but potentially "of great Emolument to Government in point of Revenue." And thus Pennsylvania's rich interior was opened to dense settlement and intense economic development.[57]

By this point, Franklin was no longer in Pennsylvania. Late in the year of the Paxton atrocities, he had gone to England, and would spend most of his remaining life in Europe as a political operator. He had stopped attending and printing treaty negotiations a decade before Fort Stanwix. Neither his continuing scientific inquiries nor his immersion in the politics leading into the American Revolution would inspire him to harmonize his warring claims: conservation would preserve a comfortable human place in the world, but the economic growth he thought had to be part of that happy condition was the greatest threat to it. He had three more stoves to invent; none would reconcile his two ideas about optimal human use of natural resources. Growth or death—they're still at war.

6

FUELED BY FOSSILS

"It has the Coffee Cups in its Belly," Franklin told his wife, "pack'd in best Chrystal Salt, of a peculiar nice Flavour." The belly belonged to a beer jug, one item in two crates crammed with things Franklin had just bought in London and was sending home to Philadelphia: yards of damask and silk and cotton (and carpeting), books and pamphlets and sheet music, multiple microscopes and one apple corer, arrays of china, glass, and silver for the table (including "4 Silver Salt Ladles, newest, but ugliest, Fashion"), swathes of tablecloths and napkins, plus that peculiarly tasty salt. Franklin didn't tell his wife what any of this cost. Except he did—in a way. He complained, in the same letter, of London's terrible air, foul with coal smoke, and he described Peter, the enslaved man he'd taken with him, as behaving "very well." And there we see the price of modern economic life over and across the Atlantic Ocean in 1758, when Franklin cataloged his London shopping while also documenting the era's two most problematic forms of energy, packed as tightly together as coffee cups inside a salt-filled beer jug.[1]

Franklin's era is distinctive, and instructive, because it had both an organic economy that still enslaved human beings *and* an emerging reliance on fossil fuels. Over the two intervals he spent in London as a colonial lobbyist, in the momentous 1750s, 1760s, and 1770s, he would reach different conclusions about each kind of energy. He refined his tentative statement of 1751 from the *Increase of Mankind* about the evils of slavery, and declared the practice irredeemable. In contrast, he thought fossil fuel consumption could be continued—with improvements, as with the technology he recommended to minimize emissions from coal. This hedging resembled his enthusiasm for colonial consumerism, which he thought needed only greater freedom from British regulation, an issue that would spark the American Revolution. His salaried obligation to represent colonial interests in this struggle is already well-known, as is his evolution from enslaver to abolitionist. These narratives are rightfully famous for their historical implications. But the parallel embrace of coal, Franklin's visible example included, is likewise momentous, the other revolution that deserves equal attention.[2]

Franklin's final three heating systems would be designed to burn coal only. His original goal of consuming minimal wood to produce maximum warmth was a legacy of the old organic economy, conserving renewable resources and living within natural limits as much as possible. He had in this way pushed an abundant "new" world forward in time, into a likely future of scarcity. From the late 1750s onward, however, living for the most part in Europe, he would adjust these speculations to a semi-industrial economy with higher rates of consumption and expectations of comfort, and a greater dependence on energy derived from coal. This led him to consider fossil fuel—where it came from, how its smoke influenced atmospheres, how it affected human welfare, and what it might reveal about the planet's past. He knew coal was one of several fossils whose ancient history, especially in North America, warned of changes in climate that could render entire species extinct. Fossils fueled both scientific knowledge and domestic comfort, though, in the end, the search for comfort would outpace the gaining of wisdom.

Franklin began to participate in Britain's unfolding industrial history while he lived in London, not as a young printer for hire (as he was decades earlier), but as the official agent of Pennsylvania's Assembly and for other colonies, in two stints: from 1757 to 1762 (when accompanied by his son and their two enslaved Black manservants) and from 1764 to 1775 (when he went alone). Both times, he enjoyed the status of a famous man of science. From rented lodgings in a townhouse owned by the widowed Margaret Stevenson on Craven Street in the Westminster area of London, he immersed himself in the capital's intellectual life and knit a broader network of correspondents, British and international. During his first residence, he toured Scotland and the Netherlands, and visited Birmingham, his wife's hometown, which fostered a community of men who, like him, were engaged in science and knew practical trades.

Altogether, it was a crash course in the human and nonhuman energies that were transforming Europe from an organic to an industrial economy. Far from the more ragged edge of that transition in Pennsylvania, with its pooling of scarce capital, ongoing appropriation of Native land, and reliance on enslaved labor, Franklin learned about the dynamics of a life fired by coal, a source of energy that seemed—for the moment—greater than anything North America had to offer.

In London, Franklin took up the question of heat again by first considering its presence in the human body. He still assumed heat was an elemental fluid, "Fire" in pure form, able to flow into material bodies that had varying ability to retain or conduct it. There was even "a certain Quantity of this Fluid, called Fire, in every living human Body." An optimal amount of it "keeps the Parts of the Flesh and

Blood at such a just Distance from each other, as that the Flesh and Nerves are suple, and the Blood fit for Circulation." The body sensed heat's distance or absence as cold; a cold body "stiffens, the Blood ceases to flow, and Death ensues." There must therefore be, somewhere in the body, "a Fund for producing" heat, and plants must, as they grow, attract both "the Fluid, *Fire*, as well as the Fluid, *Air*." Finally, "a Kind of Fermentation" or digestion within plant and animal bodies eventually released fire in its "fluid Active State again." Certain activities, like "Exercise," could quicken that process.[3]

From these conjectures, Franklin next considered combustion. He referred to "solid Fire" as what certain materials contained and released as heat when they burned. Wood was a good example, but so were "some Fossils," like "Seacoal"; "Gunpowder," he concluded, "is almost all solid Fire." Distillation, like fermentation, also released heat. The liquid in alcohol distillers' vats was, "I have been informed," about as warm as the human body, 94 to 96 degrees Fahrenheit. Given that similarity, Franklin compared combustion to human digestion, "as by a constant Supply of Fuel in a Chimney, you keep a warm Room, so by a constant Supply of Food in the Stomach, you keep a warm Body," one that, as he'd concluded elsewhere, generated its own portable atmosphere.[4]

London's much bigger atmosphere contained, Franklin was sorry to discover, much more coal smoke than he'd witnessed in the city some thirty years earlier—the city was well on its way to earning its nickname "The Smoke." It was in February 1758 that Benjamin complained to Deborah Franklin about the constant smog. She suggested he burn wood instead of coal. "It would answer no End," he replied, "unless one could furnish all one's Neighbours and the whole City with the same." He described this analogically, implicating domestic combustion in the pollution of outdoor atmospheres: "The whole Town" of London "is one great smoaky House, and every Street a Chimney, the Air full of floating Sea Coal Soot." Beggar to baroness, Londoners shared equal access to the filth: "you never get a sweet Breath of what is pure, without riding some Miles for it into the Country."[5]

With these new thoughts, Franklin began to update his fireplace design, which evolved into a different device entirely, his third model. At the start of December 1758, not quite a year and a half after arriving in London, Franklin wrote to a Boston scientific correspondent, James Bowdoin, about improvements to the fireplace in one of his rooms on Craven Street. These adjustments, as with Franklin's original Pennsylvanian fireplace inserts, were "for keeping rooms warmer in cold weather than they generally are, and with less fire." The key feature was an adjustable register, a sliding iron plate that, set into a grooved iron frame, could close the chimney entirely ("convenient when there is no fire") or open only two inches of the chimney at its back. A bricklayer had modified the fireplace itself, reducing its opening to a more heat-retaining three by two feet. (Presumably, Franklin's landlady had authorized all of this.) The new brickwork was then attractively "faced with marble slabs." The design stemmed any flow of warm air that might leak up the chimney, its expanding presence in the room creating pressure to resist cold drafts oozing in wherever they found an opening.[6]

As usual, Franklin outlined ways to monitor the fireplace's operation. First, someone could tell when the register was correctly positioned because the telltale whistling of drafts would decrease. Second, they could open the door to the room a crack, put their hand toward the top of the gap, and feel the "cold air coming in . . . but weakly, if the plate be in" place. Third, they could put something against the partly opened door, heavy enough to keep it from swinging wide when the register was in the right place, but powerless if anyone pulled the register fully out, seeing that "the door will be forced open by the increased pressure of the outward cold endeavouring to get in to supply the place of the warm air."[7]

Franklin assured Bowdoin that this "simple machine," with its articulated metalwork, was universally useful. Several visitors to Craven Street "have imitated it at their own houses, and it seems likely to become pretty common." (Visits to Franklin evidently involved generous demonstrations of his inventions.) He recommended the register "particularly" to Bowdoin, "because I think it would be useful in Boston, where firing is often dear," due to loss of trees and absence of coal.[8]

From the question of indoor heating, Franklin moved to cooling and to the circulation of air through entire houses. He would have read, in J. T. Desaguliers, of attempts to ventilate buildings, both to cool them and to refresh their air. Desaguliers had been tasked with doing that in the main chamber of the House of Commons, always hot with debate, and physically overheated whenever crowded and lit by candles. Desaguliers installed slender pyramids, six to eight feet tall, in the space above the chamber's ceiling, each vented at its bottom to receive hot air as it rose. Unfortunately, when the air in the pyramids was significantly colder than whatever was below the ceiling, it pushed down into the room and onto the unlucky members seated just below. Desaguliers adjusted by installing two small fire grates at opposite ends of the room above the chamber that, when lit, vented hotter air outward to control the drafts. (By the 1760s, after Desaguliers's death, the system was described as defunct.)[9]

Likewise, Franklin observed that, in summer, a draft of air could be felt as it passed through a room and up an unused chimney, most noticeably from about five or six in the evening to eight or nine the next morning, "the hours varying a little as the days lengthen and shorten," or when the weather changed. Franklin couldn't find any investigation of this phenomenon, though he thought it challenged "the old saying, *as useless as a chimney in summer*." To the contrary: Why not put a chimney's summertime draft to good use? (In fact, the very rich used their fireplaces in summer to display potted flowering plants.) Franklin suggested repurposing the hearth and chimney shaft as cool storage for meat wrapped in damp linen, or butter and milk in vessels covered with "wet cloths." "Evaporation" from the wetted fabric as the draft passed over it would chill the foods, otherwise prone to spoil in hot weather.[10]

Franklin attributed this phenomenon to the general tendency of heat to make air "rarefied" and lighter. A chimney insulated the air inside it, keeping it warmer than the cooling air of summer nights, then cooler than the warming air of summer days. The thermal variation created air circulation up or down the chimney. That effect might be augmented should houses be built with chimneys that maximized sun exposure on three sides, especially if they were painted black to absorb even more heat. Moreover, a rising current of air might cool an entire house. Franklin speculated that "if a house were built behind Beacon-hill" in Boston, Bowdoin (or anyone) might dig from one of its exterior doors a tunnel into the hill and then a shaft going down, gaining a current of air always insulated from the sun. The householders would then "have as much cool air passing through the house, as they should chuse." Such tunnels and shafts might also, Franklin offered, ventilate mines and protect them from damp.[11]

Geothermal air-conditioning was not unheard-of, though in the eighteenth century it was pretty ambitious for colonial consumers at the middle ranks. But so far, Franklin's ideas and devices, either to heat or cool indoor atmospheres, still belonged to the organic economy. Bricks, iron, and cellar spaces were technologies that would have been familiar within Europe, even in ancient times. This makes them interesting: the improvement of *sustainable* technology to deliver more comfort. Franklin thought this way, even though he was keeping himself comfortable in London by burning coal, not wood, a serious step out of an organic existence.

Franklin proceeded to consider one more heat-related topic: color and thermal energy. In a late 1760 letter to Mary Stevenson, his landlady's daughter, he recounted the experiment he'd done in 1729 or

1730 on "Degrees of Heat imbibed from the Sun's Rays by Cloths of different Colours." Stevenson could explore this phenomenon with a "burning Glass" that, while concentrating the sun's rays, wouldn't necessarily set white paper alight but "if you bring the Focus to a black Spot or upon Letters written or printed, the Paper will immediately be on fire under the Letters," a vivid image from a printer. Also, Franklin explained, wearing white stockings in front of a fireplace didn't amplify the heat but black stockings are "apt sooner to burn a Man's Shins."[12]

A 1759 visit to Edinburgh prompted further considerations of heat. Among other nice validations of his status in science, Franklin received an honorary doctorate from the University of St. Andrews and met members of the city's Philosophical Society, who made him a member. But if Dr. Franklin didn't already suspect his knowledge was out of date, he found out then and there. The three works on heat he'd cited in his stove pamphlet had initial publication dates of 1713, 1734, and 1735; work in chemistry, on heat specifically, had advanced significantly in the decades since. Franklin tried to get in on this discussion, writing to one new contact, physician and chemist William Cullen, for access to work Cullen hadn't yet published, promising the manuscript would "go no further than my own closet [private room] without your permission." But while asking another Scots contact for news of Cullen's work "on *Fire*," Franklin said that he himself had "been dealing in *Smoke*."[13]

This is the first sign of Franklin's renewed interest in wringing the utmost out of any fuel—with minimal smoke as proof—which differentiated his work from the increasingly specialized chemical investigation of heat. Air, for example, was being defined not as a fluid suspension of particles capable of compression and expansion (the Newtonian definition Franklin had accepted), but as having a specific chemical composition. Nor was heat anymore a mysterious and fluid "calor." Franklin knew the work of Joseph Black, who, starting in the 1750s, investigated heat intensity and capacity, and compared the chemical signatures of combustion and respiration; he defined latent heat as the energy-transfer necessary, though undetectable by a thermometer as sensible heat, to change a substance from one state to another—ice into water, for example.* Black was also the first experimenter to identify one of air's elements, carbon dioxide. (Only later was this described as a greenhouse gas, capable of trapping heat within Earth's atmosphere.)[14]

Given the ubiquity of oxygen, it's perhaps unsurprising that its discoverers seem nearly as omnipresent, though this has made distributing credit for the discovery almost impossible. Franklin knew most of the claimants. But not the very first, Swedish apothecary Carl Wilhelm Scheele, who had trustingly written to his fellow chemical investigators Antoine Lavoisier and Marie-Anne Pierrette Paulze Lavoisier about his experiments; while benefiting from his information, they filed his (unanswered) letter so deep it was only rediscovered in 1992. The Lavoisiers likewise slighted Joseph Priestley, the second investigator to isolate oxygen (who'd also unwisely shared his findings with them), and thus positioned themselves to claim priority in defining and naming this element; they also identified and named hydrogen. Franklin would meet the Lavoisiers in Paris, and, in England, he became close friends with Priestley and with Jan Ingenhousz, a Dutch physician and natural philosopher who would describe what is now called photosynthesis.[15]

But, while Franklin knew about these new definitions of chemical composition, he didn't update his work accordingly. Birmingham naturalist Erasmus Darwin would therefore dismiss his analysis of "fire"

---

* When ice melts, water forms—but the water stays at the freezing point, not getting warmer, until all the ice is gone; water becomes steam as it boils, even if the water's temperature stays the same (at the "boiling point"). Black explained these seeming paradoxes in terms of a heat that, because imperceptible during a substance's transformation, is "latent."

as "nonsensical . . . knowing nothing of Chemistry." Franklin wasn't unique in continuing to define imponderable substances, beginning to be called gases, in terms of their physical properties rather than their chemical components; this emphasis on hydrostatics, as it's now called, registered his debt to Newtonian definitions of matter and remained central to his inquiry. What's puzzling is that he ignored how chemical investigation of air was, at the time, integral to investigation of heat. The dominant theory postulated an element, phlogiston, that made certain materials catch fire; while they burned, their phlogiston passed to the air, which, when saturated, could no longer support combustion. Similarly, respiration was thought to remove phlogiston from the body.[16]

Franklin remained focused, instead, on the physics of heat, as he related in a 1762 letter to one of his American electrical collaborators. He described electricity as itself generating heat—it was its own distinct phenomenon, if similar to other kinds of "fire." Having proven (with a kite) that "artificial electricity" and "the electricity of nature" were identical, Franklin noted that both could melt metal. This was "not by what I formerly called a cold fusion." Rather, the metal not only reacted to electricity as it would to heat but could then burn other substances. Lightning and electricity were "fire" indeed, multiplying the ways to generate flame or heat: friction, fermentation, electricity, chemical mixture, and by concentrating light from the sun. The mode of production might vary, but the result was always the same. And from this insight, Franklin speculated that fire (as he still called heat) might exist in any "body, though in a quiescent state." If so, "kindling fire in a body would be nothing more than developing this inflammable principle, and setting it at liberty."[17]

The efficiency of combustion was still what Franklin found most compelling, with smoke and soot the shameful marks of waste. He'd intended the second version of his Pennsylvanian fireplace to minimize that inefficiency. But did it? Cleaning the device was tricky, so he couldn't measure the soot with any regularity. The necessary modifications to the chimney and installation of the airbox made it difficult for a chimney sweep to reach the bottom. Franklin had loosened a brick on the chimney piece's side to slide out; a sweep could, as they worked downward, brush the soot onto the hearth through this space. Swedish visitor Pehr Kalm had reported that, for the Franklins' main fireplace, "the chimney is seldom cleaned more than once a year, but Mr. Franklin was in the habit of setting fire to a sheet of paper every fortnight and let it pass through the flue leading to the stove and so burn off the wood-produced soot there also." (These pieces of paper were probably used—Franklin was sending yet more rough drafts up the chimney, never to be seen by the historian.)[18]

Coal was even sootier, though also hotter, which is why Franklin's register system at Craven Street had to be different from his Pennsylvanian fireplace. When a Scottish friend, Sir Alexander Dick, was plagued by a smoky chimney in 1762, Franklin sent him "one of my [London] Machines for your Chimney," promising "that after more than 20 Years Experience of my own Contrivances and those of others," this was the best "for the Warming of Rooms." As delivered, Franklin's device would seem too small for Dick's fireplace. A mason would need to "contract the Opening" and then "this new Brickwork may be fac'd with Dutch Tiles, Stone or Marble at your Pleasure"—consumer options not so easily recommended in the colonies. Franklin ran Dick through the theory of the register he'd devised for Bowdoin, including the "several easy Experiments" to verify its virtues. For Dick, he specified that the door to the warmed room should be opened only half an inch; if one person opened the register fully, another person with a hand to the upper part of the door's crack would feel a dramatic influx of chilly air and then, if they closed the door, they could hear "a louder Noise" hissed by the thwarted draft.[19]

The "Simplicity, Cheapness, and Easy Execution" of his Craven Street "Contrivance," Franklin told Dick, led to its widespread adoption. He claimed "many Hundreds" of them were set up in London. The device could even be "the Cure of Smoaky Chimneys" in certain cases, though not all, because that drawback had too many causes. It was a conundrum Franklin set aside for the moment. In the meantime, he told Dick he could amplify his fireplace's heat by lining the jambs supporting its sides with "Plates of polish'd Brass" to reflect warmth into the room. Two thin plates, one per side, cost 25 shillings, not something most householders could easily afford. Here again, Franklin was adapting to Britain's wealth of consumer goods, including brass, beyond what was typical in the colonies. Similarly, and on the same day, Franklin wrote another Scottish friend, the much more famous philosopher David Hume, describing the installation and function of lightning rods, substituting "a Steel Rod" in place of the iron he'd specified for colonial lightning rods.[20]

Within the same week, Franklin wrote to a third friend in Scotland, yet another philosopher, Henry Home, Lord Kames, with yet more fireplace woes. (Kames and Dick had been comparing their wretched chimneys, hence the stream of needy Scottish letters to Franklin.) Kames seems to have told Franklin of a recent work, *The Edinburgh Smoke Doctor* (1757), in which two local artisans diagnosed various problems with flues in townhouses and recommended vent designs (which they made and sold) to cure the problems.[21]

"I am griev'd that you should live in a smoaky Room at Edinburgh," Franklin told Kames, but, picking up the medical metaphor, he warned he couldn't make any firm diagnosis. One couldn't "prescribe to a Patient at such a Distance," so Franklin listed "5 or 6" possible causes. These were mostly about atmospheric conditions—whether the smoke tended to disperse depending on wind direction or ambient temperature—but also about the configurations of the house: distance from door to fireplace; height of the building relative to those around it. Franklin cautioned Kames against consulting "Workmen" because they, unlike men of science, were "ignorant of Causes," like medical

"Quacks." He instead recommended consulting an actual physician, someone trained to observe faulty respiration—in humans. If respiration was a form of combustion comparable to what a fireplace did, then bring on the medical doctors, fellow men of science.[22]

The industrious revolution's mounting energy demands were turning into the Industrial Revolution's monstrous appetite for fuel, as Franklin was learning from his new scientific contacts in Britain, particularly in Birmingham. There, Franklin knew several of the famous men who would organize the city's learned circle, the Lunar Society, including its organizer, Dr. William Small. The club was "lunar" because it met for dinner on nights of the full moon, giving everyone light to walk home—an organically sustainable plan. Other members included a Darwin (Erasmus, grandfather of the evolutionist) and a Wedgwood (Josiah, the entrepreneurial potter), plus the heart of the society, manufacturer Matthew Boulton, who financed another of the society's "lunatics," instrument-maker and chemical experimenter James Watt, in his project to invent an improved steam engine, one that would power some of Britain's first industrial concerns. It was during Watt's key effort to develop a better engine that he decided to consult Franklin, who was already famous for inventing an efficiently burning "machine."[23]

Watt was trying to improve an existing device, the atmospheric steam engine, made commercially viable in the 1720s by Thomas Newcomen's design, which dated back to 1712. In a "Newcomen," steam from a boiler fills a metal cylinder, driving up a piston, with that motion transferred, via a balance beam, to a pump. Then, a spray of cold water cools the cylinder, condensing the steam, lowering its pressure within the chamber; atmospheric pressure sends the piston back down, lifting the pumping mechanism. Franklin would have seen an image of a Newcomen in Desaguliers's *Course of Experimental Philosophy* and indeed, the technology built on scientific experiments, especially Boyle's investigation of air pressure in a closed system, which had influenced Franklin's analysis of heated air. Most Newcomens pumped water from mines, raising water by means of fire; when they pumped water from coal mines, they helped extract the very material that fired them up. They in fact required a lot of coal, and this motivated the search for more efficient versions.[24]

Watt thought the key inefficiency was reheating the chamber every time the cold water had cooled it. He believed that a separate condenser chamber would conserve heat in the piston's cylinder, thus requiring less fuel to reheat. (Actually, the mechanism operates according to principles of thermodynamics not yet identified—specifically, the greater difference the separate condenser creates between the engine's maximum and minimum temperatures.) With financial support from Boulton, Watt worked on a separate-condenser design for more than a decade, roughly 1763 to 1775.[25]

In February 1766, Watt authorized Boulton to send Franklin his most recent model (to the envy of others in the Lunar Society), and in a letter dated the twenty-second of that month, Boulton solicited advice about the coordination of the piston chamber and the condenser. Boulton asked Franklin which design of steam valve "do you Like best," and where, exactly, a jet of cold water should operate to speed the cooling. Parallel to this, Boulton wanted Franklin's help in investigating the conservation of heat, promising to send to London a piece of steel that had been made red hot, then packed in wood ashes (inflammable insulation) inside a barrel that would be wrapped in heavy wool blankets.[26]

Franklin was a very logical consultant for the steam project, precisely because of his fireplace design. His device, as its accompanying pamphlet had explained, was built to separate three kinds of air: hot, cold, and smoky. Watt's separate condenser operated similarly to keep the piston's cylinder hot, the condensing chamber cool, and (relatedly) to evacuate the residual water from the condenser and air from both it and the chamber. Correct motions of air and water at varying temperatures were crucial to the action. The stove and the engine were related contributions to the ongoing analysis of circulation of fluid substances—air, steam, heat, electricity, and water—that might be controlled to improve life: to generate wealth, multiply comforts, maintain health, and conserve resources. And it's significant that Franklin was corresponding with Boulton, often seen as the capitalizing brawn behind Watt's brains, but in truth representing the modern unity of invention and capitalism.[27]

The timing of Boulton's request was the awkward part. In early 1766, Franklin was absorbed in controversy over the 1765 Stamp Act, the British government's attempt to raise revenue in the colonies by taxing paper goods and documents. Coming so soon after the Proclamation Act of 1763, this seemed another ploy to reduce colonists' status, this time making them into a mute tax base unrepresented in any of Britain's governmental bodies. Franklin, as a colonial lobbyist, had to delay giving expert scientific advice to Boulton because he had to give expert political testimony to Parliament on February 13, 1766. The Stamp Act would be repealed, but succeeded by the Declaratory Act of March 18, 1766, stating that British Parliament had the right to make policy over the colonies, including the legal authority to raise revenue. It was a truce, not a peace.[28]

"You will I trust excuse my so long omitting to answer your kind Letter," Franklin wrote to Boulton the day after the Declaratory Act was passed, "when you consider the excessive Hurry and Anxiety I have been engag'd in with our American Affairs." (He'd returned the model and documents the week before—maybe pondering the technical problem had been a solace amidst the political acrimony.) His priorities make sense, considering his duties as a paid colonial lobbyist. Given that the steam engine has had greater impact on the world than the breakdown of British rule over North America, however, he might have acted differently. Timing was everything, as Franklin explained in his apology—though in the end his advice for the steam entrepreneurs was both timely and timeless because it was about fossil fuel pollution.[29]

Franklin said he couldn't answer Boulton's questions by inspecting a motionless model: "Experiments will best decide." What struck him most about the engine was a different inefficiency. He suggested it should "burn all your Smoke. I think a great deal of Fuel will thus be sav'd." This was for two reasons. "Smoke is Fuel," a suspension of tiny bits of combustible material, "wasted where it escapes uninflame'd." Moreover, the "sooty Crust" that smoke formed on "the Bottom of the Boiler" insulated it from heat, requiring more fuel to keep boiling the water that generated the steam. Franklin suggested developing a grate for the coal that would "make the Smoke of fresh Coals pass descending thro' those that are already thoroughly ignited," double-burning the fuel and decreasing buildup on the boiler.[30]

That idea of another way to draw down and consume smoke would not influence Watt's first condenser design. There is no more surviving correspondence between Franklin and Boulton on the matter, in any case. But Franklin kept the problem of smoke in mind, including the possibility of somehow reburning it, particularly as he adapted to coal, the default English domestic fuel. Concern over the pollution it caused, particularly in London, was no longer intermittent but persistent. The problem was growing because London's population was increasing—and because per capita consumption of coal was rising, too. It's apt and yet ironic that this is exactly when chemical experimenters were establishing that such a thing as oxygen existed in the air and was the component essential to life and health. Accepting coal smoke as a kind of tax on city life seemed a high price, as far as Franklin was concerned.[31]

A colonist's horror over soot and smoke was revealing: just as North America had greater forested areas compared to Europe, so, too, did it have fresher air. Could conserving fuel also preserve the air, creating a better atmosphere for health? England, smoke-blurred London especially, was a warning that worry about this was not misplaced. In prioritizing the Stamp Act controversy over the steam engine, Franklin was just doing his job as a paid colonial lobbyist. But in the longer term, industrialization was, arguably, the more important problem.

Indeed, the two issues intersected, as in Franklin's testimony to Parliament about the Stamp Act. The members ran him through a rather thorough 174 questions about colonists' willingness to make any contribution to British revenue. Franklin wasn't shy about presenting his fellow Americans as economic assets—if permitted to develop autonomously. As in his *Increase of Mankind*, he claimed that the continental colonists multiplied faster than their English counterparts "because they marry younger," being able to "easily obtain land of their own." These consumers with their accumulating demand for manufactures would yield plenty of revenue. Here, Franklin implied that the Stamp Act was objectionable because it taxed goods within the colonies, but that colonists would accept duties on exports from Britain as an "external tax."[32]

He also claimed that colonists were within a few years of making everything they wanted to consume, anyway. He downplayed their prowess in making iron and steel, focusing instead on the textiles that were key to Britain's industrialization, with much back-and-forth about the quality of American wool as a potential substitute. Franklin said the raw material could produce sturdy homespun and warned this would make a boycott feasible. That might be inconvenient for colonists, but not for long. The last two questions, and his responses, suggested as much:

[173] Q. What used to be the pride of the Americans?
   A. To indulge in the fashions and manufactures of Great-Britain.

[174] Q. What is now their Pride?
A. To wear their old cloathes over again, till they can make new ones.

This defense of industrious colonists was slightly at odds with Franklin's support of his industrializing British friends. In his letter soliciting Franklin's advice, Boulton had affirmed his and his friends' support "for the rights, of your oppress'd Countryman." Neither he nor Franklin would have been unaware, however, that an American market for manufactures, including whatever steam engines and coal might produce, would have massive implications for British industrialization, and for American consumers' right to reject British exports and taxes accordingly.[33]

From his interaction with Boulton and Watt, Franklin may also have learned caution about profiting from a new technology. He had already gleaned that lesson from Desaguliers's *Course of Experimental Philosophy*. There, Desaguliers said he'd declined to seek a patent for his devices to improve the heating and circulation of air, which he'd introduced to the Houses of Parliament, also comically relating how someone had impersonated him to apply for a patent. Desaguliers particularly deplored the steam engine inventors who sought patents: "*There are several Persons who have Money, that are ready to supply boasting Engineers with it, in hopes of great Returns; and especially if the Project has the Sanction of an Act of Parliament to support it—and then the Bubble becomes compleat, and ends in Ruin.*" The possible financial wreckage was only one difficulty—the human desire to profit from natural phenomena in the first place was, for Desaguliers, the worse impulse. "*Our Legislators may make Laws to govern us, repeal some, and enact others, and we must obey them,*" he admitted, "*but they cannot alter the Laws of Nature; nor add or take away one ιώτα [iota] from the Gravity of Bodies.*" He would repeat this warning even when he praised a man who had patented a grate for burning coal. Desaguliers noted that it merely provided heat indoors, as the sun did outside—nature being the ultimate benefactor, not any tinkering human.[34]

Indeed, patenting inventions was not yet standard practice. Franklin was a fellow of London's Society of Arts (founded 1754), which promoted useful devices as tools of philanthropic improvement, not economic gain. In 1764, London physician Alexander Small lamented to Franklin that this ideal hadn't influenced the famous British competition to develop a way to determine time at sea, making it possible to pinpoint a ship's position. This longitude question would, in Britain, be mostly solved with John Harrison's chronometers, though Harrison had multiple competitors. As Small warned, their feuding was a nefarious business, and he gave Franklin gossipy details about who was busy sandbagging whom. "What Candide [candid] Patrons we are of the Sciences," Small said, with maybe an intended implication that sunny views of science and improvement were what Voltaire, in his *Candide* (1759), called "Panglossian." All the same, the period after 1760 was a boom time for British patents, a crescendo of inventions, improvements, and entrepreneurship.[35]

Franklin knew that the scramble for preeminence in technology and science typically ignored people in the Americas, most of whom (he being the obvious exception) were considered competent only to select natural specimens for the real experts to study. Knowledge and inventions were supposed to radiate from imperial centers, a prejudice that in particular excluded anyone who wasn't white. This cultural hierarchy persisted despite industrialization of the colonial economy. The Schuyler copper mine in Bergen County, New Jersey, is a case in point. There, on land claimed by Arent Schuyler (ancestor of Elizabeth Schuyler, the philanthropist who married Alexander Hamilton), an enslaved Black man found copper ore while tilling a field. Schuyler began to extract the ore in 1715 and, when his mine flooded, imported what is likely the first steam engine in British America. (The second seems to have been in Jamaica, in 1768.) Fueled by wood or charcoal, this Newcomen could pump seven hundred thousand gallons of water a day. Like the Pennsylvanian fireplace, the New Jersey steam engine teetered between organic and modern energy regimes.[36]

As did Franklin's household in Philadelphia. The two crates of British goods he'd sent home were just the beginning. In spring of 1765, Deborah Franklin sent letters at a brisk clip, reporting on the new house they were building. In his absence, she was a deputy husband, legal contractor for the project, endlessly patient with her partner's tendency to second-guess and micromanage at a distance. When he learned that the new kitchen stove was cranky, he wrote, "I could have wished to have been present," telling his wife "it is a mere Machine, and being new to you, I think you will scarce know how to work it. The several Contrivances to carry off Steam and Smell and Smoke not being fully explain'd to you." (Was this slightly patronizing? No. It was very patronizing.) Franklin also fussed about the "furnace," a firebox that supported a large built-in metal pot, useful for cooking or heating water for other tasks, such as laundry. Franklin said if an iron furnace hadn't been installed yet, "I shall bring a more convenient copper one" from Britain, land of shiny metal goods. Meanwhile, Deborah entertained guests in the only warm part of the house: "I am much oblig'd to my good old Friends that did me the Honour to remember me in the unfinish'd Kitchin," Benjamin joked, "I hope soon to drink with them in the Parlour."[37]

The completed house was both well-heated and fireproofed, featuring both of Franklin's main contributions to modern comfort. While installing protections against fire (no flammable wood trim leading from room to room; multiple lightning rods on the roof), the Franklins made sure they'd never be cold. They had some pretty grand fireplaces, as much for decor as actual heat, defying Franklin's disapproval, in his stove pamphlet, of such preening waste. Insurance records reveal that the ground-floor dining room had "A Rich Chimney piece" with columns and a decorated mantelpiece made from imported marble. Two other rooms on the upper floors had chimney pieces modeled on the "tabernacle frame" design, with columns supporting a decorated triangular pediment, like the Parthenon in Athens. Benjamin sent the fireplace surrounds from London to display his family's status: owners of a grand house filled with costly consumer choices, a place to live and entertain, no longer combined with printshop or post office.[38]

And yet the heating was necessary, not just ornamental. The mantelpieces arrived during another unusually cold Philadelphia winter. Deborah Franklin noted that the snow was too deep even for sleighing and that the Delaware River froze thick enough for pop-up taverns to serve customers and even for an ox to be roasted—on the ice.[39]

Little surprise that the chilly 1760s were boom times for Pennsylvania's stove makers. Warwick Furnace, Rebecca and Robert Grace's establishment, where the second version of the Pennsylvanian fireplace had been produced, was still making stoves (though the Graces would sell their half interest in the operation to their son-in-law in 1764). But several newer ironworks offered stiff competition. Mark Bird's Hopewell Furnace in the Schuylkill Valley, near Warwick Furnace, was a prime example. Hopewell made six-plate stoves—freestanding metal boxes on metal legs, radiating heat from all sides—which flourished in the region from 1761 to 1772. From 1769 onward, however, these began to be replaced by ten-plate stoves with interior spaces to use as ovens, thereby consuming fuel more efficiently, for both cooking and heating. (Meanwhile, the old German fireback was vanishing, as wood became too expensive for this modest assist to make good sense.) The ten-plate stoves had richer ornamentation—urns, tulips, scrolls— than Franklin's fireplace insert. They were sometimes decorated on all four of their vertical sides, and often bore the manufacturer's name.[40]

The Pennsylvanian fireplaces had been an interim solution, never needed by the rich (as the Franklins now were) and no longer sufficient for those at the middling ranks. As his new house went up in 1765, Franklin was trying to find two surviving examples of his model for an unidentified French contact, telling Deborah to pay whatever it took. Franklin knew "many People [had] laid them aside," so assumed they could be easily found, especially those "cast by our Friend Grace, when the Moulds were good" and the plates' outlines crisp. But the Philadelphia friend who went on the hunt, Hugh Roberts, came up empty-handed. Roberts reported that both stove iterations—either "the first impression" Potts and Potts had made of the firebox model or the Warwick version with "the Sun in front and Airbox"—were "much out of use." Many had been cannibalized to form firebacks or other items; "I have not found a second hand one Compleat." The Potts family did have some "with the sun in front" but not the airbox—a version they may never have made.[41]

In fact, it's impossible to pin down what exactly "the" Pennsylvanian fireplace was at any given time. The one surviving example, and part of one plate of another, deepen the mystery. The six-plate version at the Mercer Museum, which you met at the very start of this book, was found in 1910 at a house in rural Montgomery County, north of Philadelphia. It had been disassembled, with some plates set atop a chimney in the parlor and the rest stored outside. Only a fragment of the airbox remained. And the firebox is significantly different from Franklin's first or second models: each of its side and back plates has a slot cast into it to evacuate smoke and, in case those three vents weren't enough, there is a round hole cast into the top plate to accommodate stovepipe. These modifications are significant enough to suggest a different model (version 2.5?), matching Franklin's renewed efforts to banish smoke, though it's unknown whether he recommended these new elements to the forge himself.[42]

The other surviving plate fragment came from John Bartram's garden just outside Philadelphia. Archaeologists found it at the site of a greenhouse built circa 1760 to 1761. This front plate, with sun and foliage (miraculously making it identifiable), is intriguing because it lacks the Latin tag, *Alter Idem*. Was it the first model, done by Potts and Potts and, destined for a greenhouse, never put into a chimney, making an airbox superfluous? Maybe it was from a later period and once used indoors—in 1770, Bartram pleaded with Franklin to write to him from London: "Squeese out a few lines . . . to comfort thy ould friend in his new stove room." The earliest forge records don't always distinguish between the first two models, anyway. It's only when customers needed replacement plates, or specified a preference for with-or-without airbox, that the records make any distinction. In September 1747, for instance, Grace ordered specific firebox plates versus airbox plates, also separate airboxes, for people who wanted to add those—again, a hint of DIY for ongoing heating upgrades. And to complicate things further, there were knockoffs of Franklin's device, especially the firebox model; the Winterthur Museum displays one (from circa 1790) with the BE THINE LIBERTY design instead of foliage and sun. Even these, however, became scarce. Twenty-five years after its invention, the Pennsylvanian fireplace was obsolete. Of course, the fact that Franklin kept working on new stove models indicates that he knew very well the quest for an optimal indoor atmosphere wasn't over.[43]

Indeed, newer modes of indoor heating carried more prestige. Coal-fired "warming machines" were England's most luxurious designs. The best were probably the stoves of Abraham Buzaglo, a Jewish Moroccan man who had emigrated to England and established himself as an inventor, particularly of patented, highly decorated iron heaters. These ascended in layers (more than seven feet tall) like tiers of a fossilized wedding cake. While he was governor of Virginia, Norborne Berkeley, fourth Baron Botetourt, bought "a Buzaglo" for the governor's palace in 1770, and, liking the results, ordered another for the House of Burgesses. This one went to the state's post-independence capitol building in Richmond and was used into the nineteenth century. Although Buzaglo designed his stoves for any fuel—including wood or peat—the ones in Virginia burned imported coal. Coal grates had proliferated in the colony since the 1730s. Coal discovered around Richmond was beginning to be mined and sold (including to Pennsylvanians), though its industrial extraction lay in the future.[44]

Given a taste of coal, American consumers wanted more. Philadelphia continued to welcome imports from England, principally for the city itself. Commercial operations, particularly smithies, were the earliest and steadiest customers. But households also began to use coal, a more compact fuel than firewood, given its greater energy density. One ship's master advertised his cargo of "Sea Coals" as "extraordinary good and pleasant for Gentlemens Parlour Fires, as also for Smith and Sugar baker Use."[45]

So far, consumption of fuel had smoldered within a general consumer culture, but was now threatening to blaze out of control. This was apparent in colonial protests over the next British attempt to raise colonial revenue, the Townshend Duties. Passed in 1767 and 1768, the Townshend Act levied export taxes on some consumer items that colonists could not easily produce themselves, including glass and paper. This followed the logic of Franklin's testimony that colonists might accept "external" taxes, also taking stock of his bold talk about colonists eventually making their own consumer items. But the duties incited new outrage. In Boston, local merchants wrote up a plan for a multiyear boycott, even of goods not subject to the new duties. They exempted coal, however, alongside subsistence necessities (like fishhooks and salt), a telling admission of the colonies' rising energy expenses outside heavily wooded areas. That factor persisted even after partial repeal of the Townshend Duties in the spring of 1770.[46]

Like medieval English folk before them, colonists were entering an age of fossil fuel dependence, exploiting a commodity that took aeons to create but mere hours to incinerate. Investigations of heat and improvements in heating would therefore, in a very real way, examine time as well as space, the full dimensions of the natural world, on scales that grew to encompass the whole planet and its most remote past, all of its global atmospheres included.

Franklin edged into these bigger dimensions by reconsidering heat and atmospheric circulation on a hemispheric scale. In the 1760s, he wanted to know what happened when, pushed by thermal forces emanating from the Gulf of Mexico, the Atlantic storms he'd described in 1750 blew out to sea, forming a powerful air current that, as it turned out, mirrored a powerful current in the ocean. Franklin knew that eastbound ships could cross the Atlantic faster than when going west. One of his New England cousins, a mariner, explained why: "The long passages made by some Ships bound from England to New York" resulted when their captains sailed, obliviously, in the middle of the "*Gulph Stream*," where they had to fight the current (and wind). Franklin and his cousin, Timothy Folger, did the first chart of the Gulf Stream in 1768 to show ships' captains how to avoid it when sailing west—or find it when headed east. Franklin would contribute to two more charts of the Atlantic, each centered on the Gulf Stream. These represented the atmosphere as a complex system, with wind moving both longitudinally and latitudinally.[47]

Franklin's expanding vision of atmospheric phenomena is evident in updated editions of his *Experiments and Observations on Electricity*. Although the very first edition had exactly matched its title, over time, Franklin added different writings, increasingly not on electricity. That was most dramatically the case with the fourth edition, published in 1769. This 496-page work tripled the length of the previous edition, a slim 154 pages. The fifth edition (1774) changed little from the fourth, essentially the author's final compendium, his accelerating pace of political work preventing further updates. In these final two collections, Franklin reprinted his pamphlet on Pennsylvanian fireplaces. At last, he affirmed the essay as a contribution to science, leading into the electrical works that came later. This was an indication that, while the fireplace itself was no longer in demand, the science that underpinned it was, if anything, even more so.[48]

Franklin accompanied the nearly twenty-five-year-old essay with new thoughts on heat in a letter he'd written in 1759 to Mary Stevenson. Publication of the letter, among several to Stevenson (including the one on thermal experiments with a snowbank), reflected Franklin's belief that women could do science. But in focusing on fireplaces, the letter implies that heating is a domestic comfort, therefore associated with women, whose place within European and colonial culture was supposed to be at home. Franklin's letter responded to Stevenson's "very sensible" question about making a chimney narrower without choking the draw of air needed to keep coal alight; he agreed with her that it was better to shrink the mouth of the fireplace, not the funnel at the chimney's top. This referred to an earlier conversation with several unnamed people in which Franklin praised Stevenson both for her being right and for giving up the point, with a diffidence well-suited to a spirit of inquiry.[49]

In describing Stevenson's question as "very sensible," Franklin may have intended a triple pun. He was praising her intelligence, but also her thoughts on something felt by the senses: heat. Furthermore, he noted that her conversation had been sensitive to the feelings of others. As such, the letter reinforced Franklin's own sensibility about domestic space (and conversation), noting white women's presence in scientific debate—if their discourse shed more light on a subject than it generated hot tempers over it. The exchange also revealed that, whereas Franklin's Pennsylvanian fireplace had been designed for firewood, in London any talk of domestic heat involved coal.

By including white women in fireside chats about heating, moreover, Franklin admitted his increased debt to women's household labor, including their moral labor. This is another way of saying that

Franklin at last repudiated domestic slavery. During his first political posting (1757–1763), two enslaved young men, Peter and King, had served Franklin and his son. (The Stevensons did not have slaves.) King fled London, opting for another employer, possibly without being formally manumitted. Franklin tracked him down and tried to persuade him to come back, also advertising (anonymously) for his return, but decided against using force or the law. King's actions may have enlightened Franklin about the injustice of their connection, as did Franklin's wife, Deborah, who was advocating for a Pennsylvania home without enslaved people. The timing of these events, starting in the early 1760s, is significant within the longer history of energy use. King and Deborah Franklin could imagine a better use of human energies, even before fossil fuel would be a common substitute for mechanical labor, not just for cooking and heating.[50]

When Franklin returned to London in 1764, he came alone. Revealingly, that worked just fine. Franklin wrote a comic description of the household, "The Cravenstreet Gazette," in 1770, while his landlady, Margaret Stevenson, was briefly away. This series of mock news items document the pleasant microstate of Cravenstreet, in near-collapse when "Queen Margaret" must delegate authority—"it is impossible Government can long go on in such Hands"—an obvious satire on what was going on in the British Empire. However hilarious, the reports are in fact the best evidence of how Franklin lived in London in the latter part of the eighteenth century, as the British Empire reached a crisis point, and when the metropolis was fired by coal but still relied on human labor for everything else.[51]

The gazette discloses the household's minimal human labor—and its only visible energy. Craven Street has one servant, "Nanny" (Ann Hardy), who does pretty much everything but the cooking, which Stevenson ordinarily does; her daughter fills in while she is away (and mends some of Franklin's clothing). So two women, persistently impeded by their ridiculous American lodger, "Dr. Fatsides" (he is untidy; he joggles the tea table and sloshes the teapot; he misplaces everything), keep the place going. Food is the major fuel and constant preoccupation, but keeping warm isn't a worry, even during the late September of the gazette's "publication." This indicates both that the doctor's adjustment of the fireplaces paid off in their low maintenance, but also that the household's energy needs were otherwise outsourced, with coal and kindling delivered and chimney sweeps (and maybe other cleaners) servicing the premises.[52]

All of this hints that Franklin had, in 1757, separated Peter and King from home and kin in Pennsylvania not for labor, but out of vanity, to display his status, maybe overcompensating for his memory of being a colonial nobody on his first sojourn in the capital. On his third London residence, he no longer wanted that. And yet as he, at this phase of his life, quietly distanced himself from the starkest use of human energy, he was moving fully toward fossil fuel dependence.

His life at Craven Street reveals Franklin's multiple adaptations to a world different from Pennsylvania, ones more materially complex and oblivious of how material conditions in overseas colonies might make that possible. Both places had enslaved people, though more obviously in America, and Franklin's rejection of that labor in his two trans-Atlantic homes after 1767 was one of his most powerful adjustments in relation to the organic economy, on a par (if unconnectedly) with his acceptance of coal. So in Philadelphia, white people read and sewed and talked within human-scaled atmospheres heated by wood that they knew came from land taken from Indigenous people, and that this (whether they admitted it) was an unresolved problem, newly obvious given the recent Seven Years' War. Meanwhile, in London, women and men debated the science of heat in rooms warmed by coal, which, unlike the trees of the Lenape nation, seemed unproblematic. Which was itself a problem, as Franklin's fretting over coal smoke revealed.

Warm comfort was one of modernity's signal temptations during the Little Ice Age, whatever its costs. It's striking that the English saw coal, in a kind of trickle-up economics, as the everyday convenience even of the super-elite. Caricatures of the royal family, George III and Queen Charlotte, show them settled before coal-burning parlor fireplaces, preparing hot snacks; she is frying sprats and he is toasting muffins. The comedy works because the snug domesticity, in some cozy corner of some grand palace somewhere, is somehow believable.[53]

In two winter letters from February 1768, Franklin and one of his Scottish friends, Lord Kames, exchanged philosophic thoughts on indoor heating as integral to a good life. Kames reminded Franklin of their earlier correspondence about smoke, calling his friend "an universal Smoke Doctor." He'd just moved into a house poorly sealed against "what is commonly called neighbour smoke"; he wondered if he should adopt "your Philadelphia Grate, as it promises to save Coal, and to diffuse an equal heat through the Room." It was a small, specific problem but with greater consequences. "I begin to perceive in my decline of life that happiness," said Kames, "comfort at least, depends more upon what a Philosopher would call slight conveniences than a young man in the ardour of his studies is apt to imagine." Kames was seventy years old. Franklin, aged sixty-two, agreed: "I have long been of an Opinion similar to that you express," he told Kames, "and think Happiness consists more in small Conveniences or Pleasures that occur every day, than in great Pieces of good Fortune that happen but seldom to a Man in the Course of his Life." Small comforts adding up to a state of happiness—that was the great promise of the consumer revolution, which, as always, required burnable energy.[54]

And it's notable that Franklin's work on heat, given its originally colonial context, could be so easily adapted for British consumers at more mature phases of the industrious and industrial revolutions. The 1769 London edition of his scientific works had notable impact on the first edition of the *Encyclopaedia Britannica* and, through it, on an even wider reading public. But this was mostly because of his efforts to minimize pollution, which coal-burning Britons needed even more than wood-burning colonists.

In the *Encyclopaedia Britannica*, Scottish political economist James Anderson composed the entry on "smoke" as a domestic problem, an "extremely disagreeable" vapor "prejudicial to health" that humans generated while trying to "enjoy the benefit of fire." Anderson, who wrote a separate *Practical Treatise on Chimneys* (1776), had no interest in smoke that might occur naturally. He made the encyclopedia entry instead into an analysis of chimneys—their function, construction, drawbacks. But like Franklin, he linked human comprehension of smoke (and heat) to the natural world beyond the rooms equipped with chimneys. "The earth is every where surrounded with a great body of air called the atmosphere," Anderson explained, a concept necessary to understand the movement of heat, air, and other vapors in any space, domestic or cosmic. He then focused his essay on the problems of smoky chimneys within the mini-atmospheres where humans tried to keep warm but needed fresh air.[55]

The fresh air was, meanwhile, for Franklin, a new concern—and goal. Since he'd first declared that heating a room was a war against drafts, in recommending ways to make a room "tight," several theorists had urged better ventilation for interior spaces, and also tried to identify how much air, and what parts of it, benefited humans and other living creatures. After Joseph Black had, in 1754, isolated the first chemical element of air, carbon dioxide (which he called "fixed air"), Joseph Priestley found that this gas could be fixed within water, thus creating carbonated water. Priestley declined to profit from his invention, offering it as free fizzy refreshment to the world: "If this discovery . . . be of any use to my countrymen, and to mankind at large, I shall have my reward." (Cheers to that.)[56]

Priestley next did experiments that identified oxygen. Though he didn't name it or describe it fully, he established that it was the part of air necessary to life. In 1771, he found that a sprig of mint, set in water to re-root itself and then put (with its flask) into a sealed glass enclosure, could keep candles alight and a mouse alive, restoring "air made thoroughly noxious, by mice breathing and dying in it." While he was creating that noxious air, Priestley established a grim baseline: the very few minutes it took a rodent to expire if no plant was present. In contrast, he discovered, it took fourteen minutes for a mouse to become visibly "uneasy" if given a vase of greenery. And this time, Priestley amended his scheme to investigate life-giving air for mice (and men): he removed and saved his fellow mortal.[57]

Let us now praise that mint-sniffing mouse in its little glass house, the small surviving inhabitant of history's first (almost) self-sustaining artificial atmosphere. The experiment proved the possibility of enclosed life-support systems, later realized in submarines and spaceships that carry or generate oxygen. Franklin rebuked Priestley for his abuse of "honest harmless Mice," but he loved the discovery, marveling "that the vegetable creation should restore the air which is spoiled by the animal part of it." He hoped it would "give some check to the rage of destroying trees that grow near houses" on the grounds that they generated "unwholesome" air when the contrary was true: "We Americans have every where our country habitations in the midst of woods, and no people on earth enjoy better health, or are more prolific"—explicit endorsement of the American environment's ability to sustain white bodies. (Priestley quoted Franklin in his printed account of the experiment.) The new significance of the air changed the calculus for creating artificial environments, no longer carefully protected from the bigger atmosphere but, to a more recognized extent, needing to be part of it.[58]

The invention of devices to circulate air was, for this reason, connected to the invention of stoves to warm it. Both kinds of machine could improve health, striking a balance between warmth and fresh air, even in winter. Physiologist Stephen Hales had led the way with two volumes (1743 and 1758) on "ventilators" for hospitals, prisons, and ships, where lack of fresh air was a medical threat to confined people. Other inventors followed with a wealth of other ventilators, along with new exhortations to avoid overheating, especially while sleeping—a problem earlier generations hadn't needed to worry about, at least in winter. Simpler mechanisms also helped keep rooms cool and airy, including sash windows that could be lowered a crack for quick refreshment and gadgets like the aeolus, a round metal insert for a window, cut and bent into vanes to assist air movement—another tweak to boost consumer comfort at home.[59]

Accordingly, by the 1770s, Franklin acted as a consultant for efforts to heat—healthfully—buildings for large groups of people, with artificial atmospheres bigger than his original sitting room. He corresponded with the Reverend Samuel Cooper, minister of Boston's Brattle Square Church, who wanted to heat his new meetinghouse. Franklin said such a large space couldn't be fully warmed if a "Machine" were fired just one day a week; that would benefit only the people right around it while launching "cold Currents" of stagnant air to plague everyone else. If Cooper insisted on a heater, Franklin recommended "those now used at the Bank [of England], or that in Lincoln's-Inn Hall," one of England's professional organizations for judges and barristers. Such a warming machine was centrally positioned, with its smoke descending to pass through a flue beneath the floor and out a chimney placed on an exterior wall. Franklin may have been twitting the Boston clergyman a bit by saying such a stove could be styled to resemble the classical temples or altars where ancient pagans had worshipped. More practically, he questioned whether heating large spaces was always worth it. People never caught colds or other illnesses from chilly rooms; they could bundle up for church and go back home to thaw out.[60]

Slightly later, Franklin was drawn into an effort to keep even more people warm and yet not stifled indoors. The fellow mortals he sought to benefit were, surprisingly, a set of British men who were increasingly his political enemies. In 1769, three years after they'd grilled Franklin about the Stamp Act, members of the House of Commons consulted him about a very different matter. Many members had been complaining of the "great Inconvenience from the heat of the House of Commons when it is full, and from the Cold when it is thin." "Doctor Franklin" and other "learned Mathematicians" (what we would call scientists) toured the affected rooms and the spaces above them, inspecting "the present Ventilators" and declaring them useless. The House needed "to erect new ones, with some Stoves." The estimated cost of the renovations was £300, very roughly the equivalent of £52,580 today.[61]

But isn't human health priceless? Because indoor air could be "mephitic," one of Franklin's scientific contacts explained—in relation to the miserable House of Commons—indoor spaces needed "a current of air," most easily supplied by a chimney, which generated "a kind of tide in the flue" to keep the interiors fresh. So, too, might "outlets" in the seats of the Commons carry off any "personal atmosphere surrounding the Members," an image that's maybe too evocative. Given Franklin's mounting criticisms of Parliament and impending break with Britain, his eleventh-hour attempt to clear the literal air in Britain's main political body is oddly perfect. Equally, it was an important historical measure of the shift away from a concern to preserve a warm atmosphere around the body and to the creation of a comfortable and ventilated indoor space that could accommodate multiple bodies—or only a few, if they didn't show up to speak or vote or interrogate their local colonial expert.[62]

Even as the government fretted about air quality indoors, some of the worst air was still to be found outside. That was the price of a growing population combined, in London (and other parts of Britain), with a reliance on coal. True, Londoners were at least unburdened in knowing that burning coal would lead eventually to catastrophic levels of atmospheric CO_2, because no one knew that yet. What was coming to light, however, was that coal, among other fossils, revealed Earth's abyssal past, when vast waves of destruction had crashed across the face of the planet again and again, a disaster scenario uncomfortably at odds with the domestication of coal fire as a nice source of toasted muffins.

On both sides of the Atlantic, scientific analysis of American fossils promised the massive availability of fossilized fuel—power— but also the prospect of natural extinction—death—with unresolved questions about the optimal distribution of either. Charles (II) le Moyne de Longueuil, Baron de Longueuil, had in 1739 recovered fossils of vertebrates along the Ohio River, at the spot Lewis Evans marked on his 1749 map. The site was a salt lick, known to Native hunters as a place to find deer; Native men must have told the French about it. In 1765, Irish-born Indian agent and land speculator George Croghan, whom Franklin knew from several treaty negotiations with the Haudenosaunee, extracted fossils from what the English renamed Big Bone Lick. Whatever Croghan's intellectual interests, his action was part of his mission to remove Indigenous nations from their land; he'd taken a bribe from Virginia land agents to engineer, against Lenape interests, a white settlement at the Forks of the Ohio River.[63]

Croghan sent some of the fossils to Franklin and some to William Petty, the scientifically inclined (and politically powerful) second Earl of Shelburne, grandson of the William Petty who'd advised William Penn on colonial political arithmetic. Each man got a package of tusks and teeth, with a jawbone for Shelburne and a vertebra for Franklin thrown in. Franklin thanked Croghan for the specimens, which were "curious" for three reasons. "No living elephants" had ever been described by white people anywhere in the Americas. Nor were they "remembered in any tradition of the Indians." But at the least, they were proof of climate change. Living elephants were found in "hot countries where there is no winter." Yet fossils resembling their tusks were now being unearthed in Siberia, Peru, and the Ohio country, "which looks as if the earth had anciently been in another position, and the climates differently placed from what they are at present."[64]

Only one of Franklin's ideas turned out to be entirely true. He was correct that no living person had seen the creatures that bore those tusks, teeth, and bones. But he was only partly right about climate change. Like many people, including his friend Bartram, he was using Thomas Burnet's theory of a degraded, postdiluvian planet to explain American fossils. As he put it rather cheerfully to one correspondent, "'Tis certainly the Wreck of a World we live on!" But Franklin somehow assumed that while climate might change, animals might not. Long-ago climatic differences had in fact supported cold-adapted mammoths and mastodons in non-elephant-friendly zones, for instance. Meanwhile, Franklin was completely wrong about what Natives knew. They did have oral accounts of ancient beasts, maybe based on their discovery of the fossilized remains long before Europeans got involved. But their knowledge mattered to Franklin only if it was immediately useful to settlers and colonial officials, not as a more abstract point about natural history.[65]

In his 1767 assessment of the specimens Croghan sent to Franklin and Shelburne, plus a set of fossils held in the Tower of London, anatomist William Hunter called the creature in question "the American *incognitum*," something unknown, distinct from modern elephants. In Siberia, Indigenous people said that comparable fossils were from "the mammouth, an animal of which they told and believed strange stories," though "modern philosophers have held the mammouth to be as fabulous as the centaur." Hunter agreed with Burnet that "some astonishing change must have happened to this terraqueous globe" in remote times. Thinking the *incognita* had been carnivorous, and knowing they were huge, he said that "though we may as philosophers regret it, as men we cannot but thank Heaven that its whole generation is probably extinct," stating a hypothesis about annihilation of species that was, at the time, extraordinary.[66]

More typical was Hunter's indifference to what Indigenous people might think about the specimens. Yet again, scientific investigation of fossils excavated from American land aligned with the project to depose the people native to that land, remapping it in terms of colonial interests and European expertise. These were yet more steps out of an organic economy—once common around the world—toward excavation and identification of fossilized materials, including coal, and away from cultures that continued to conserve the natural world as it existed.

A few years after his assessment of American fossils, Franklin discussed fossil coal with a French contact, Jacques Barbeu-Dubourg, a botanist and physician. The Frenchman had begun corresponding with Franklin on politics and science in 1768. He suggested a French translation of Franklin's latest *Experiments and Observations on Electricity*, with the expanded content that the only French translation, based on the first edition of 1751, didn't have, plus some even newer material. The result was the most ambitious collection of Franklin's writings to appear in his lifetime, in any language. It included the original stove pamphlet, plus relevant letters Barbeu-Dubourg and Franklin exchanged while editing the collection, published in 1773.[67]

During their correspondence, the two men pondered coal, the fuel of the future from a distant past. Barbeu-Dubourg thought it not so very old, all things considered: "We regard it as a fossil of plant origin, and not from the earliest antiquity of our globe," which (investigators were realizing) hadn't sustained plant life. Franklin agreed "that coal is of vegetable origin, and that it was formed towards the surface of the earth; but as previous convulsions had buried it deeply before in several places, and covered with several very thick layers, we are obliged to the following convulsions to have put the ends of the coal beds before our eyes, in order to put us within reach" of it. He spoke from experience, relating his 1771 visit to a coal mine in Whitehaven, England. He had walked down the vein of coal until he was "80 fathoms" below sea level, "and the miners assured me that their works extended up to a few miles beyond, always descending by degrees below the sea." Franklin saw that "the slate stone which forms the roof of the coal mine is imprinted in several places with leaves and branches of ferns, which no doubt grew on the surface ... Thus it appears that this coal suffered a prodigious subsidence," some past cataclysm dragging it down into the earth.[68]

As he'd done in comparing indoor atmospheres to the entire natural atmosphere, so Franklin connected Earth's most remote chapters to everyday energy use. When he returned to Craven Street in November 1764, after his two-year visit back to America, he started to monitor his landlady's fuel purchases. Their coal merchants' elegantly engraved trade card—"Tuckwell & Cooper"—survives among Franklin's papers in Philadelphia. (He also treasured the card for the supplier of Craven Street's tastier energy source: cheese.) As it turned out, Franklin was watching the household's coal consumption for a reason.[69]

In a 1773 letter to Barbeu-Dubourg, Franklin revealed that he had developed a fourth heating device, fired only by coal. He'd evidently been working on this as early as 1768, though his earlier Craven Street invention of the 1750s (the sliding vent system, his third heating device) was probably background research for this latest iteration. The new design, like the three before, resulted from experiments within a specific room, this time Franklin's combined sitting room and study in London. But, for the first time, Franklin tested his heater with instrumental readings of temperature—though using an unusual instrument.

In 1766, after the frenzy of work around the Stamp Act controversy, Franklin had gone on holiday, touring the German-speaking parts of Europe with a friend, a scientifically minded English physician. The two men marveled on seeing a pulse glass, a thermometrical device that shows how heat increases vapor pressure: a glass tube is partly filled with a volatile liquid—alcohol or ether—which is boiled to reduce air in the tube before it's sealed. Under these conditions, even the warmth of a hand makes the liquid bubble. Franklin brought several of the glasses back to London, where the instrument-maker Edward Nairne made even more sensitive versions. Franklin wanted to use them to compare indoor and outdoor atmospheres. "I bored a very small hole through the wainscot in the seat of my window," he said, "through which a little cold air constantly entered, while the air in the room was kept warmer by fires daily made in it." When one end of a pulse glass was in the draft and the other out in the warm room, "philosophical spectators" who visited Dr. Franklin could marvel at the bubbles fizzing up the tube.[70]

Pulse glasses charmed Franklin, who'd shunned thermometers in his previous heat experiments, into instrumental verification of temperature changes. He considered other possible uses for the glasses and, given his recent contemplation of Watt's separate-condenser steam engine, his thoughts logically turned to generation of power. "Our engines for raising water by fire," he reasoned, relied on "water reduced to vapor by heat." Maybe "a much greater power may be obtained with little expense," using the principles of pulse glasses. Maybe, attached to a moveable beam (of the kind steam engines used) or a wheel (as Franklin had sparked into motion with electric batteries), the motion of their contents could serve some "mechanical purposes." Reusing a piece of paper on which he'd drafted a phonetic spelling system, Franklin sketched a circle with two pulse glasses laid across each other at right angles, each with their bulbs bent down ninety degrees, all in the same direction. In theory, as the bulbs went from cold to hot in succession, with their weights shifting (as heat modified their contents), they might, just might, spin the wheel.[71]

With this small sketch, as in his earlier design for a one-dogpower electrical motor, Franklin pondered one of the era's big quests: turning a stationary energy source into a rotative device to power some system of mobility. No one had managed to do it yet. And engines that used heat transfer (instead of heat-generated steam pressure) to produce power wouldn't be the first to succeed at the task. Steam would do it. Not until 1824, with Nicolas Léonard Sadi Carnot's *Reflections on the Motive Power of Fire*, would anyone analyze how a boiler's heat might itself generate mechanical work, as related to the second law of thermodynamics: heat transfer or conversion has motive force but always entails some loss of energy. Again, Franklin was considering this problem much as he'd done when telling Boulton that steam engines should consume their own smoke, maximizing their efficiency by that means and by preventing buildup on the boiler that impaired its heat absorption.[72]

As he considered Boulton's model, the pulse glasses, and his proposed heat engine, Franklin decided to invent yet another heating system. A bill he paid to a London carpenter, Josiah Day, indicates he did almost three years of fresh research in and around his day job of monitoring the collapse of the first British Empire. Day did many tasks at Craven Street: adjusting Franklin's desk, making bookshelves, putting casters under furniture that Franklin wanted to be able to slide around. And, in May 1769, Day built what his bill laconically describes as a "Modell for Grate." Wood would have been cheaper than metal for this, more easily modified, and—like the steel Franklin had used for his prototype in Philadelphia—easier to shift about. Still, billed at 15 shillings, requiring ten feet of "clean yellow Boards" and four days of work, the grate model was one of the most expensive items on Day's bill. The amount of wood suggests this may have been an update of the sliding register Franklin had installed previously, with "grate" a synonym for fireplace. That this model was made shortly after the pulse glass experiments suggests the two projects may have been linked. As they were on Day's bill: he built the model grate in May and repaired Franklin's long-suffering landlady's hole-bored wainscot in December.[73]

When did Franklin have a working prototype? In April 1771, he paid a mason for unspecified work, likely to a fireplace or chimney, the only masonry within his rooms. Another mason's bill, from July 1771, lists charges for "Stoping Holes in ye back of a Chim.y" (This was on top of any alterations Franklin had made to his sitting room in 1758, or else for a second fireplace, maybe in his bedroom.) And just under a year later, in February 1772, Day (the busy carpenter) built a second "model for a grate." This one cost less, 3 shillings and 3 pence, so was smaller than the first, maybe just a key component. A receipt from a London ironmonger, James Buttall, is unhelpfully undated and lacks any detail, but may have been related to the stove work; Buttall, "at the Gridiron," was around the corner from Craven Street, at Charing Cross on the Strand, not ten minutes away on foot. Though what Franklin may have commissioned there couldn't easily be carried home: this time, he'd invented a round heater of the type people usually think of as a "Franklin" stove. (Maybe he burned his wooden prototypes in it to celebrate—more rough drafts up the chimney.)[74]

Franklin had little hope of the new device's being widely adopted. "It demands too much attention in its operations to be governed by ordinary servants, which is why I cannot recommend its use, although I use it for myself." It was "strictly a curiosity, or a philosophical experience," to tackle the nagging problem of smoke. Within *his* device, Franklin said, "the smoke imperceptibly changes into flame, instead of soiling the fireplace; this flame descends, and serves to heat the plates and the chamber; and at the same time it prevents the coals which had begun to catch fire from being consumed" too quickly because it restricted their air supply.[75]

In a 1773 letter making final corrections to Barbeu-Dubourg's translation of the original stove pamphlet, Franklin enclosed an

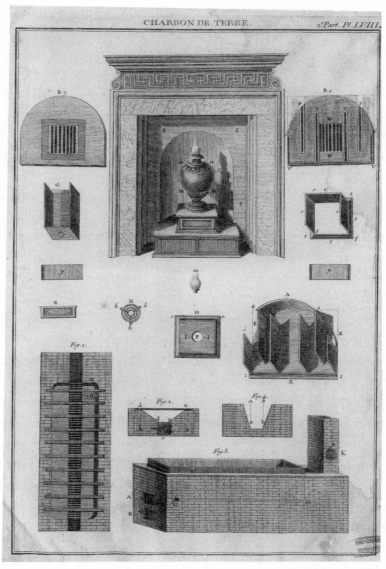

FRANKLIN HESITATED TO PUBLISH AN IMAGE OF HIS FOURTH STOVE (THE GRACEFUL URN AT TOP), GIVEN ITS PERSNICKETY NEED FOR SPECIALIST ATTENTION.

(Jean-François-Clément Morand, L'art d'exploiter les mines de charbon de terre. Seconde Partie. Suite de la quatrième section *[{Paris}, 1779]. CNAM-BIB Gd Fol Ky 35 [P.1] Res, Cnum-Conservatoire numérique des Arts et Métiers)*

engraving of his new device. But he declined to write instructions for its installation or use. "The Description must be postpon'd for want of Time," he said, "and I had rather it should not be published at present" anyway. The diagram indicated that this, like the Pennsylvanian fireplace, would sit in a hearth and chimney, a house's existing heatproof and vented space. Although its design makes it seem like the mythical "Franklin stove," it's smaller (and fancier) than the potbelly stoves most people have in mind, being modeled on a classical urn or vase. A small grate (H), with three concentric iron circles, set within a firebox or "chamber," constituted Franklin's key design, the part Day may have rendered in wood. A set of plates (B^1 and B^2) with channels would, in addition, slow airflow and maximize heat absorption. Franklin apologized to Barbeu-Dubourg: "It is fit only for burning Pit-coal, and therefore can be of little or no Use with you who generally burn Wood."[76]

Barbeu-Dubourg illustrated his edition of Franklin's essays generously, with two new pages of plates for the stove pamphlet adapted from the 1744 originals. He wanted versions whose quality would match the volume's other plates, permitting greater detail. But without instructions for Franklin's new stove, Barbeu-Dubourg was reluctant to include its illustration, which he set aside. Meanwhile, Franklin was discussing his "vase stove" with London friends, at least one of whom, Francis Dashwood, Baron Le Despencer, had one made and installed in 1774, along with a "Pensylvania Stove" for his library. (That might have been Franklin's London register system, not the full Pennsylvanian fireplace.) The bookish Dashwood is thought to have founded the Order of the Friars of St. Francis of Wycombe, a "Hellfire Club" that performed erotic variations on Christian rituals. Clearly, staying warm indoors appealed to all kinds of people. Though Dashwood wasn't totally godless. He and Franklin produced an abridged, quick-reading Book of Common Prayer, so that "pious and devout Persons" who were in poor health would no longer have to skip long religious services held "in a cold church."⁷⁷

Finding new ways and reasons to burn coal represented a universal gravitation toward the solution the British (and Dutch) had already accepted, as translation and circulation of Franklin's work continued to reveal. And yet the raw iron that constructed his London stoves and grates may have followed him from Pennsylvania. By the 1770s, colonial furnaces would be outproducing their English and Welsh counterparts, turning out one-seventh of the world's total supply, reforged, in Britain and the colonies, into myriad useful things—if permitted to do so. Franklin satirized Britain's imperial overregulation in his "Edict by the King of Prussia" (1773). In this spoof, Frederick II declares an ancestral claim to Britain and a plan to resume Prussian rule, laying out new laws for Britons, forbidding them to do more than smelt iron from ore, though he was "graciously pleased to permit" them "to transport their Iron into Prussia, there to be manufactured, and to them returned." Abandon the monopolistic regulations, Franklin was warning—American iron and British coal did not need excessive supervision to power up an age of steam.⁷⁸

In 1775, the famous wizards of steam, now incorporated as Boulton and Watt, had won an extension of their 1769 patent for their separate-condenser steam engine, with exclusive rights to the design, after a long lobbying campaign in Parliament. With particular cunning, Watt claimed intellectual property in the entire *concept*, not the technical specifics of a working model. Within Britain, therefore, use of any separate-condenser steam engine was illegal unless negotiated through Boulton and Watt. As with Franklin's heating designs, their engine was designed to minimize the use of fuel—in this case, coal. So, anyone who adopted their device paid a set fee plus an estimated percentage of the value of the fuel saved, typically around a third. The patent was guaranteed for twenty-five years; it would outlive Franklin, running through a thicket of lawsuits, yielding Boulton and Watt £30,000 in damages, worth as much as £3 million today. Still, pirated separate-condensers popped up beyond the reach of British law, as in France.[79]

The transition from organic to fossil fuel energy was by no means complete. Watt didn't coin "horsepower," but he used it, evocatively, to measure the force of steam, not least to calculate payments to him and Boulton. Indeed, horses had powered earlier mining pumps, and their effort was easier to visualize than the bushels or acres of fodder that powered them. But Watt was coy, if not in denial, about the era's other great source of mechanical power: human beings. Although he would eventually support abolition of the transatlantic traffic in human beings, the Watt family had built its wealth on the profits of that very trade. "Manpower" was a later term, coined in 1825, to describe the human body as "a self-moving, self-supplying, steam-engine," eventually specified as one-eighth to one-tenth of one horsepower. Significantly, this designation happened before the end of enslavement in the Americas, where human bodies were still forced, crudely and cruelly, to make profits.[80]

A world dependent on fossil fuel is ugly. The human exploitation that came before it was ugly, too. The one attractive fact about Franklin's observations, at the cusp of two flawed energy systems, is that he didn't see coal as a substitute for slavery, or vice versa. He followed the lead of other people in his household who criticized enslavement: Deborah Franklin, who hadn't witnessed anyplace dependent on coal, and King, who didn't seem to use that British reality to explain his desire for freedom.

Franklin connected enslavement to coal only once, in a (rather forced) fictional dialogue on slavery between an Englishman, a Scotsman, and an American. He wrote this 1770 piece to refute the opinion that enslaving was something only colonists did. The "American" of the conversation protests, scoring his best point with the obvious fact that a great many Britons profited from the slave trade. He also argues that some forms of labor within Britain are kinds of enslavement, as with the status of colliers: "All the Wretches that dig Coal for you, in those dark Caverns under Ground . . . are absolute Slaves by your Law, and their Children after them." True: until 1779, Scottish coal miners were chattels, bought and sold with the mines they worked, a condition their children inherited. But the "American" concedes that colliers aren't entirely comparable to enslaved Black people, and the "Englishman" tells him to drop the point.[81]

What no one suggests is that the colliers might eventually free themselves and others because of what they mine. Emancipating enslaved people wasn't imperative because there was at last a substitute for their labor—it was just the right thing to do, coal or no coal. Franklin's hopes for that fuel operated independently and, in contrast to his evolution in relation to enslavement, in a direction that's harder to praise. His optimism that some technical intervention would mitigate coal smoke seems questionable. He of all people might have known that burning coal, even while trying to design better ways to do it, defied Poor Richard's advice that "To err is human, to repent divine, to persist devilish."[82]

7

THE AMERICAN REVOLUTION?

It may not have been the very coldest time in Franklin's life, but the winter conditions of 1776 were bad enough to nearly kill him. After his return to America in late 1775, he helped invent the United States and joined a US commission to Canada, inviting Canadian settlers to join the new nation. The commissioners headed north in March 1776, just in time for a freak April snowfall and decidedly un-spring-like conditions. Franklin broke out in "Boils"—probably psoriasis, a skin condition that flares with winter's cold and lack of sunlight. As the rash inflamed his head, he acquired a Canadian souvenir: a warm cap of marten fur, softer than a wig or tailored hat. He wore the cap when Congress next sent him to France on October 27, 1776, for a rough thirty-day Atlantic crossing that, Franklin said, "almost demolish'd" him. At least he kept his head warm—and his headgear renewed his fame. Fur hats were prime accessories for men of science who ventured into cold climates (Scandinavia, Scotland, Canada) or atmospheres; balloonists would favor them, too. French rapture over Franklin's cap was just the beginning. His work on heating would go into its widest circulation yet, though the notorious US insurgency wasn't necessarily the reason for this.¹

In the final third of the eighteenth century, war rejoined the era's other two crises, climate change and resource scarcity, and the famous American inventor's ideas and devices gained new authority. The Little Ice Age wasn't over, after all. Any advice about keeping warm was still welcome. Nor was fuel scarcity a problem solved, except in places with ample forests or else commercial coal mining, the latter still rare beyond Britain (and China). And the war meant that Franklin's international visibility, as an emissary of the upstart United States, cast borrowed light on his inventions and science.

This light wasn't always flattering. One month after Franklin sailed for France, Philadelphia poet Hannah Griffitts denounced his politics—and his stove. Her short poem's long title, "Inscription, on a Curious Chamberstove, in ye form of an Urn, contriv'd in such a Manner, as to Make ye flame descend, instead of rising, invented by ye Celebrated B F," describes the urn-shaped device—stove model number four—that evidently accompanied Franklin from London the preceding year. (Griffitts was not a confidante of Franklin or anyone in his family, but his latest stove was evidently the talk of the town.) The poem opens with praise for Franklin's investigations of the atmosphere:

> Like a Newton, sublimely he soar'd,
> To a summit—before unattain'd
> New Regions of Science explor'd,
> And the Palm of Philosophy gain'd;
> With a Spark that he Caught from ye Skies,
> He display'd an unparraled wonder,
> And we saw, with delight & surprise
> His Rod Could defend us from Thunder.

Flattering, so far, but, like her fellow Quakers, Griffitts deplored the decision to establish the United States by military force. Her disapproval extended to Franklin's complicity in the war effort.

> O, had He been wise to pursue
> The Tract, for his Talent design'd
> What a Tribute of Praise had been due
> To the Teacher and friend of mankind,
> But to Covet—Political fame
> In Him—was degrading ambition
> A Spark—that from Lucifer Came
> And kindled the flame of sedition.

Hints of Prometheus here, overlaid with Christian belief in fire's hellish power. God would judge Franklin after death; Griffitts offered the latest stove as a funeral urn to adorn his grave:

> Let candour then write on his urn,
> "Here lyes—the Renowned inventor;
> Whose flame to ye Skies ought to burn,
> But Inverted—descends to the Center."

Pretty damning, but Griffitts was not quite done. She went to the trouble of footnoting her use of the word "defend" to praise Franklin's lightning rod, which Kant had thought made him into a modern Prometheus. "'Defend' was, I supose intended," Griffitts said, "but I don't believe such virtue was lodg'd in ye man, or his art." So, too, her dismissal of his ongoing project to defend humanity from cold and smoke. That did not excuse the state-sponsored violence for which Franklin was soliciting international support, once he'd fled London, become a delegate of the Continental Congress, and, as its most famous member—indeed, as an international celebrity—agreed to serve as US emissary to the body's most coveted potential ally, France. Coveted, because France was at the time not merely a chic purveyor of cultural prestige, but a bona fide superpower: a rich, well-armed, and formidable global colossus. The infant United States sought help from France much as a weaker nation might today seek help from . . . the United States.[2]

Franklin represented his new nation at the court of Louis XVI for about a decade, from late 1776 into 1785. At the same time, his scientific work was circulating internationally, and his investigations of the atmosphere in particular were supposed to float peacefully above national divisions, nature and science as universal unifiers. Before formal hostilities in early 1775, his work on heat had been available in three European languages: English, Dutch, and French. From 1775 to 1780, the Pennsylvanian fireplace and its successors would gain further fame in Italian and German. The translations show that Franklin's work had practical appeal to multiple nations, even those that fought on opposite sides of the war.

That there was interest in his work at this particular time is itself interesting. During the war, most European powers, and American colonies still controlled by those powers, would not recognize the United States. Until 1778 and formal French recognition, France's foreign minister, Charles Gravier, comte de Vergennes, refused to meet with Franklin or even acknowledge that he was living in France. Not that Franklin could complain of neglect. The lieutenant general of the Paris police, Jean-Charles-Pierre Lenoir, was most attentive. Lenoir used his extensive network of spies and censors to collect and control information about the celebrated US envoy, so famous that his face (Franklin quipped) was "as well known as that of the moon," even if, for the moment, kept in the very darkest of lunar phases.[3]

Nor did most people dare to officially express enthusiasm for the United States or anything it or its formal delegates had written. British officials only printed the Declaration of Independence alongside a rebuttal, for example. The Declaration and the constitutions of the independent US states were translated into French starting in 1776, but until the formal alliance of 1778, these were printed outside France—and anonymously. Even seemingly innocuous relics of the American rebels could seem compromising. When the Russian Minister in the Hague (Netherlands) forwarded to St. Petersburg a portrait of George Washington he'd received (maybe from Franklin or John Adams), Catherine II sent it right back and ordered him to return it. Within this atmosphere, in which everyone knew about the war but had to be careful what they said about it, the circulation of Franklin's scientific works, including on heat and heating, are a distinctive test of how a political insurgent's ideas can be deemed innocuous, maybe even beneficial.[4]

In the end, fuel scarcity, not politics, predicted receptivity to Franklin's work on heat: warmest in the northern Italian states, somewhat cooler in Britain and France (though equivalently, despite the war dividing them), much less so in German-speaking territories, and not at all on the Iberian Peninsula and in Russia, despite their different stances toward the United States. The varying responses reveal the uneven transition to new fuel regimes, with a desire for energy autonomy at least as powerful a deciding factor in the age of revolutions as revolutionary ideology.

For most of his Paris mission, Franklin lived in the suburb of Passy, midway between the city's center and the court of Versailles, his two main sites of business. In Passy, he rented part of the Hôtel de Valentinois, the home of Jacques-Donatien Le Ray de Chaumont, a supporter of the American cause. This was the grandest residence Franklin would ever occupy, with a main house, two additional wings, and extensive gardens. Grand it may have been, but Franklin thought it didn't quite live up to its Italian motto, *"Se stà bene, non si muove"* (if one is well, no need to budge). So, he made modifications, mostly to improve light and heat, two essentials of modern domestic life, and to augment his comfort, as his health declined even after recovery from his grueling journeys.

He left his landlord a helpful list of the alterations, intended to offset any damage he and his resident grandson, William Temple Franklin, might have done to the place. He detailed the enlarged windows in his study and in the kitchen, for example—he and his cook having greatest need of light—and got a mason to correct the kitchen chimney's "intolerable disease of smoke." He'd bought two stoves (*cheminées*), one for his bedroom and a second for his grandson's, and possibly two other stoves (it's not clear if these were different from the pair he'd already listed). The bedroom stoves were French chest-of-drawer models, possibly for burning coal on grates above pull-out drawers to collect falling ash that, if kept clumped around the coal, smothered the fire. Franklin modified his bedroom stove with an iron plate and drawer, the plate most likely the kind of register he'd developed in London, the drawer another concession to coal. He noted as well that he was leaving behind "a great deal of coal," though other records show him buying firewood, definitely for the foundry he operated to cast lead type for his printing press, possibly for the kitchen, and for ordinary fireplaces. Finally, he put a streetlamp out in front of the house and erected a lightning rod atop its roof, both practical measures, though the rod was also his equivalent of running up an aristocratic banner to show who was at home. The greatest expense was for modernizing the kitchen, 172 livres (in 1785, the year of Franklin's reckoning, a livre was worth between 2 and 12 US dollars today), and the next, the lightning rod, 150 livres, with the stove in his grandson's room in third place, around 48 to 120 livres, Franklin didn't quite recall.[5]

Circulation of Franklin's scientific work within Europe obviously benefited from his visibility in public affairs, but it didn't hurt that there was another spate of colder weather. The period from the late 1770s onward, maybe extending into the second half of the nineteenth century, represented a cooler trend in the Northern Hemisphere on both sides of the Atlantic, spiked, in some places, with droughts. The renewed cold crushed hopes that things might be improving. In late 1770, for example, just in time for the weather to contradict him, North Carolina physician Hugh Williamson had presented to the American Philosophical Society "An Attempt to Account for the Change of Climate, Which Has Been Observed in the Middle Colonies in North-America." Winters and summers were becoming milder, Williamson insisted. According to him, anthropogenic change—from colonizers' cutting of trees and plowing of land—contradicted the ancient theory of climatic fixity: "The face of a country may be altered by cultivation."[6]

Whatever trends Williamson (and others) may have been sensing, they didn't hold. Although the winter General George Washington kept his troops at Valley Forge (1777–1778) is fabled for its miseries, the winter of 1779–1780 spent at Morristown, New Jersey, was worse, inciting mutiny and prompting Washington to begin the first weather diary he's known to have kept. On March 18, 1780, he wrote despairingly to the Marquis de Lafayette, "The oldest people now living in this Country do not remember so hard a Winter as the one we are now emerging from. In a word, the severity of the frost exceeded anything of the kind that had ever been experienced in this climate before." This winter prompted the first systematic collection of US weather data; the American Philosophical Society asked "the curious in every part of the continent" to send information, either from instruments or observation of precipitation and the effect of cold on plants, animals, and humans. In France, the cold was particularly apparent in 1783–1784 and 1788; both times, harvests shrank. In other parts of Europe, too, extreme weather exacerbated poverty and social unrest.[7]

But if anything, and yet again, the extreme conditions (plus the war) did not discourage Promethean prophecies of human control over the material world. In 1778, one of Franklin's scientific contacts, naturalist Georges-Louis Leclerc, Comte de Buffon, predicted in his *Époques de la nature* that humans, armed with science and technology, would create a seventh and final epoch for the planet, commanding nature with its own laws but according to their preferences. Buffon narrated the history of an initially molten Earth slowly cooling and forming a nontoxic atmosphere. People should moderate the continued cooling trend, especially in the Americas, other parts of the world having already been terraformed, Europe in particular. Not unlike Hannah Griffitts, Buffon deplored war's malign distractions—"Let us suppose thus a world at peace, and see more closely how much the power of man could influence that of Nature." Buffon's book would not be sold until early 1779, a year after France had entered the American Revolutionary War; he'd made clear his preference for a global colonization of nature, an effort that cut across Western nations while excluding Indigenous people.[8]

Debate about North America was exactly what prompted Buffon's confidence that humans could modify the climate. He cited Hugh Williamson's 1770 essay in the *Transactions of the American Philosophical Society* on the climate of the middle colonies becoming more moderate. "These opinions of M. Williamson are very just," Buffon concluded, "and I do not doubt that our posterity will see them confirmed."[9]

His enthusiasm for terraforming included the use of fire. "The habitual use that man has made of fire adds much to this artificial temperature in all those places that he inhabits"—cutting down trees and burning them had a doubled benefit. In fact, "man" had a distinctive tool in fire: "It is more difficult for him to cool the Earth than to heat it." Too true, but in an era when cold was the foe, a happy prospect. Buffon rejoiced that Earth's ancient vegetation had been transmuted into rich seams of coal, "the treasures that nature seems to have accumulated in advance for the needs to come of great populations." Humans had a forever fuel supply "because a single coal seam perhaps contains more combustible matter than all the forests of a vast country," an idea of cornucopianism transposed from the new world. Buffon also linked coal to the extinction and fossilization of ancient creatures in a long discussion of the artifacts Indian agent George Croghan had sent to Franklin and other experts.[10]

Buffon's cheerful faith in anthropogenesis forecast the industrialized nineteenth century and the mounting extraction of natural commodities onward into some kind of an Anthropocene. Franklin would have disagreed, having repented of his earlier enthusiasm for making North American colonialism visible from outer space, skeptical that humans could fine-tune nature to their specifications. But this did not prevent his continued improvement of artificial climates for human comfort, so long as they conserved fuel.[11]

It was while he was in France, for example, that Franklin finally agreed to publish a description of his urn-shaped chamber stove, and the engraving of it Barbeu-Dubourg had commissioned, thanks to the pleading of Jean-François-Clément Morand. (It's possible that Barbeu-Dubourg had introduced Morand to Franklin.) As early as the summer of 1777, Franklin had corresponded with Morand, a medical doctor, scientific experimenter, and member of the Académie royale des sciences, to which Franklin had been elected. Morand was busy writing what was eventually a four-part study of coal mining. While discussing this work, Franklin gave Morand a copy of the stove engraving. In late 1778, Morand asked Franklin for permission to publish it. "If the charming coal heating stove" were better known, he argued, it might be widely adopted. Morand also requested "an abridged notice, explaining the plate." Franklin obliged, and a copy of the engraving appeared in Morand's *L'art d'exploiter les mines de charbon de terre* (1779), along with three engravings based on drawings Franklin provided of coal mines and mining equipment in Newcastle.[12]

Essentially, Morand was advertising an improved way to burn coal within a work promoting commercial extraction of that very material. It may have been the possibility of coal's greater availability that persuaded Franklin to discuss his newest invention publicly. As well, he owed an enormous debt to France, which had recognized the United States in early 1778 and offered it the financial and military assistance for which Congress had dispatched Franklin. Behind the scenes, his scientific contacts had been easing Franklin's access to France's power circuits since 1776—why not favor them with his famous name and works?[13]

Indeed, both the cold of winter and the heat of the American Revolution likely solicited new interest in Franklin's work. His stove pamphlet was only at this point translated into German, for example, appearing in a three-volume collection of his works drawn from the Barbeu-Dubourg edition and from a 1779 collection in English, edited by British politician and physician Benjamin Vaughan, Franklin's friend and supporter (even during the war, when this edition was published). The German compendium came out in Dresden in 1780, consisting mostly of Franklin's electrical experiments, with some of his political writings. The presence of the latter is notable, given that the German states, including Dresden's Saxony, were either neutral during the war or supported the British. Franklin's essays on science were for that reason presented as politically neutral, and the stove essay certainly supplied a full sense of Franklin's intellectual arc, despite its disdain for the "German" stoves in Pennsylvania. More significantly, the volume reproduces both Franklin's *Increase of Mankind* and a letter he'd received from British politician and scientific observer Richard Jackson, commenting (favorably) on the population essay, and, like Franklin, predicting the rise of colonial consumers and demise of Indigenous people.[14]

By including the pieces on heating and on population growth, the compendium astutely conveys Franklin's own irresolution over the problem of deforestation: improved stoves play David against the growing Goliath of heat-seeking American settlers. Whether intended in those opposing terms or not, the overall effect was to present the inventions as hypotheses, not proven solutions. And for this reason, the Dresden edition is distinctive in putting those four pieces together; the essay plus letter on heating came from Barbeu-Dubourg's French edition, and the essay plus letter on population from Vaughan's English edition.

Franklin's various fireplace inserts may have seemed quaint to German readers because they and their ancestors had been conserving wood and using stoves for a longer time. Rising population and intensifying industrial activities had been driving up demand for fuel for centuries. Efforts to preserve wood and woodlands had therefore begun in the late Middle Ages. Forestry became a distinctively "German" occupation, one that was skilled, specialized—and state-sponsored. While individual British and British-American consumers followed commercial prompts to make decisions about energy use, government was more important within the German-speaking lands. Stoves were a particular concern in the eighteenth century, resulting in large iron stand-alones, often highly decorated, sometimes fitted with compartments for cooking, developed to control smoke, and made to maximize heat from minimal fuel. These were more readily available than elsewhere in Europe, let alone colonial zones. They were expensive, however, and critics said they couldn't solve the problem by themselves. In any case, Germans were already shifting from wood to coal, favoring a new fuel at least as much as new technologies.[15]

A quick glance farther east, into Russia, is a useful comparison. There, as in Sweden, the Franklin stove's invisibility signals that energy availability was a major factor in its reception—or lack of reception. Franklin's work on electricity was famous among educated Russians, whose elite was interested in science, as with the empress herself, Catherine II (r. 1762–1796), who kept an eye on Western Europe's technological and industrial improvements, particularly as they might affect Russia's economy. Russian timber and iron exports, for example, were rapidly increasing, including to Britain. But crucially, Russia had a wealth of firewood in its huge forests; burned into charcoal, the wood helped produce the iron. In the eighteenth century, Russia was the world's biggest smelter of iron. The nation had further energy abundance in its rivers, tapped for hydropower. Finally, Russia's large population of serfs meant that forced human labor was another significant source of energy (and revenue). These factors were linked; after emancipation of serfs in 1861, landowners would compensate for lost income by selling off wooded property. This was the largest phase of deforestation in Russian history: millions of cherry orchards' worth of trees, put to the axe.[16]

Before that point, however, conservation of fuel wood was not a high priority except for poorer consumers, especially in cities. Russians had already developed heating technologies that maximized fuel use. For ordinary people, especially in the countryside, a large stove (*izba*), made of stone, clay, timber, brick, tile, or some blend of these, was their house's centerpiece. They used it for cooking, for heating, and for sleeping snug above it. Wealthier Russians had upright stoves of the kind found elsewhere in Europe; the Winter Palace in St. Petersburg, for example, has fireplaces and Dutch-style tiled stoves; the Pavlovsk Palace outside the city has fireplaces plus masonry and plaster stoves. Abundance of fuel and existing heating technologies, not politics, constituted two reasons for indifference to Franklin's stoves. Catherine II had a policy of neutrality during the American War, if tending strategically to question Britain's imperialism more than US independence. Only in 1789, after the war, would the Imperial Academy of Sciences in St. Petersburg elect Franklin an honorary member. Russia would recognize the United States in 1803, the year (maybe not coincidentally) several of Franklin's nonscientific works were translated, although his "Speech of Father Abraham" and his autobiography had already appeared in 1784 and 1791, respectively, as if they seemed safer (or more saleable).[17]

Translation of his work may have been unnecessary, in any case. Russians interested in science could probably read Franklin in English, French, or German. For example, the Russian National Library holds the *Oeuvres de M. Franklin*, also volumes of the American Philosophical Society's *Transactions* that would include his late thoughts on science. The Russian State Library holds eighteenth-century editions of his scientific compendia in English (the 1773 edition), French, and German, each of which includes the stove pamphlet. It's not certain those works arrived at these libraries during Franklin's lifetime. But a contemporary catalog of the private library of Yekaterina Romanovna Vorontsova, Princess Dashkova, Franklin's main learned contact in Russia, includes his political essays in English (maybe the 1779 edition from London) and a volume of *"Transactions of the American Society,"* probably *Transactions of the American Philosophical Society*, and most likely the volume Franklin would send Dashkova in 1788, which includes a later essay of his on smoky chimneys. Dashkova's library also had an *Histoire de l'Électricité*, probably the French edition of Joseph Priestley's work, which covers some of Franklin's work on the atmosphere.[18]

Russian interest in his works seems less, on the whole, when compared to other European nations and language groups. The contemporary Russian periodicals that discussed science don't include anything by him. And a search of amalgamated library catalogs for Franklin's name (in Cyrillic) as author yields no Russian translation of his scientific works during his lifetime. Unpublished letters or scattered published references may eventually reveal something. But there's nothing comparable to the circulation of his stove works in other languages that, as in Britain, fostered manufacture of the actual devices. With their huge land empire and energy reserves, Russians found advice about fuel conservation interesting, but not compelling.[19]

Elsewhere, lack of firewood wasn't itself always sufficient to generate translations of Franklin's work on heat. His stove pamphlet seems never to have been translated into Spanish, Portuguese, or Catalan, for instance; certainly, for those languages, he lacked the kind of acolyte he had in Barbeu-Dubourg. And, as in Russia, people in Spanish or Portuguese territories who kept up with science could probably read French (or English). But Portugal and Spain had shortages of wood that were unknown in Russia. One observer in central Spain said that food cost less than fuel to cook it. Here, Iberian caution about backing the United States may have been relevant. Portugal and Spain, as politically conservative and imperial powers, weren't motivated to endorse a revolt against monarchy and imperialism, preferring neutrality, though, in Spain's case, with some surreptitious assistance of the Americans. Although Spain entered the war in April 1779, it did so as France's ally, without recognizing the United States.[20]

This caution contrasts with Britain, where, somewhat oddly, given the war, Franklin's example remained popular. An Englishman, James Sharp, published a rhapsodic account of Franklin's stove in London that, from 1780 to 1785, ran through *ten* editions. Sharp was an iron merchant who made and sold multiple versions of the metal fireboxes that had been Franklin's basic fireplace design, testifying that he'd had one in his dining room for the past twenty years. His pamphlet's standard title, *An Account of the Principle and Effects of the Air Stove-grates... Commonly Known by the Name of American Stoves*, was in one of its editions (1781) augmented with the more explicit *Pensilvanian Air-grates*, the old fireplace updated in light of newer goals about circulation of air. Sharp provided illustrations of the fireplace's variations (large, small; plain, fancy) and gave a French translation alongside the

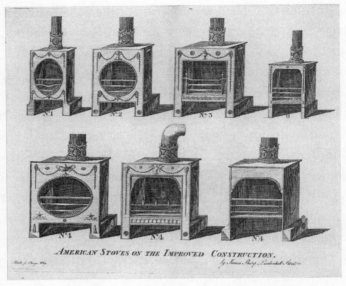

AMERICAN STOVES, MADE IN ENGLAND
DURING THE AMERICAN REVOLUTION.
(James Sharp, An Account of the Principle and Effects of the Air Stove-grates . . .
Commonly Known by the Name of American Stoves *[London, 178–?],*
call number *60–283, Houghton Library, Harvard University)

English. The wartime, bilingual publication indicates that Franklin's outlaw status in Britain as a rebel against a constitutional monarchy did not cancel the validity of his work in science.[21]

Sharp's credit to Franklin was, if anything, more dutiful than it needed to be. What he produced was more of a freestanding stove, with a metal stovepipe that could fit into any existing flue in a fireplace or elsewhere—a blend of Franklin's second and fourth models, or, like the surviving example, cast with a top plate that accommodated a pipe. Still, he insisted that his overall concept was an homage to Franklin: "These Stoves are called *American,* because the first Patterns in cast Iron upon this Principle were the Invention of the celebrated and ingenious Dr. *Benjamin Franklin,* who then resided in *Philadelphia.*" (The italicized words made explicit the wartime praise for an American enemy who hailed from the Declaration of Independence's birthplace.) Sharp quoted extensively from Franklin's stove pamphlet, placed sunburst medallions on the fanciest of his stoves, and emphasized Franklin's point that less fuel burned meant less smoke made: "a very strong Argument for their general Use in this great Metropolis." In one stone church outside London, two of Sharp's stoves, he claimed, easily drove off both cold and damp.[22]

Unlike Franklin, Sharp gave exact dimensions for the spaces his stoves heated and of the degrees Fahrenheit they achieved. Rooms could be warmed 20 to 30 degrees above outdoor winter conditions, keeping steady at 60 degrees with a moderate fire; in a room that was 48 by 24 feet, with 30-foot ceilings, a Sharp stove generated 54 degrees indoors when it was a challenging 22 outside. At the time, an influential English medical text recommended an artificially heated indoor temperature no greater than 58 or 60 degrees, a range associated with refreshing outdoor air. This prescription, plus Sharp's degree readings, indicate that, in wintertime, the middling ranks still expected to wear more clothing indoors than many people do today, but nevertheless wanted a significant difference between indoor and outdoor temperatures. Also, because medical thermometers to determine body temperature didn't exist until the second half of the nineteenth century, regular thermometers were increasingly recommended to gauge ambient temperature as a proxy measure for health, one step closer to thermostatic control of indoor atmospheres.[23]

Across the Channel, Franklin renewed discussion of his stove designs with a widening circle of European contacts. Fellow man of science Étienne-François Turgot, Marquis de Soumont, requested information about Franklin's vase stove in a letter in the spring of 1781. Turgot was specifically interested in "the method you have used to ignite the smoke" in order "to reduce the consumption of wood." Barbeu-Dubourg had reproduced Franklin's refusal to explain this stove fully "because its success depended on attention" most servants could not give, but Turgot wanted to try it "in our kitchen fireplaces which consume an enormous quantity of wood." Franklin replied, supplying a copy of the plate he'd had engraved in London to describe the device. It worked because it had two "legs" of different length, modified from the earlier tubing with equal legs that Franklin had first used to investigate "the Pressure of the Atmosphere." In this case, the stove was the shorter leg and the chimney, as the longer, had greater pressure, which pushed the smoke down into a grate of coals. This could indeed have "Advantage in our Kitchens," if "the Repugnance of Cooks" to "new Instruments & new Methods" could be overcome.[24]

Encouraged by these exchanges, Franklin found time amid his diplomatic duties to read new research on heating. From Vienna (Austria was neutral during the war), his friend Jan Ingenhousz sent extracts from sources Franklin couldn't read in their original German or even get his hands on. This included an account of what Ingenhousz called a "smokeless furnace." While he didn't name the author, it was French inventor and physicist André Dalesme, whose device had been described in both the *Journal des sçavans* (Paris) and the *Philosophical Transactions* of the Royal Society of London for 1686. Franklin knew of this invention from Martin Clare's work, cited in his original stove pamphlet. The device, made from iron, was shaped like a pipe for smoking tobacco. When the top of the pipe's stem filled with hot air, it generated a down draft to fan the fire and keep it actively reconsuming any smoke that failed to rise. It was supposedly so effective that no odor was detectable from even the most redolent of fuels, including coal soaked in cat piss.[25]

Surviving books from Franklin's personal library indicate four other relevant works, two published in Stockholm and two in Paris, with publication years extending from 1772 to 1777, suggesting he acquired them in Paris. The short 1772 work, *Poële hydraulique, économique et de santé* (*A Healthy and Economical Hydraulic Stove*), was an affidavit by two doctors on the Faculty of Medicine of Paris that an anonymous inventor's portable device, resembling a bain-marie, generated moisture and heat. This defined yet another measure of an artificial atmosphere, ambient humidity, in addition to temperature and fresh air. The two Swedish works together described and illustrated multiple large heating systems that, installed within fireplaces, used baffles to collect and distribute heat.[26]

The fourth work was a French translation of an Italian essay, *Lettera del Conte Cisalpino . . . in cui si descrive un cammino, e stufa du nuova invenzione*, republished in a scientific periodical, *Observations sur la physique*. Cisalpino (likely a pseudonym) gave a description of his "new invented stove" as another example of the strategy to propose a design as a modification of Franklin's—invoking the famous name to gain credibility. In this case, the inventor praised "*la stufa di Franklin*," but offered improvements. He reduced the size of the fireplace's mouth (which everyone was doing), but his better modification was building a channel into the back of the fireplace that (unlike the Pennsylvanian fireplace) evacuated the smoke well above the mantelpiece, with less risk of it leaking into the room. The channel could be opened or shut with a sliding plate operated by an iron rod with a handle at the mantel's lower edge, easily reached. There was no airbox to concentrate the heat, but the count recommended a conical metal tube to convey smoke up the chimney, surmounted with a metal cap to prevent wind from sending the exhaust back into the house. Aside from the iron, the count's fireplace was made of Italian marble (one would hope so), though he said more economical versions could be made with terracotta tile. Whatever its refinements, the "cis-Alpine" fireplace, whose name may have emphasized its merits in an Alpine and chilly part of Italy (or else, more daringly, referred to a historically autonomous Italy), validated Franklin's design and theories about heat and smoke.[27]

Franklin's expertise came to even greater attention in France because of emergency conditions during yet another historically terrible winter. This occurred in 1783 to 1784, when persistent haze occluded the sun, making conditions darker and colder. The timing of the gloom was ironic, descending as negotiations to end the war were yielding peace, one problem swapping in for another. (It was also ironic that, during the previous year's unusually warm winter, Antoine Lavoisier and Pierre-Simon, Marquis de Laplace, had struggled to get ice for a historic experiment conceptualizing basal metabolism with a "calorimeter," a device that measured how long it took a heated object—or a miserably shivering guinea pig—to melt the ice.) Because he was preoccupied with the peace talks, Franklin would speculate on the cause of the haze only later. In the meantime, authorities in Paris declared a state of emergency, cold having made firewood scarce and disastrously expensive. The government announced that baking, not heating, was the priority. Bread was notorious as a measure of public confidence in France's ruling order. Clueless French noblewomen who airily declare that hungry folks can go and eat cake are mythical, but bread riots in eighteenth-century France were all too real. So, during the awful interval as 1783 turned into 1784, public officials decided that the people might have to be cold but—*mon Dieu!*—let them eat bread.[28]

Compounding the problem was the fact that the French were running out of wood to burn, running out of wood for everything. The monarchy had become more stringent in regulating royal forests and even private woodlands for shipbuilding. Alongside this conservation, done to protect state interests, available trees for charcoal, construction, or firewood were dwindling in number. One estimate indicates that across France, wood would cost 91 percent more in 1789 than it had in 1726; less than 13 percent of the nation's lands remained forested. Nevertheless, like other Europeans, French people were adopting the ideal of indoor "comfort" and (like the English) adapting their word *"confort"* accordingly. Consumerism was, moreover, now linked to civic identity and the possibility of equality—you were what you ate, drank, and wore, and you shared status with fellow subjects accordingly. On the eve of the French Revolution, Parisians burned an average of one ton of wood per person each year. The tally was even higher in the countryside; only city folk, like Franklin and his grandson, might burn coal in stoves. Unsurprisingly, given the 1783 climate emergency, France's *savants* offered advice about forest conservation; equally unsurprisingly, the resident American man of science was consulted, giving him a new and collaborative role within the diplomatic and intelligence networks that had been woven—if antagonistically—into his mission all along.[29]

The instigator of this exchange was Antoine-Alexis Cadet de Vaux, the chemical experimenter and apothecary who was Paris's health inspector. Cadet de Vaux had informed France's foreign minister, Vergennes (Franklin's official counterpart), of the vase stove Franklin had described in Morand's 1779 work on coal, complete with engraved illustration. Because cast iron was more expensive in France than in Britain (or Pennsylvania), the vase stove was not an optimal solution to what might turn out to be a short-term problem, as everyone hoped it would be. This was the exact difficulty Franklin had faced originally, in Pennsylvania, when he decided to modify fireplaces rather than replace them with an entirely new technology.[30]

So as he'd done before—four times before—Franklin invented a device to sit within a fireplace, generating and circulating more heat than the fireplace itself could muster while using less fuel. This machine was also designed to use minimal metal, a first for him, though entirely following the logic of his project.

The proposed device, documented in a sketch Franklin provided, was his response to Lenoir, head of the Paris police, whose spies had been keeping tabs on him since his arrival. Lenoir was charged with keeping order in the city during the climate emergency. On the advice of either Vergennes or Cadet de Vaux, Lenoir asked Franklin for a device to accomplish his signature feat: "To destroy as you do, Sir, the smoke from the Coal and turn it to the benefit of the heat." Franklin's obliging design resembled a portable space heater with a closed, circular basket to hold coal, in essence an improved version of the brazier Europeans had often used to heat small spaces, as Franklin had described in his stove pamphlet decades earlier. This new metal stove (*poêle* or *poële*) could nest in a fireplace, whose chimney or flue would vent the smoke. Its key feature was the circular cage for the fuel. When the embers were almost out, someone could open and refill the cage, then close and flip it, making the smoke and flame sink into the fresh coal, maximizing fuel consumption.[31]

In a subsequent letter to Cadet de Vaux, Franklin affirmed that this stove was designed for coal only, following Lenoir's request for something that spared wood and charcoal. (Though Franklin complained that there was little coal to burn.) Relatedly, Lenoir had asked Franklin about making bread from maize, maybe fearing harvest failures due to the wintry weather (though that didn't happen). Maize was cheaper than wheat, but the French resented it as a hardship food; they fed American corn to livestock. Franklin noted that cornmeal took longer to bake, requiring more fuel, but offered a recipe parboiling and then blending it with wheat flour, as was common in North America. The recipe reduced baking time while cutting the alien corn with some reassuring French wheat. To the problems of keeping people warm and fed during unusual cold weather, Franklin proposed hybridity: a Pennsylvanian fireplace insert stripped down radically and adapted to coal, and an American grain eking out traditional daily bread. The debt to the generations of Native women who had domesticated maize and taught settlers what to do with it went unacknowledged.[32]

This final Franklin stove was the smallest of them all, though equally bold in addressing a climate crisis by using colonial resources (American grain; Franklin's brain). The American diplomat's knowledge was American know-how, white settlers' adaptation to places once rich in resources, including trees for fuel and land for maize. The obvious new factors were the notoriety of the American War and Franklin's heightened visibility in representing his nation. As his friend Ingenhousz put it in late 1782, "Every thing coming now from you gets an additional importance." There was an unacknowledged feedback loop here. France's effort to rebuild its fleet, after defeat in the Seven Years' War, put yet more pressure on the nation's forests—worsened by their military aid to the United States. The merchant ship built in 1765, renamed for Franklin's almanac, the *Bonhomme Richard*, and deployed in the new war was a case in point, as was the fleet of 228 ships Admiral de Grasse led to victory at Yorktown in 1781, a flotilla of hundreds of thousands of butchered trees.[33]

Franklin's collaboration with Lenoir was an emergency measure, responding to a sharp convergence of the era's three huge crises with one small, hot device. Similarly, in 1779, he had brokered a US passport for some French coal-mining proprietors who wanted to import a Boulton and Watt steam engine from Britain. The French would not definitively switch to coal until the 1830s. But addressing the winter crisis of 1783–1784 was one of the episodes that went into that historic transition, infused with anxieties over wood and forests, as demands on them intensified. (Not even the French Revolution solved the problem.) Notably, this was the only time Franklin's conservation efforts were part of a government plan, not just pitched as a consumer option. In his fifth stove, American autonomy merged into France's Ancien Régime, a political system far more centralized and rigid than Britain's rule over the American colonies, another reminder that conservation has often been contentious because, when hierarchically organized, it does not automatically align with all human interests.[34]

Franklin's original wood-burning stove, as a consumer option, remained a more salient design in France once the immediate winter crisis eased. While writing an analysis of the Pennsylvanian fireplace, mathematician and military engineer Charles-Louis François Fossé went to Passy to discuss it with Franklin himself. In 1783, he presented Franklin with a draft of his work. Franklin at some point gave him a copy of the engraving of his coal-burning vase stove. Fossé could not replicate that, not directly, given that France's reliance on coal lay in the future; his exchange with Franklin was about a wood-burning present, as the wood ran out.[35]

The work Fossé published in 1786, *Cheminée économique [Economical Fireplace], a laquelle on a adapté la méchanique de M. Franklin*, adapted Franklin's mechanism in one of the most ingenious ways yet, by making it from sheet metal. With improvement of rolling technologies in the late eighteenth century, that material was more readily available, no longer reserved for expensive prototypes, as with Franklin's first steel fireplace inserts. From sheet metal, a heating unit could be made to measure, whatever the room or fireplace. This didn't necessarily make it cheap. What Fossé described was the equivalent of being fitted for a suit or dress—not possible for everyone. And just as a French consumer who could afford it might have garments tailored from luxury imports—Italian silk velvet, say—Fossé recommended Swedish sheet metal (the best), though French stuff would do. The results, based on Fossé's two years of experiments at home to heat at least four rooms in his house (including two he described as large), were the usual gains in comfort, economy, and health. Like the *poële hydraulique*'s inventor, Fossé also emphasized an optimal humidity, of particular benefit to plants, this new measure of an indoor atmosphere reflecting new consumer desire for interiors with many living things. A good balance of heat, air, and moisture would result only if wood was burned, however. Coal would scorch sheet metal and make it buckle, requiring frequent replacement, in which case cast iron was better.[36]

Likewise, in 1789, Lyon architect Joseph-François Desarnod would publish an analysis of "the celebrated M. Franklin" and his "Pennsylvanian fireplace," inspired by the now chronic wood shortages near any urban center of consequence. Desarnod had studied Franklin's design from the stove pamphlet reproduced in Barbeu-Dubourg's translation and came up with some improvements (as usual), making a successful prototype in 1786. He retained the firebox-plus-airbox design from Franklin, removing the bellows insert in the hearth's bottom plate, thus giving more space for fuel and (he thought) augmenting the draft that channeled smoke, eliminating even the strong odors of peat or charcoal. His text cites affidavits from the Académie royale des sciences and the Société royale de médecine, plus his fifteen-year royal patent on his model of Franklin's fireplace in three sizes. (If Desarnod had read James Sharp's bilingual account of his similar London model, he doesn't admit it.) He gives instructions for installation and operation, with variations for wood, charcoal, or mineral coal. He concludes with pictures of his different models. It's a nice touch that a small sun adorns the front plate of the very first.[37]

Did Desarnod's design take off? Maybe it already had. One of Franklin's scientific contacts in Paris uncovered a smoke-consuming stove design at the Académie royale des sciences from a century earlier (this may have been Dalesme's cat-piss-defying device, yet again) and pointed out that a similar design was common for kilns that fired faience, a tin-glazed pottery that requires high heat. And it's possible that, once the weather warmed back up, there was less immediate need to conserve fuel in Paris, anyway. (France's bread supply was a different story.) But the French solicitations of advice from Franklin reveal yet again the circulation of information about conservation between old and new worlds and, likewise, emerging consensus that coal was the answer, the hot and sooty deus ex machina.[38]

Franklin's heating designs had their greatest impact in a place he never visited: the northern parts of what would become Italy. His fame there originated from his work on electricity; two important electrical experimenters, Giovanni Battista Beccaria and Alessandro Volta, corresponded with and cited him. But Italian readers also learned of Franklin's experiments with heat. In 1767, the Grand Duke Leopold, ruler of the Grand Duchy of Tuscany, told Filippo Mazzei, his agent in London, to find and buy him two Franklin stoves. Mazzei contacted Franklin himself, who somehow found time to sweep him around the likeliest ironmongers, carping amiably about their stoves, but picking one man to visit Craven Street and copy what he'd installed. The grand duke may have possessed the first Franklin stoves on the European continent.[39]

Franklin's stove pamphlet first appeared in Italian in Milan, translated by an admirer, the editor and cleric Carlo Giuseppe Campi, who in 1774 produced an edition of Franklin's essays derived from Barbeu-Dubourg. Campi didn't include any of the electrical works, saying they were too familiar to attract readers. But he did include the stove pamphlet. He requested further essays from Franklin, also explaining his writing him in French, "because I don't know if you understand Italian." That must have amused Franklin, who, after learning some French in his youth, next acquired Italian as part of a chess-playing contest with a Philadelphia friend, in which the victor set an Italian exercise for the loser: "We thus beat one another into that Language." Franklin was flattered over the translation, bragging about his Italian fame in his autobiography, and sending Campi some essays that hadn't yet appeared in Italian.[40]

Campi's work set off a cascade of reprints of the stove pamphlet in northern Italy. His translation reappeared in a multivolume compendium of noted men of science, *Scelta di opuscoli interessanti tradotti da varie lingue* (*A Selection of Interesting Pieces Translated from Various Languages*), that Giuseppe Marelli published in Milan from 1775 to 1777, for a total of thirty-six volumes. (Another editor would continue it until 1803 under a slightly different title.) In fact, the very first piece in the first volume was Franklin's *Descrizione della stufa di Pensilvania*.

Next, the stove pamphlet was cited in Giuseppe Vernazza's *Lettera del Conte Cisalpino* one year later—1776, when the American revolt might have encouraged discussion of the American invention, also reported in the French translation of this work, discussed above.[41]

Whether with permission or not, the Venice printer Antonio Graziosi would republish Marelli's translation of Franklin's stove pamphlet no fewer than three times, in 1778, 1788, and 1791. Why this lavish attention to a device that would, by 1791, be half a century old? Both politics and material conditions were factors. Like the United States, Venice was a republic, where expressing sympathy for the American cause was safer than in a monarchy. (In contrast, Milan was part of the Hapsburg monarchy, and Turin belonged to the Duchy of Savoy, though Franklin's work had appeared in these places before the American War.) Still, Venetians played it safe by maintaining an ambassador in Great Britain through the war, and none of the Italian states allied with the United States.[42]

It was a careful balancing act: to publish one of Franklin's writings not overtly linked to the revolt against Britain indicated some sympathy for the Americans, but not too much. In his introduction, Graziosi strategically thanked the Venetian ambassador to France, Daniel Dolfin (who'd formed a friendship with Franklin), for introducing the stove—"from the bosom of America"—into his own house, with no harm, only benefit. Still, the authorities worried the stove pamphlet might have some kind of covert radical message, "death to tyrants" encoded in American thrift. The Holy Inquisitor, Venice's office for the detection of heresy, therefore examined the text (whose author, after all, wasn't Catholic), but found nothing to offend. In his third edition, Graziosi printed their approval.[43]

Graziosi did not print any radical works. Like Franklin before him, he published a newspaper and almanac, bread and butter for a printer of the time. He sold a wider variety of artistic works than Franklin could have done in Philadelphia, including plays, romances, and opera libretti, in French, Latin, and classical Greek, plus Italian and, occasionally, Venetian. He'd printed some other practical applications of science, as with a 1784 account of ballooning. And midway through his run of three Franklin editions, in 1785, Graziosi printed a work on the fossil coal found near Bergamo, Giovanni Maironi da Ponte's *Delli carboni fossili o antraci bituminosi di Gandino nella provincia bergamasca*. The closest he came to sympathizing with revolt against the established order was in 1775, when, as the rumblings from America were getting quite loud, he printed Carlo Goldoni's libretto for *Il mondo della luna*, a comic opera about a woman-ruled utopia on the moon.[44]

Still, it's not easy to determine whether Franklin's celebrity or his heating expertise was the greater reason for someone to buy his stove pamphlet in Italian, though Graziosi was probably counting on both. The work didn't interest everyone. A surviving copy, from the Massachusetts Historical Society in Boston, has several uncut pages; some long-ago reader gave up. But a very different level of engagement appears in a surviving copy from the John Carter Brown Library in Providence, Rhode Island. This 1791 *Stufa di Pensilvania* is annotated, in Venetian, with instructions and a hand-sketched diagram of how to reduce a fireplace's size in order to install the American invention. The number of Venice editions is itself unusual—there hadn't even been that many in Philadelphia. This is for good reason: Venice is a very revealing case of fuel shortage during the Little Ice Age.[45]

Venice had been conserving wood since the late Middle Ages. In the premodern period, the city's name described both a republic, radiating from the lagoon into the mainland (*terraferma*), and an empire, with territories in Eastern Europe and the Mediterranean. All the lands beyond the city were managed for their economic value, including as sources of wood. Much of the city is built on wooden pilings, used since the Middle Ages to create ground beyond the islands' original footprints. That use of timber continued into the early modern era,

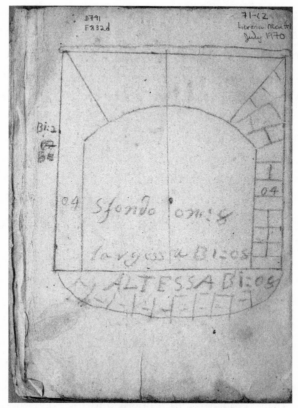

THE FRANKLIN STOVE IN THE PALAZZO, WITH INSTRUCTIONS FOR
ADAPTING A VENETIAN HEARTH TO A PENNSYLVANIAN FIREPLACE.
*(*Descrizione della stufa di Pensilvania *[Venice, 1791], call number D791.F832d,
courtesy of the John Carter Brown Library)*

though the greater strategic need was for warships to defend or extend the empire. Venetian authorities thus developed one of Europe's most elaborate regulation of forests. The republic's 1476 forestry laws designated oak and pine trees on the *terraferma* for shipbuilding; meticulous tree-by-tree surveys began in 1569.[46]

Wood was also essential for ordinary construction or repairs, but above all for fuel. Households burned wood or charcoal every day, as did bakeries, cookshops, forges, and the many workshops in this semi-industrial city, with its valuable glass products and dyed silks. Charcoal often had the edge over firewood, given its greater portability and heat. Communities in forested areas in the sub-Alpine regions north of Venice cut wood, burned at least some of it into charcoal, and transported it down to the city. Such forests were managed by Venetian authorities beginning in the late sixteenth century, and not until the nineteenth century would coal replace wood and charcoal.[47]

Northern Italy was not an outlier in the timing of its adoption of coal; Britain, in its early shift, was the exception. This was why Franklin's example remained salient in Venice, warm comfort when little else might have been.

Metal versions of his stoves, complete with inner baffles, were therefore imported and installed in several of the northern Italian states. Builders in Venice imitated them in both metal and terra-cotta. The learned brothers Alessandro and Pietro Verri, active in Milan, imported a version of the famous fireplace from England in the late eighteenth century. Leopold II, Holy Roman Emperor, King of Hungary and Bohemia, Archduke of Austria from 1790 to 1792, and Grand Duke of Tuscany from 1765 to 1790, bought and installed several Franklin fireplaces in his Italian residences. Franklin's status as a master of heating systems made him a byword in Italy. When the writer and revolutionary Ugo Foscolo endured a cold December in 1808, in Pavia, he said he "sighed a Franklin" (*sospiro una Franklin*), his use of the feminine "una" referring to the stove, *la stufa*, that bore the American man's name, the invention (not the inventor) being the feminine solace for which the shivering poet heaved his sigh.[48]

Finally, the Franklin fireplace was hybridized into the famous ceramic stoves of the Castellamonte factory near Turin, which moved the design out of the princely palazzo and into middle-class and public spaces. Most of these stoves resemble those found elsewhere in Northern Europe: closed, fire-proof boxes from which heat could radiate on all sides. The Castellamonte manufacturers adopted some of Franklin's design by making parts of their stoves' interiors metal, giving them a greater capacity to retain heat than ceramic could achieve. These stoves, as well as metal Franklin fireplaces (*caminetti Franklin*) closer to the original Pennsylvanian design, were common heating systems in Northern Italy until the adoption of steam heat, radiators, and a decisive shift to coal after around 1880. Several such *caminetti* survive in the ducal castle of Agliè, near Turin. Castellamonte Franklin stoves heated Italian public buildings and private residences even into the early twentieth century. One was listed on Italian eBay in 2022, with a front plate showing the American inventor in noble profile.[49]

FRANKLIN AND HIS STOVE (AND HIS FUR CAP, FOR EXTRA WARMTH).
(Giovanni Antonio Sasso, after Jean François Bosio, Beniamino Francklin *(Milan, 1815), Philadelphia Museum of Art: Gift of Mrs. John D. Rockefeller Jr., 1946, 1946–1951–149)*

Signor Beniamino Francklin, *il dottore di Pensilvania*, who'd learned Italian by being beaten at chess, who'd described distant Venice in his stove pamphlet, had an afterlife in Northern Italy he'd probably never dreamed of. More than in any other part of the world, his invention represented him. An 1815 posthumous portrait, printed in Milan, shows Franklin seated at a table, his freestanding rectangular stove behind him.

It's the image made nearest his lifetime that includes both him and his stove—in a place where the stove was almost as famous as he was. Thirty copies of the *Stufa di Pensilvania* survive in libraries today, for example, but only one copy of the first *Poor Richard's Almanack* in English. Northern Italy existed at a sweet spot for adoption of Franklin's design. It depended on shrinking supplies of wood, had not yet shifted to stoves or modified fireplaces, lacked easy access to coal, and, given its northward location, was vulnerable to cold during the Little Ice Age.[50]

And yet the international reception of Franklin's stove project is eclipsed by the global proliferation of *The Way to Wealth*, another of his works that ignored politics in favor of economics, though in this case, an economy of abundance. This essay began as the 1758 preface to his expanded almanac, *Poor Richard Improved*, in which Franklin recycled many of the proverbs he'd used in earlier almanacs—"*'tis easier to build two Chimnies than to keep one in Fuel*," and so forth—and compressed them into "Father Abraham's Speech," as if delivered by a wise village elder, "a plain clean old Man" speaking pithy truths to rapt rustics gathered round. By the end of the eighteenth century, the essay had been extracted and reprinted over a hundred times in seven European languages, compared to the stove pamphlet's five. And that was just the beginning. Retitled *The Way to Wealth* circa 1771, the speech was recast as primarily economic advice and eventually circulated worldwide, representing new confidence in what by the nineteenth century would be called "capitalism."[51]

The stove pamphlet never had that kind of afterlife. But note that it faded, and *The Way to Wealth* took off, as coal consumption was rising. As they moved into fossil fuel dependence, people preferred to believe in wealth, not dearth. Coal may have killed Franklin's stove project, in its original wood-conserving form, but it would be the lively fuel of industrial capitalism, the revolution that, far more than resource conservation, and more even than the American revolt against British rule, marks the modern world.

8

NOVUS ORDO SECLORUM

On September 3, 1783, Franklin emerged from a tense meeting at the Hôtel de York on Rue Jacob in Paris with only two massive crises to worry about, not three. The war was over. With some covert French aid, the Continental Army had defeated British troops at the Battle of Saratoga in 1777, leading to formal French recognition and an alliance, and serious aid, the following year. With that help, and similar validation and assistance from other European powers, the United States fought its way to a final victory at Yorktown, Virginia, and to peace talks. Franklin was one of three US delegates to represent the United States and to sign the final treaty at Rue Jacob that early autumn day. At last, the United States had a fully acknowledged independent existence, which the treaty mapped in words, starting at the "northwest angle of Nova Scotia" and laying out the nation's original territory between the Atlantic and the Mississippi, about eight hundred thousand square miles, not quite 10 percent of North America.[1]

Novus ordo seclorum, as the Great Seal of the United States proclaims: the republic inaugurated a "new order of the ages." Adopted in

1782, and much like the effort to rename our era as an unprecedented Anthropocene, the Latin motto makes an extraordinary claim, decreeing the beginning of an entirely new epoch. Similarly, the hypothesis of anthropogenic climate change, expressed as a hope that US settlers would swarm across North America and warm its climate, would prevail through the early nineteenth century. But was the creation of the United States epochal in that same way? The Treaty of Paris did not just end the war—it distributed material assets: land (those 800,000 square miles), money (who owed what to whom), and access to resources and to markets. While people have quibbled ever since about who got the most out of the concessions, the continuation of Pennsylvania's industrialization, and its continued prosperity, suggests that independence from Britain left more than enough of the key materials to build the United States' eventual economic core, mapping a course toward a natural world deeply affected by human actions.

Released from the emergency defense of his nation, and despite an accumulation of maladies (psoriasis, gout, and a bladder stone so large and jaggedly painful he could feel it even when turning over in bed), Franklin plunged into one of the most intense periods of scientific engagement in his life, mostly to examine atmospheric phenomena. His work reaffirmed that climate and atmosphere constituted irregular patterns that changed over time, ranging from days to millennia—yes, epochs. As usual, he related these phenomena to the everyday human creation of heat and emission of smoke, the generation of substances that, he conjectured—distinctively for his era—affected the bigger atmosphere, one chunk of firewood or lump of coal at a time.[2]

In the air is often a metaphor, evoking zeitgeist or cultural atmosphere—unless it's describing real things up in the air. There was a superabundance of that aerial matter in and around Paris during

Franklin's residence (even as revolution and independence were parts of his zeitgeist, too), showering him with a wealth of phenomena to revive his war-deferred work in atmospheric science.

First up (in the air) was the aurora borealis, which had dazzled Parisians with a vivid display on December 3, 1778. *Magnifique*—but what was it? Franklin reasoned, yet again, that convection was at work. Hot air would always rise: "If in the middle of a Room you heat the Air by a Stove, or Pot of burning Coals near the Floor, the heated Air will rise to the Ceiling." Just so, "a Similar Operation is perform'd by Nature on the Air of this Globe," with air from equatorial regions spreading up and out, "thus a Circulation of Air is kept up in our Atmosphere as in the Room above mentioned," as anyone could verify by holding a lit candle in a doorway between two differently heated rooms, first up high, then down low, seeing the flame waver first in one direction, then the other. Given that electricity also circulated in air in the form of lightning, might it not be affected by temperature extremes? The aurora borealis might be an electrical presence within a polar atmosphere, unable to discharge within a dense layer of cold air.[3]

Franklin posed this idea as a question—most of the sentences at the end of his essay conclude in question marks. (The aurora borealis does involve electricity, specifically yet-to-be-identified electrons, though Franklin didn't grasp the significance of Earth's magnetism in its light display.) He translated his essay into French and, after a friend made corrections, it was read at the Académie royale des sciences. When it appeared in the *Journal de physique*, a note politely marveled that a scientific genius could also be engaged in political affairs, as if declining to judge the results of either. It wouldn't be the last word on the northern lights—and Franklin didn't care. He thought it enough that his thoughts might "produce a better Hypothesis," so "not be wholly useless." And it's notable that, in this latest leap from the atmosphere of a room to that of the planet, he refers to "Nature" deploying the equivalent of either a stove or "a Pot of burning Coals," the latter the size of his small revolving coal grate.[4]

Not long after the shimmering light show, another wonder rose above Paris: balloons. Hot air made possible the first artificial flights: the first successful ascent of a globe-shaped, inflatable balloon in spring 1783, and the first manned flights (tethered, then untethered) in fall that same year. Hydrogen turned out to have even greater lifting power, as demonstrated at the Champ-de-Mars and Jardin des Tuileries in 1783 as well. (It was quite a year.) Franklin saw one of the hydrogen balloon ascents from his terrace at Passy, seven days before the treaty signing at Rue Jacob. Jacques-Étienne Montgolfier invited him to the tethered, manned liftoff of his and his brother's eponymous hot-air balloon, eight days after the treaty signing; Franklin RSVP'd, but may not have attended. He did observe the first untethered Montgolfier flight from Passy in November, excitedly relating that "when it went over our Heads, we could see the Fire" that kept it aloft—fueled by straw! As the balloon's "rarified Air cools and condenses," it begins to sink, and then the passengers can be seen "throwing in" more straw, as if flying globes, like horses, eat hay.[5]

But within this soaring drama, our main character wouldn't have deserved the name "Benjamin Franklin" if he hadn't tested this atmospheric invention himself, this time using an assistant. (The seventy-five-year-old may have been ill and bed-bound, as was increasingly the case.) His grandson got him one of the mini-balloons being hawked to sky-watching Parisians and filled it (carefully) with hydrogen. The small globe rose "to the Cieling in my Chamber," Franklin exulted, "and remained rolling about there for some time." At last, there was a possible way to escape from Earth, with balloons offering a plausible technology to explore and maybe exit the atmosphere—no more fanciful schemes for getting to the moon by harnessing flocks of birds, and so on. Though balloonists quickly discovered that the higher they rose, the colder the atmosphere got and the more labored their breath, confirming what had been observed in ascents of high mountains. As a life-support system, the atmosphere's direct benefits for humans had a physical boundary; some obstacles to unearthly travel remained.[6]

Amid the heavenly displays, Franklin had a flood of British visitors and letters, making up for silences necessitated by war. (In one case, a balloon made the reconnection: a French aeronaut and a British doctor made the first aerial crossing of the Channel in 1785, delivering the world's first airmail, a letter from Franklin's estranged son to his grandson, William Temple Franklin.) These renewed conversations, too, were invitations back into science. Franklin's work on atmospheric phenomena was of course already renowned. His work representing the Gulf Stream as a complex system of heat and motion, for example, circulated along with that of William Gerard De Brahm (a rival) and Charles Blagden (an admirer). The war had kept Blagden, who was British, from traveling to France, but he arrived in 1783 and presented Franklin with a copy of his essay from the *Philosophical Transactions* of the Royal Society of London on the temperatures of water in the Gulf Stream. Franklin's own meteorological works had been circulating since the first edition of his *Experiments and Observations on Electricity* (1751), with expanded content in both the English and French editions that followed. To the former, he had added maps illustrating his work on storms and on the Gulf Stream. But Blagden's visit may have reminded him that he still had several things to say.[7]

First, Franklin reviewed the basic nature of heat. Aware of the chemical "philosophers" who were "in the way of finding out at last what fire is," he maintained his essentially Newtonian view that heat existed "in the state of a subtile fluid," defined by its scattered rather than dense particles. When abundant, the fluid generated the sensation of heat; when lacking, cold; and when it emitted "vibrations," light. It existed within anything flammable, and when these burnable substances returned to an "original fluid state," they manifested as ordinary fire. Most interestingly, Franklin proposed that this subtle fluid "is attracted by plants and animals in their growth, and consolidated." His statement reflected recent theories of plant respiration, as with Joseph Priestley's discovery that plant life could effect "purification of the atmosphere," restoring air mice had breathed within a closed container. That same year, Jan Ingenhousz showed that plants required exposure to the sun to absorb carbon dioxide ("fixed air") and emit oxygen, a process later called photosynthesis; he reported to Franklin his continuing investigation of "the influence of the vegetable kingdom upon . . . our atmosphere," but the rudiments of an essential atmospheric cycle were now identified.[8]

Franklin next composed some "Loose Thoughts on a Universal Fluid" in 1784, a speculative essay he sent to Ingenhousz and to Pennsylvania scientific investigator David Rittenhouse. By reinterpreting Newton's hypothesis of a universal fluid, he sought to explain both light and heat—and therefore life. His conjectures anticipate the wave theory of light: "Universal Space . . . seems to be fill'd with a subtil Fluid, whose Motion, or Vibration, is called Light." This property might also explain heat, and here Franklin's conjectures express a theory of energy, if without using that word. For this he drew, at least partly, on recent chemistry that described transformation of different material states:

> the Power of Man relative to Matter, seems limited to the dividing it, or to mixing the various kinds of it, or changing its Form and Appearance by different Compositions of it, but does not extend to the making or creating of new Matter, or annihilating the old: Thus if Fire be an original Element or kind of Matter its Quantity is fix'd and permanent in the Universe. We cannot destroy any Part of it, or make addition to it. We can only separate it from that which confines it & so set it at Liberty, as when we put Wood in a Situation to be burnt.[9]

This elementary fire was essential to life. When human bodies lacked it, they were "frozen" (dead); when they had the normal level, they were "in Health." Fittingly, given his bold attempt to define life itself, Franklin phrased his hypothesis as questions. "Is not the natural Heat of Animals," he asked, "thus produc'd, by separating in digestion the Parts of Food, and setting their Fire at Liberty?" In positing this heat exchange, he still found it likely that, above Earth's atmosphere, there must be a "Sphere of Fire," a zone of this fluid surrounding the planet, set into trembling motion, ultimately, by the sun, with vibrations that heated Earth by day though "discontinu'd in the Night, or intercepted & reflected by Clouds."[10]

That realm of fire does not exist—not as a place of sensible heat, at least. Nor, as modern thermodynamics have determined, is energy stable—it is constantly dissipating. But Franklin interestingly posed a final question about what people in the past had "suppos'd" to be true about nature: "May it not have been from such Considerations that the antient Philosophers suppos'd a Sphere of Fire to exist above the Air of our Atmosphere?" His question was a way to think about how the atmosphere diffused solar energy, with the sun's fire converted into light, sensible heat, and organic life, diminished from a force that, if unmediated, would have damaged the planet considerably.[11]

Franklin used atmospheric circulation to explain one final climatic phenomenon: the cause of the bitter winter of 1783 to 1784, for which he'd developed his rotating coal grate. The cold, he pointed out, had been a secondary phenomenon. First, in late June of that year, there had been a "constant Fog over all Europe," yet more stuff up in the air, about five years after the aurora borealis and right before France's historic bouquet of balloons. This oddly "dry" haze occluded the sun. It drew much comment and speculation about its origin. But not everyone connected it to the succeeding event, the unusually cold weather, which prolonged winter conditions and created shortages of fuel. Abnormally thick ice in the Seine River, once it melted in spring, had flooded parts of Paris and surrounding areas. At the time, Franklin was preoccupied with the peace negotiations. He gave French authorities emergency advice on coal and corn in borrowed time, though they of course had him over a barrel: he needed their goodwill exactly then. Luckily, his widely reprinted work on lightning rods guided a spate of adoptions related to the haze, whose tiny particles sparked electrical storms, without his direct intervention.[12]

The next year, Franklin had time to ponder the odd fog and the cool conditions, and to interpret them as cause and effect. His resulting work, "Meteorological Imaginations and Conjectures," published in the *Memoirs of the Literary and Philosophical Society of Manchester* in 1785, was in essence an analysis of his old enemies, climate change and bitter winters.[13]

He began by conjecturing "a Region high in the Air over all Countries, where it is always Winter, where Frost exists continually." This lay beneath any celestial zone of fire, an atmospheric layer where heat could not be retained. It was Franklin's sharpest deviation from interpretations of Earth's climates as determined by the sun and differentiated by latitude, proposing instead a realm of ice above them all. He noted that hail could fall even in summer. Rain might actually be thawed hail or snow; the large hailstones that sometimes survived the descent were proof of an intense cold above. Cold precipitation and solar heat were counterforces, though sometimes operating together. Franklin suggested that rain, falling on sun-warmed ground, carried the heat inward, maybe to a depth of thirty feet. This was why a winter's first snows thawed quickly, and, with the removal of a white coating that reflected the sun, Earth's surface retained warmth. For that reason, in winter, "the extreme degree of its Cold is not always at the time we might expect it," when the sun was farthest away and the days shortest. The old English proverb got it right: "As the Day lengthens, / The Cold Strengthens." With "the Causes of refrigeration continuing to operate," Earth's surface became colder and retained snow, which reflected the sun's heat, generating a feedback loop.[14]

Franklin was here drawing on the insights he'd gleaned, decades earlier, from Newton's and Boerhaave's analyses of heat absorption. But he scaled up the phenomenon he'd used to explain swatches of cloth sinking into a snowbank, now visualizing the whole Earth—as when he'd imagined colonists "scouring" North America of its forests to give it a brighter appearance from afar. These descriptions anticipated the concept of an entire planet's reflectivity, an effect first called "albedo" in 1860, in which snow and ice help keep Earth cool, unless climate change forces a breakdown of the feedback effect.[15]

The strange fog during the summer of 1783 was prime evidence for Franklin's conjecture that it was not just correlated with the unusual cold, but its cause. The dry haze could not be dissipated by the sun in the way "moist Fog" could be. While the murk lasted, trying to burn a sheet of paper by focusing sunlight through "a Burning Glass" barely worked. No wonder the summer was ominously cool, "the Surface [of the earth] was early frozen," the first snows stayed on the ground, and the air and wind were "severely cold"—"Hence perhaps the Winter of 1783, 4, was more Severe than any that had happened for many Years."[16]

But what was the "Cause of this Universal Fog"? A dry haze resembled smoke, Franklin reasoned. Maybe a meteor, one of those "great burning Balls or Globes which we happen to meet with in our rapid Course round the Sun," had exploded upon entering the "Atmosphere," generating smoke. Or maybe a volcano was to blame. Mount Hekla in Iceland had recently erupted, along with a "Volcano which arose out of the Sea near that Island, which Smoke might be spread by various Winds, over the northern Part of the World." Franklin noted that this lasted a long time, a "vast Quantity of Smoke long continuing to issue during the Summer." (Right idea, though wrong volcano: the disastrous Lakagígar or Laki Fissure event of 1783–1784 continued for about eight months; its choking, poisonous fog and the subsequent famine killed at least a fifth of Iceland's population and more than half of its livestock, with enough of its plume surviving to suffocate people in Ireland and Britain.) Franklin recommended study of "whether other hard Winters recorded in History were preceded by similar permanent & widely-extended Summer Fogs." Such knowledge could help people prepare for extreme winters, anticipating fuel shortages and flooding when ice thawed.[17]

The "universal Smoke Doctor" was worthy of his name—he'd formed an important hypothesis about climate and volcanic activity, now recognized as a factor in short-term cooling, including some of the Little Ice Age's coldest intervals. Accustomed since the 1750s to thinking of North America and the Atlantic as swept by crescents of northeasterly winds, Franklin had developed a way to understand the spread of fine particulate matter across oceans and continents as a climatic variable. If his visualization of that circulation had begun with his first fireplace experiment, his fifth stove, the small device fashioned for the conditions of 1783–1784, was yet another doubled contribution to climate science and to climate-change adaptation.

Franklin also speculated that some climatic changes might persist for very long spans of time. While most of his contemporaries compared their cold years to warmer ones recorded in the Middle Ages, Franklin looked further back. He'd followed the work that gave Earth a long, slow geological history, punctuated by violent upheavals. He was "reconciled" to "those convulsions which all naturalists agree this globe has suffered," resulting in uneven and broken strata of different minerals: "clay, gravel, marble, coals." The climate, too, must have changed. He conjectured that the planet's early seas had been larger. Their shrinkage had exposed the European continent, whose original extent was proof of an "antient Climate, which seems to have been a hot One." Nor was that the last great shift. If it were true, as "Historians tell us, that old Greenland once inhabited and populous, is now render'd uninhabitable by Ice, it should seem that the almost perpetual northern Winter has gained ground to the Southward." "If so," he continued, "more northern Countries might anciently have had [grape] Vines than can bear them in these Days."[18]

Franklin postulated a world in which a person's life and experience (including his own) was but a small moment in time and space, a fragment of a pattern that extended far beyond what anyone could see in the world around them. This perspective defined climate science as the expert and collective analysis of long-term and complex processes in the larger atmosphere. When he sailed back to America in 1785, Franklin used his time at sea to elaborate that science while also, yet again, emphasizing how improved understanding of heat could be applied to everyday life; by connecting the phenomenon back to one individual's personal experience, he made the science believable.

He was aware that, aged seventy-nine, he was probably doing his last sea voyage, so he took a farewell look at the Gulf Stream. He'd last addressed it during his 1775 (westward) and 1776 (eastward) Atlantic transits, when he'd taken location-specific water samples, recording their temperature differences. He had already explained the current in relation to wind set into motion by the heating of air down in the Gulf of Mexico. Proving that the Gulf Stream's water was warmest where it ran fastest would help confirm the phenomenon, hence the water samples. He compiled a final set of thermometric data from water collected at different points during his 1785 crossing and presented the results in an essay of 1790, along with a third chart of the Gulf Stream. The data, Franklin's analysis, and his chart show the current as a crescent that trended north and east, connecting air and water in a pattern both longitudinal and latitudinal—climate as complex system.[19]

Franklin also used his homeward journey to summarize his

thoughts on heating systems. He assembled these in a letter to Ingenhousz, resulting in his longest explanation of his final fireplace's principles. This "Letter from Dr. B. Franklin, to Dr. Ingenhausz, Physician to the Emperor, at Vienna, on the Causes and Cure of Smokey Chimneys" would appear in the *Transactions of the American Philosophical Society* for 1786, the second volume of the journal of the scientific society Franklin had founded in Philadelphia in 1743. "Smokey Chimneys" was later reprinted in London, both in a 1787 collection of Franklin's works and in freestanding editions. These late thoughts also appeared in a French scientific compendium in 1789, and in Italian in a Milan science periodical—volume nine reproduces Franklin's letter to Ingenhousz; volume ten has a companion essay from the *Transactions of the American Philosophical Society* on Franklin's London vase stove. Finally, though German readers hadn't shown that much interest in Franklin's stoves, they, too, were plagued by smoky chimneys; Franklin's letter was translated and printed in Hamburg twice in 1788.[20]

In the letter, Franklin frames domestic heating in terms of the problem he felt he hadn't solved: emissions, including their indoor impact on human health and welfare (his correspondent, after all, was a physician), and their atmospheric impact (his correspondent had given the first description of photosynthesis). Significantly, his text reflects new confidence that heat might be easily generated and retained. So he welcomes the drafts he'd abhorred before, his appreciation of fresh air accompanying a sense that, on reflection, it was relatively easy to keep people warm, but at the risk of polluting the air they breathed. "The doctrine of chimneys appearing not to be as yet generally understood," Franklin begins, as, with a seriousness maybe genuine, maybe slightly self-mocking, he resumes his long quest to separate heated air from smoke, to circulate the former within a house and banish the latter from it. Who, after all, would not want uncontaminated air to breathe?[21]

He begins with the smoke itself, the vestige of inefficient combustion. People wrongly assumed smoke was a light substance, able to whisk itself up a chimney. Or they thought a chimney had some occult power to suck smoke upward. Wrong in both cases—it was convection at work, as always: warm air expands and rises, optimally lifting the smoke with it. A chimney's shape, including its top funnel, predicted little about its drawing power. Only its height might be implicated, with taller chimneys able to exit smoke somewhat more easily, though this didn't explain why some tall chimneys didn't do that while some shorter ones did just fine. As far as Franklin was concerned, chimneys were smoky for nine other perfectly obvious reasons, each of which could be rectified once diagnosed.[22]

Nine reasons may seem rather thorough—recall that Franklin had posited only five or six in 1762 when trying to keep his Edinburgh friends from being smoked alive. But he was, after all, reminding everyone who burned something to keep warm that they were probably doing it in some wrong way or another. Their mistakes resulted in greater fuel cost per household and greater pollution for everyone—a smoke footprint, long before the concept of a carbon footprint. Correcting faulty modes of combustion was worth it both for microcosm and macrocosm. Franklin's detailed advice might resemble counsel that could be given to Western consumers today if they had to live in direct, cough-inducing contact with the filth their energy needs generate, or if tasked with reducing their tailpipe emissions themselves, tinkering away in their garages and driveways.

Franklin's first possible chimney fault fingered a development that was otherwise positive, this being the trend toward draft-resistant housing. New homes were so well crafted—"each room is as tight as a snuff-box"—that no passages for air existed, "except the key-hole, and even that is sometimes covered by a little dropping shutter." This even tighter "tight" room admitted little fresh air to help push the smoke up the chimney. People who "stop every crevice in a room to prevent the admission of fresh air, and yet would have their chimney carry up the smoke, require inconsistencies, and expect impossibilities." Franklin recalled the woes of "a nobleman in Westminster" who spent £300 on multiple fixes to a smoking chimney in a brand-new house, none of which worked, "for want of understanding the true principles," something money couldn't necessarily buy.[23]

Franklin confessed that he'd evolved to be more accepting of drafts. He'd earlier had an *"aerophobia,* as I now account it," but came to regard "fresh air as a friend: I even sleep with an open window." (This had triggered a tiff with aerophobe John Adams when, in 1776, the two revolutionaries shared a New Jersey bedchamber with one window and two opinions about it.) Even damp air was welcome and, given Franklin's new habit of bathing twice a week entirely "covered with water," he no longer thought damp air or clothing could make him catch cold. "I find it of importance to the happiness of life," he told Ingenhousz, "the being freed from vain terrors." His new opinion reflected the evolving medical definition of contagion. Franklin's doubt that cold air made people ill was now vindicated by new analysis of air. It did not impart disease itself, but was a medium for "corrupted animal matter," either from "putrid marshes and stagnant pools" or by being "breathed over and over again by a number of assembled people." It wasn't "by *going out* of the room" that someone caught a cold—"it was in fact by being in it."[24]

Modern indoor health, therefore, required a room to be heated, though this meant determining "how much is *absolutely necessary.*" Again, Franklin encouraged people to use their houses as laboratories: go to the main door of a room where a midsized fire is burning, shut the door until smoke begins to enter the room, then open the door, slowly, to find the width that admits just enough air to shoo the smoke up the chimney. Several "easy and cheap methods" could assist with this, including sash windows whose top halves, when lowered, could admit fresh air. On the same principle, narrow crevices could be cut along wainscot or plastering where it met the ceiling above the chimney, too narrow to chill the room but wide enough to produce a smoke-purging draft.²⁵

Altogether, Franklin was continuing to modernize heating systems, adjusting them to new technologies and consumer amenities (wainscot, sash windows), while counteracting some of that gadgetry's drawbacks, as with those little metal shutters over keyholes. For the first time, he mentioned the utility of a household thermometer, registering how their cost had dropped. The instrument could verify "the difference of climate between the upper and lower parts" of a room, for instance. But he quickly added that someone could explore their room's upper atmosphere with much simpler equipment, a stepladder, which they could ascend "till their heads are near the ceiling" and able to sense the temperature difference, if any.²⁶

Moreover, Franklin now had wider experience with heating systems than he'd had in the colonies in the first part of the century. He described what the French called a *vasistas*, a pane of glass within a hinged frame that, mounted in a sash window, admitted a small draft if opening the whole window seemed excessive. Franklin used the spelling "*Was ist das?*" indicating a possible German origin for the mini-window, or maybe German speakers' fascination on seeing it in France. (Though the name may have been a corruption of "aeolus," the English term for a similar device—or else the German question punned on that.) Franklin also mentioned that some English people had aeoluses in the form of round tin plates with vanes cut and bent into them, which augmented the dispersal of fresh air, though their "noise, only, is a little inconvenient." Indeed, while we might have found the dwellings of Franklin's era cold or smoky, their inhabitants might have found our HVAC systems unbelievably noisy.²⁷

A second cause of smoky chimneys, Franklin explained, was the continuing preference for fireplaces with huge, ostentatious maws. The legacy of great roaring fireplaces to announce wealth was partly to blame, but also the architects who made fireplaces and chimneys bigger than they needed to be for perceived "symmetry and beauty" within rooms. "Our fathers made them generally much too large" and reductions were jarring, "the human eye not being easily reconciled to sudden and great changes." The remedy was to ignore what your eyes told you and hire a bricklayer or mason to fill unnecessary space, "imaginary" aesthetic principles be damned. Anyone "bigotted" in favor of "a large noble opening" was, after all, risking "damaged furniture, sore eyes and skins almost smoked to bacon." Likewise, the third cause of smoky rooms, too short a funnel at the chimney top, could also be adjusted (as Mary Stevenson had noted years ago at Craven Street) by contracting the mouth of the fireplace, thus generating greater heat within the chimney to lift the smoke.[28]

Franklin's seventh and eighth causes of smoky chimneys (poor placement of doors relative to fireplaces; unlit fireplaces whose cold chimneys welcomed smoky air from outside) revisited his original focus on the circulations of heat versus smoke within single rooms. But the remaining four causes addressed bigger problems of the construction and placement of entire houses. This was a more complicated scenario, reflecting Franklin's late-life mingling with people who, like that luckless smoked nobleman in Westminster, had big complex dwellings only a lot of money could buy. For problems within single rooms or chimneys, Franklin made recommendations based on his original Pennsylvanian fireplace design (adjust the door, in this case by switching the side that was hinged) or on the sliding register he'd developed in London, useful in sequestering an unlit fireplace from unhelpful atmospheric conditions.[29]

In quick succession, Franklin considered how multiple chimneys within a house, let alone a room (this also a problem, obviously, only for the rich), could overpower one another, with the weakest fireplace wheezing out fumes because pressure was building up elsewhere. Luckily, the adjustments Franklin had recommended for rooms or houses with only one fireplace and chimney typically worked where there were several. Though another problem was chimneys overtopped by taller buildings or hills, with wind spilling into them "like water over a dam" to drive smoke back inside. The opposite situation could also be a problem, with a "commanding eminence" unshielded from wind, which eagerly poured down the chimney. Finally, very strong wind could be a nuisance whatever a house's position and elevation. Franklin thought a metal "turncap" affixed above the chimney and turned with a vane might shelter a chimney from wind, unless the whole funnel needed to be raised with iron bars—or unless neither a turn cap nor additional height worked, which Franklin admitted was possible. (Imagine your heating or HVAC system roaring greedily away, spewing acrid haze everywhere—and nothing to be done about it, except maybe move.)[30]

To decode these mysteries, further scientific investigation was needed. The experts who did "experimental philosophy" should do smoke experiments, Franklin said, and present them in public lectures, using small glass houses. In these, people could see smoke circulate through adjoining rooms and via "moveable glass chimneys" with different configurations. He specified that a glass house would be best situated within the lecture room's circulatory system, with a working fireplace and a window that opened, one artificial atmosphere surrounding another. "By the help of such lectures," Franklin said, "our fumists would become better instructed," using glass houses too small to live in, let alone throw stones in (even at each other's theories of smoke).[31]

In describing this scenario, Franklin again showed that his fine-tuning of domestic heating had gravitated away from the humbler ranks. In three anecdotes, he traced his social ascent. He recalled renting a room in London in the 1720s with a smoky chimney whose funnel turned out to be cracked; the "landlord" stopped up the chimney rather than pay for repairs, leaving young Franklin to keep himself warm some other way. Contrast that scenario to his later familiarity with a "friend's country house near London." Franklin tried to figure out why smoke from the fireplace in the "best room" stayed right there. Not until the friend clambered up a ladder to inspect the chimney from above did he find it clogged with an abandoned bird's nest. Finally, at the really deluxe tip of the social pyramid, Franklin admired an iron fireplace he'd seen in Paris, which pivoted in the wall between a bedroom and a study. For the convenience and privacy of the unnamed "master," a servant kindled the fire in the study, and the master swiveled it into the bedchamber "with a touch of his foot," then swung it back to the study when desired. (An iron backplate diffused heat to whichever room didn't face the fire.)[32]

Luckily for anyone who had any indoor heat, whether for a small chamber in a city or a huge house in the country, "smoke is a very tractable thing." Franklin reminded Ingenhousz that he had "made it *descend* in my Pennsylvania stove," the fume bowing to his will, then forced by his sliding register to retreat up the chimney, "another instance of the tractability of smoke." (For the published version of his letter to Ingenhousz, Franklin also reproduced his 1758 letter to James Bowdoin describing his London invention.) Here, again, Franklin was turning his attention to what seemed a much more promising fuel— yes, coal. The "Staffordshire Fire-Place," someone else's invention for which Franklin provided a diagram, was a simple brazier extended from a chimney bricked up except for an opening the width of the brazier. Even as coal smoldered in the device, right out in the room, its smoke "bends and enters the passage above it, the draft being strong" in a space so efficiently heated.[33]

The whole problem, really, was chimneys in the first place. Franklin described them as one of the "modern improvements of living" that emerged, he thought, in Tudor times. They're in fact ancient and had been used in European houses since the Middle Ages. Franklin was correct, however, that their proliferation was a modern phenomenon.

"Such is now the growth of luxury," he chided, "that in both England and France we must have a chimney for every room," sometimes even for *servants*, he marveled, maybe remembering his shivering sojourn in that London room without a fire. The new expectation of heat for all, spread over multiple (and private) chambers, required multiple exits for smoke, so each chimney might contain several flues, some of which might fail. Also, "this change of manners soon consumed the firewood of England" and was threatening the same for France. To which Franklin did not add any comparison to Pennsylvania or any other part of North America.[34]

He instead focused on stoves and fireplaces in select parts of Europe. "The prosperity of a winter country," by which he meant anyplace with a substantial cold season, "depends on the plenty and cheapness of fuel," whether it was wood or coal or peat. Where fuel was expensive, "the working people live in miserable hovels, are ragged, and have nothing comfortable about them." This was the case unless they'd adapted technology to improve their circumstances. He noted that Venice, crowded with people but not trees, did not have multiple chimneys in its houses, but rather flues that serviced fireplaces at different levels of a building. Venice prospered, bustling with activity. Elsewhere, without adequate warmth, people worked less and more slowly, with impaired contributions to the industrious revolution. This was a final, econometric evaluation of fuel, heat, and life, though Franklin's comment about comfort registered a moral imperative: basic well-being should be available to all social ranks.[35]

Only coal could accomplish that. Franklin cited attempts in the 1600s to regulate industrial use of coal in London, on the grounds its smoke was unhealthful. "Luckily," the smoke was now considered "salubrious," given that there had been no great pestilence in the city since coal became the general fuel. (This was a correlation, not cause and effect; coal smoke does not, in fact, ward off plague.) As far as Franklin was concerned, coal had only one drawback: its smoke generated even more soot than wood did. Hence the importance of finding a way to burn the smoke. While admiring the ingenuity of Chinese underfloor heating, produced with "sulphurous pitcoal," he worried that the underside of the floor must be thickly fouled. "I conceive that burning the smoke by obliging it to descend through red [hot] coals" would solve the problem, and he sketched a diagram of how the heated air could be drawn upward, forcing the smoke to form a layer beneath it, nearest the flame.[36]

Did smoke-bending technology really work? While still at sea, alongside his other writing, Franklin composed a long-deferred analysis of the vase-shaped stove he'd developed in the 1770s, because he thought its downdraft design most likely to eliminate emissions, his most ambitious goal. His correspondence shows that at least two of these stoves existed, made for Baron Le Despencer and himself; he'd thought his own worth packing back to Philadelphia. An image and brief instructions had appeared in Morand's work on coal. The new essay revealed that he'd used his design for four years in London, acquiring detailed knowledge of its efficacy, including its ability to reduce smoke.

Franklin began with "an ingenious French philosopher, whose name I am sorry I cannot recollect"—in fact, the long dead but ever-recurring André Dalesme, whose brazier supposedly eliminated smoke (and smell) from anything burned. Franklin next reproduced an extract from Johann Georg Leutmann's *Vulcanus famulans . . .* ("Vulcan's Assistant"), which promised a similar result with another design. First published in Wittenberg (1720), Leutmann's essay describes a device to give maximum warmth using minimal wood while eliminating indoor smoke—the usual trifecta—modeled on Dalesme's 1680s bowl-and-pipe device. (Although Franklin said he'd translated the extract, he probably had help from Ingenhousz.)[37]

Franklin described this smoke-banishing action, yet again, as that of a "syphon," not one that raised water, but a metal tube bent into unequal, upward-extending lengths. To vent away smoke, the longer "leg" of Leutmann's stove had to be heated to trigger its drawing power—the exact opposite of what Franklin wanted. (Given that Leutmann specified four feet of iron pipe plus an inverted funnel above it to draw off the "unwholesome vapour" from coal, and that the whole thing had to be taken to the kitchen and fired up there before being toted back into whatever room needed the heat, Franklin's warning to Barbeu-Dubourg that servants might find this setup tiresome makes sense.) Franklin concluded that Leutmann's device was merely for showy experiments, whereas his had a practical goal: burning coal smoke, which required an even stronger downdraft, pushing smoke back down into the fire. Franklin said he'd used his stove for three years in London, starting in 1771, and then a fourth year following, in Philadelphia, before he decamped for Paris. (He seems to have counted in units of twelve months; his use of the stove in London extended over four calendar years.).[38]

Franklin's vase stove derived, as well, from an analysis of how coal burned—"pitcoal" appears in the essay's title. Within a steady flame, he said, coal would, paradoxically, burn slower than if only partly exposed to it. "That flame should be a kind of pickle, to preserve burning coals from consuming, may seem a paradox to many," he admitted. He'd discovered the phenomenon when testing a metal prototype (maybe after an initial mock-up done in wood, with puffs of smoke to check the device's airflow). The metal version was an inverted pyramid made from four triangular sheets of iron. He'd cut a small door in front of the grate for removing cinders, and through which he could view (and time) the combustion. When surrounded by flame, a piece of coal burned more slowly. (Reduced access to oxygen retards a fuel's conversion into heat and pollutants.) He gave exact measures for this, referring to the largest of his three rooms at Craven Street, a combined study and parlor that faced southwest, onto the street, with windows that caught the afternoon light:

> Two fires, one made in the morning, and the other in the afternoon, each made by only a hatfull of coals, were sufficient to keep my writing room, about sixteen feet square and ten high, warm a whole day. The fire kindled at seven in the morning would burn till noon; and all the iron of the machine with the walls of the niche being thereby heated, the room kept warm till evening, when another smaller fire kindled kept it warm till midnight.[39]

His device's efficiency was visible even to his neighbors. "The effect of this machine, well managed, is to burn not only the coals, but all the smoke of the coals, so that while the fire is burning, if you go out and observe the top of your chimney, you will see no smoke issuing, nor any thing but clear warm air, which as usual makes the bodies seen through it appear waving." (Refraction: light seems to move as it passes through different densities of air.) True to form, Franklin wanted to see the stove's cheerful fire, so he replaced the metal of a sliding door with either a pane of glass or (even better) of "Muscovy talc" (heat-resistant mica), so "that the flame might be seen descending from the bottom of the vase and passing in a column through the box C, into the cavities of the bottom plate, like water falling from a funnel, admirable to such as are not acquainted with the nature of the machine, and in itself a pleasing spectacle." The chimney into which such a stove vented wouldn't need cleaning, soot being fuel that Franklin's stove did not waste. The reburning of smoke would reduce fuel costs, as would recycling cinders, which could be cast on top of fresh coal and reincinerated.[40]

Franklin's vase stove was strikingly different from his Pennsylvanian fireplace for two reasons. First, it was a solitary rather than sociable machine, with indoor heating continuing its association with privacy. Franklin explained his stove as best suited to a "studious man who is much in his chamber, and has a pleasure in managing his own fire." (Which was presumably the case also for Baron "Hellfire" Le Despencer, famous for his bookish interests and his indifference to conventional family life.) Servants would find the device vexing, and it wouldn't suit families gathered in larger rooms. Second, the vase stove was not designed to conserve wood, because only with significant alterations could it burn wood. It was best lit by laying down some sticks of charcoal, then "brush" (commercially available bunches of twigs), then paper (inky scraps from the studious man's desk, presumably), the reverse of laying down a typical fire. But by lighting the top layer of paper from a candle, a downdraft would ignite the twigs and charcoal, which, when fully burning, could then be fed a "hatfull" of sea coal.[41]

Reducing heating costs and reducing emissions were excellent goals, but Franklin was achieving them by switching to another fuel: coal to the rescue, yet again. And while the solitary nature of the vase stove was new, it extended the egocentric logic of the patriarch's hearth, cultural icon in Europe and its colonies, so different from the Indigenous council fires that united people across households and nations. The net result would engrain yet more deeply into colonial North America a culturally specific way of life. Finally, the recurring measure of coal by the "hatful" beautifully summarizes the historic transition from clothing that supplied a warm zone around the body—products of the needle—to a less-swaddled body basking in room-sized warmth, triumph of the flame.

The essay he was writing at sea in 1785 also gave Franklin the chance to explain his fifth stove, the rotating device he'd sketched out in Paris. He admitted that it didn't consume smoke as well as the vase stove, though it was still better at that than an ordinary fire, and "fitter for common use, and very advantageous, compared to the vase stove." For instance, "it gives too a full sight of the fire" and, turned horizontally, the grate would support a teakettle, two jolly solaces for anyone housebound in bitter weather. An engraving displays the little brazier alongside the vase stove (also a version of Leutmann's fussy brazier, based on Dalesme's), with details of the baffles cast into the interior plates for the base of Franklin's iron vase. To give a sense of scale, the rectangular base for the urn was about two feet wide, so the height of the entire device was a little more than that; the rotating grate was about a foot wide. Note the iron instruments at the bottom for handling either coal or else a stove made very hot by coal.

Each of his last two heating devices represented the shift to coal,

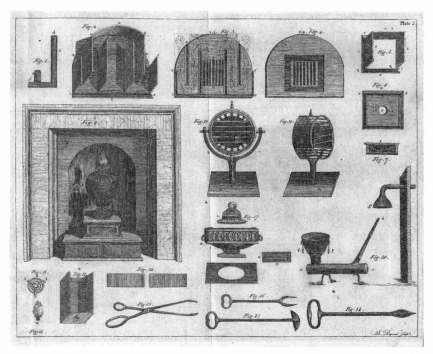

TWO FINAL FRANKLIN STOVES: THE LONDON URN (MODEL FOUR)
AND THE PARIS ROTATING GRATE (MODEL FIVE).
*(Benjamin Franklin, "Description of a New Stove for Burning of Pitcoal,
and Consuming All its Smoke,"* Transactions of the American Philosophical Society 2 [1786],
courtesy American Philosophical Society)

though Franklin was also associating fuel adaptation with emergency circumstances. Multiple emergencies: he would have learned that the British officers who commandeered his Philadelphia house during the war pulled down its fences and exterior bathhouse, most likely to secure fuel, as supply lines to the wooded countryside and foreign coal fields failed.[42]

The ethos of Franklin's last two stoves was more urban than rural, more cosmopolitan, more luxurious, with a new priority for individuals to be warm in rooms scattered through a house. Franklin thus bequeathed the fruits of his decades of heating experimentation mostly back to the old world, validating its consumer-driven industrious revolution and its adaptation to fossil fuels. He did this with a warning, having learned the hazard of inventing things adopted hither and thither without his feedback. The English "workmen" who'd read about his first fireplace in 1744, for instance, had tried to implement his design without including its airbox, declaring that this yielded more space on the grate for coal, but this unwisely abandoned the efficiency of the original. "On pretence of such improvements," Franklin said, "they obtained patents for the invention, and for a while made great profit by the sale, till the public became sensible of that defect."[43]

Caveat emptor: anyone buying a stove called *Franklin* or *Pennsylvanian* should compare it to what the inventor described in print. He knew he would never return to Europe. After 1785, he existed there mostly on paper, sometimes in iron, never in the flesh. Meanwhile, he was beginning to think of himself, again, back in Philadelphia's green country town. Or was it still?

In 1782, Francis Hopkinson, son of one of Franklin's collaborators in the original electrical experiments and a fellow signer of the Declaration of Independence, shared with Franklin his worry about trees in a city with streets named for them.[44]

The Pennsylvania Assembly had ordered Philadelphia's trees to be removed along public streets and alleys, on the grounds they were fire hazards and impediments to firefighting. Hopkinson protested. He wrote a parable, published in Franklin's old *Pennsylvania Gazette*, done from the perspective of an indignant wooden column, Silvester, who helps hold up the ceiling of the State House. Insisting on his right to speak as "a *standing member* of that House," Silvester chides his human colleagues for their chauvinism and ingratitude. Trees are living beings, too, and civically responsible ones. Having kept up with modern science, Silvester cites the discovery, popularized by "*Linnaeus*," that trees, like people, are male or female and, loving each other, reproduce sexually. Conversely, "we [trees] employ none of our powers in devising means for the more speedy and effectual destruction of our species," a pointed comment on the late war. Even after death, Silvester notes, trees are sweet and useful, unlike humans.[45]

The best proof of trees' goodness is their living contribution to human comfort and health. If deciduous, they shed their leaves in fall, maximizing Earth's absorption of warmth from a dimmer winter sun, then grow back the greenery to offer shade in summer. Most thrillingly, they are humanity's "safest Physician": they breathe, and in breathing, "they purify the ambient air," as "PRIESTLY" and "INGEN-HOUSZ" had proven. Even when trapped in Philadelphia under the occupying British, they had done their atmospheric duty, though whole "groves" of them on the far banks of the Schuylkill had fallen to the invaders and breathed no more. The State Assembly should spare any veterans and survivors. "A few hours are sufficient to execute this fatal law," Silvester pleads, "but it will take years to repair the damage when you shall have discovered your error." Knowledge of trees' healing respiration was a "modern discovery," perhaps, "unless we may suppose it to have been known to the Indians of America," who used a leaf to signify a healer.[46]

Yes, Indigenous people appreciated trees. But they considered them kin, comparable to humans, whereas Hopkinson was telling a pretty fable. He made an alarming elision of Natives when he had Silvester insist he belongs to "a numerous race, descended in a direct line from the *aborigines* of this country; those venerable ancestors who gave the name of Penn's-*sylvania* to this State," as if, by this point, white settlers could more easily think of trees as aboriginal than humans. Within those very terms, Hopkinson's plea was successful. The Pennsylvania government conceded "that trees planted in the streets thereof conduce much to the health of the inhabitants, and are in other respects of great public utility," and amended the proposed act. And thus, a city whose first streets were named for Pennsylvania's trees—chestnut, walnut, spruce, pine, and so on—took on the task of preserving its actual trees, an early example of atmospheric science affecting public policy.[47]

Franklin thanked Hopkinson for his "ingenious Paper in favour of the Trees." True, they might be fire hazards, but Franklin knew Philadelphia had adequate protection, having helped create the city's fire brigade—"a Tree is soon fell'd; and as Axes are at hand in every Neighbourhood, may be down before the Engines arrive." Meanwhile, the trees' ongoing benefit massively outweighed their drawbacks, as Franklin had noted to Priestley in 1772, praising their ability to maintain healthy air. Now, Franklin made a remarkable speculation about what are now called microclimates— that they were created by trees: "I wish we had two Rows of them in every one of our Streets, The comfortable Shelter they would afford us, when walking, from our burning Summer Suns, and the greater Coolness of our Walls & Pavements, would I conceive [add to] the improv'd Health of the Inhabitants, [and] amply compensate the Loss of a House now and then, by Fire."[48]

This was a new vision of a city whose artificially cultivated nature had distinctive value. The human equilibrium with the natural world was being recalibrated. Protection of urban nature was, accordingly, part of the management of Philadelphia's local climate, in which people needed trees to breathe and to keep cool. (Given that Franklin wrote on Christmas Eve in Paris, his memory of summer heat in Philadelphia was impressive.)[49]

And yet, by century's end, Philadelphia's city fathers no longer followed Penn's plan to name the streets for Pennsylvania's natural things. The first exception, designated around 1770, was Callowhill, family name of the colony's other original proprietor, Hannah Penn. Nor were tree streets necessarily named for Pennsylvania species. Appletree Street appears in the 1791 street directory, after the introduced Eurasian cultivars that lined it at the time. Cherry Street appears around 1809, commemorating the era's craze for cherry trees, with their ex post facto grafting onto George Washington's boyhood. Cherry was another Eurasian fruit tree that, along with apples, reinforced settlers' European heritage, a legacy they saw in the trees as they walked in their shade, and picked and ate their fruit.[50]

It was more than slightly ironic, as well, that while trees retained value in the city, the immediate countryside was almost wholly deforested as the settler population grew. On the eve of the American Revolution, Pennsylvania had been second only to Virginia in population and Philadelphia was the largest colonial city. The first US census (1790) confirmed that Pennsylvania remained the second-most-populous state with 434,373 people, still fewer than Virginia (747,610) but more than Massachusetts (in third place with 378,787), and Philadelphia was still the largest urban center.[51]

The state's official population was overwhelmingly white, with a minority of Black people, mostly in cities, and an even smaller number of Natives. The earlier episodes of settler violence, with Teedyuscung's murder and the massacre at Paxton, had erased Pennsylvania's history of pacifism, with the related effect of extracting yet more land from Natives, now justified in terms of their reluctance to support the American cause. At the start of the war, when the Haudenosaunee were urged to choose a side, they couldn't agree. Instead, they "covered the fire" that united them in council at Onondaga, suspending their confederation. Most Haudenosaunee warriors who entered the conflict fought for the British; some supported the United States; none officially represented their confederacy.[52]

The Haudenosaunee decision released Lenapes from any obligation, as their honorary nephews, to choose a specific alliance, US or British. At this point, some Lenapes remained in New York, New Jersey, and Delaware. But most from Pennsylvania had already been forced west, then into the Ohio Valley. (New Jersey Lenapes declined a 1771 invitation to join them.) There, as war had crept into the valley, Lenapes tried to position themselves so that no one would "stain their *Councel-fire* with blood," as they explained to a Moravian missionary. This proved difficult.[53]

By the 1778 Treaty of Fort Pitt, Lenapes allied with the United States and against the British, making them the second Native nation to sign a US treaty. In this, US forces gained the right to pass through Lenape territory to attack the British. The treaty declared that "all offences or acts of hostilities by one, or either of the contracting parties against the other, be mutually forgiven, and buried in the depth of oblivion" and that they would henceforth "hold fast the chain of friendship." No mention of a council fire. The treaty acknowledged Lenape territory and their authority "to invite any other tribes who have been friends to the interest of the United States, to join the present confederation, and to form a state whereof the Delaware [Lenape] nation shall be the head, and have a representation in Congress." This prospect of a fourteenth state, designated for Indigenous people—a homeland—was one of the most radical propositions to come out of the early republic. But it obliged Lenapes to confirm their renunciations of pacifism and antislavery; the treaty required them to return runaway "servants and slaves" to US citizens who claimed them as property.[54]

The Continental Congress never accepted the treaty. Lenapes may not have known that, at least immediately. One of the Native signatories later said the written treaty didn't match the terms spoken at the negotiations; another signatory died on the 1778 US-Lenape campaign against Fort Detroit, possibly assassinated by his ostensible allies. And not all white settlers respected Natives as military allies or neutral parties. In 1782, at a Moravian mission called Gnadenhütten, Pennsylvania militiamen attacked pacifist Lenapes and Mohicans who refused to fight in the war, systematically murdering twenty-eight men, twenty-nine women, and thirty-nine children, as their victims prayed, sang hymns, and begged for mercy. The massacre convinced Lenapes that alliance with the United States was pointless; the Treaty of Fort Pitt was broken, and a Native state in the United States—an arresting what-if—did not come into being.[55]

The Treaty of Paris (ratified 1784), in which Great Britain recognized the United States, ceded territory while disregarding Indigenous sovereignty. Some Lenapes stayed east, particularly in New Jersey, but continued settler invasion and armed conflict sent many Natives from the Ohio territory into what would become Indiana and Wisconsin. Some Lenapes joined Oneidas in New York and in Canada, where the

A LATE-EIGHTEENTH-CENTURY ONEIDA WAMPUM BELT (TOP), REPRESENTING THE SIX COUNCIL FIRES OF THE HAUDENOSAUNEE. A MODERN ONEIDA FLAG WITH BELT DESIGN SHOWN BELOW.
(Oneida Language and Cultural Centre, Oneida Nation of the Thames)

Haudenosaunee reconstituted itself after the war; to commemorate the event, the Oneida wove a treaty belt with wampum, with diamond shapes representing the council fires of the six nations then in the confederation, and an open end for additions, if ever needed.

From the US Midwest, surviving Lenapes were forced out again, into territory in Kansas and Oklahoma in 1867. In the latter state, the Delaware Nation of Oklahoma has made its official home. State-sponsored Indian removal fulfilled Franklin's prophecy that white people would displace Indigenous people. While rapid growth of the white population played a part in this, formal political action was the necessary mechanism of expulsion.[56]

Confident that white settlers would fill the land through natural increase, Franklin was reluctant to encourage emigration. In his "Information to Those Who Would Remove to America," written in 1784 (as he was moving back himself), Franklin favored his theory, from his *Increase of Mankind*, that ample land for young white people guaranteed population growth in what was now the United States. "Land being cheap in that Country," he claimed, "from the vast Forests still void of Inhabitants . . . the Propriety of an hundred Acres of fertile Soil full of Wood may be obtained near the Frontiers." He had paused, however, and deleted the phrase "easy to be obtained" about the land, the one possible trace of consciousness (or conscience) that he was describing Native land, still forested and contested. He simply stated that increase resulted from the "Encouragement to early Marriages, by the certainty of Subsistance," not specifying that the "Marriages" and "Subsistance" only applied to white people, naturalizing a process in which political decisions did the real work. Finally, he praised "the Salubrity of the Air," perfectly suited to the American-born settler body.[57]

This justification of Indigenous removal had a global afterlife in the genesis of growth economics. In *An Inquiry into the Nature and Causes of the Wealth of Nations*, published in 1776, the first year of the American War, Adam Smith used Franklin's *Increase of Mankind* to distinguish between two kinds of economies, "thriving" versus "stationary." China was the latter. Smith describes it as an ancient and rich country, but with such a large population that wages were low; it no longer thrived or developed, "but it doesn't go backward." The opposite occurred in "the colony of a civilized nation which takes possession either of a waste country, or of one so thinly inhabited, that the natives easily give place to the new settlers, [and it] advances more rapidly to wealth and greatness than any other human society." In such a place, people married early and had ample resources to sustain children. British colonists therefore "double in twenty or five-and-twenty years," making them "much more thriving." In these details, Smith shows his debt to Franklin, either to his essay (he owned a 1760 edition) or via texts that cited Franklin.[58]

And so, from this American source, Smith derived one of his central and influential economic principles: it was not "the richest countries" that were "the most thriving . . . [but] those which are growing rich the fastest." Rate of growth is the best measure of a nation's value, the main premise of all growth economics that followed Smith. He was adamant, as well, that consumer demand was central to growth: "Consumption is the sole end and purpose of production," with industrious consumers, even the working poor, committed to buying soap, candles, tea, sugar, tobacco, and the like.[59]

To the accumulating qualms about growth economics, not least for their environmental impact, add its originator's dismissal of America's Natives as thinly settled, easily removed, and possessing no legal sovereignty. Smith also aligned with Franklin's assessment of slavery, dismissing it as economically feasible only for high-value crops like sugar and tobacco, but otherwise inferior to free labor. This ignored the immense economic impact of enslaved peoples' work in all colonies and in Britain itself, an inconvenient truth. Whatever his theory's flaws, Smith powerfully endorsed the aggregate, expanding, nonconservationist economy that had become the United States of America.[60]

Franklin himself thriftily reused the paper on which he'd drafted "Information to Those Who Would Remove to America." The preliminary sketches of his coal-burning rotating grate, recommended to Paris authorities in the cold winter of 1783–1784, appear on the back of one sheet. The two-faced document records North America's and Europe's opposing material circumstances, wood remaining in the one place but barely in the other, though coal (possibly) could fill the gap—anywhere. It was as if his long residence in Europe had made Franklin forget his old conservationist impulse, while coal somehow became available in the United States, as was happening in Europe.[61]

Certainly, getting firewood remained a problem in older US cities, including Philadelphia, a scarcity that threatened Pennsylvania's maturing semi-industrial economy. As "Silvester" had said, the war had spoiled landscapes and disrupted supply lines. Iron forges and furnaces were both culprits and victims of the damage. Prisoners of war had sometimes substituted for the ironworkers who joined the military effort or ran away—many manufacturers leased prisoners of war from the Continental Army. Sometimes, they paid the army in iron products. Meanwhile, military action gouged the countryside. Valley Forge, scene of the Continental Army's survival in the bleak winter of 1777–1778, was named for the Mount Joy Forge owned, after 1757, by the Potts family who had cast the first version of Franklin's Pennsylvanian fireplace. (General Washington rented a Potts family house as his headquarters and residence: George Washington slept there.)[62]

A shortage of fuel continued to loom. Yet again, the impact was worst for poor people in areas far from woodlands, who faced emergency conditions during periods of unusual cold. "Some knowing Ones here in Matters of Weather predict a hard Winter," Franklin wrote his beloved sister, Bostonian Jane Franklin Mecom, in early autumn of 1786. "Permit me to have the Pleasure of helping to keep you warm. Lay in a good Stock of Firewood, and draw upon me for the Amount. Your Bill shall be paid upon Sight, by Your affectionate Brother B Franklin." Brotherly love, indeed.[63]

But where was any of this wood to come from? New Englanders were steadily deforesting parts of what would become New Hampshire and Vermont, but it wasn't clear that the mid-Atlantic states had equivalent hinterlands. A German visitor, Johann David Schöpf, commented in 1783 that one New Jersey iron plantation had "exhausted a forest of nearly 20,000 acres in about twelve to fifteen years, and the works had to be abandoned for lack of wood." Coming from a place where forest conservation had existed since the Middle Ages, Schöpf was appalled: "Nobody is concerned about forest preservation, and without an uninterrupted supply of fuel and timber many works must go to ruin." Without the discovery of "rich coal mines," Schöpf said, the region's iron industry would surely die.[64]

And yet it carried on. A Swedish visitor (and industrial spy), Samuel Gustaf Hermelin, who'd met Franklin in Paris, was sharper-eyed in detecting surviving woodlands. True, "timber had been used up" in Valley Forge, for example. But farther away from the longest colonized areas, available trees and iron ore mapped onto each other, forebodingly for the trees. So the postwar future seemed bright for the recovery of the iron industry, already making an estimated ten thousand tons of bar iron each year. The discovery of ore fields near Blue Mountain (a ridge of the Appalachians) promised further increase, though Hermelin deplored the costs, economic and ethical, of making charcoal, which still relied on timbering and on exploiting enslaved workers. In 1809, Benjamin Henry Latrobe would be the latest foreign visitor to prophesy, yet again, that the iron industry would die from *"the scarcity of Charcoal."*[65]

But it didn't die. Instead, removal of Natives from western Pennsylvania transferred landscapes from people who had conserved them to people who would intensively exploit them. That transfer would sustain industrial development on a massive scale, with eventually devastating environmental consequences unfurling across Pennsylvania and then beyond. At first, these transformations were not regarded as forms of damage. Franklin may not have thought humans could remake the climate to suit themselves, but many other people took that as gospel. Thomas Jefferson would be one of the most famous; the ordinary farmers who spilled over the Appalachian border that the prewar Proclamation of 1763 had drawn against them likewise assumed they could, by bringing down forests and breaking into new land, create a North America with a more moderate climate, pioneering the last and best of Earth's epochs: *novus ordo seclorum*.[66]

9

KEYSTONE STATE

American independence, the Civil War, full-blown industrialization—Pennsylvania was the heart and hearth of it all. "One [is] inclined to think that Pennsylvania set up the Government in 1789," said Henry Adams, John and Abigail Adams's great-grandson, then "saved it in 1861 . . . developed its iron and coal power; and invented its great railways." The state's list of industrial firsts and greatests is long indeed (including its having what was then the largest railway system in the world), all depending on the fact that it was the first in the United States to have an economy dependent on fossil fuel and waged labor. That transformation began while Philadelphia was the new nation's capital and cultural center—First City; it reached logical extremes during and just after the Civil War, establishing a template for modern economic life throughout the world.[1]

Franklin witnessed this history's first phases without comprehending where they would end, a founder of something he never saw or foresaw. This is obvious from the last project he set into motion. In a codicil to his will, he outlined a philanthropic scheme to run into the future, maybe forever. His goals reveal, however, that he thought he lived in a final stage of economic development, a stable hybrid of fossil fuel and organic energies, including the labor of human bodies. He didn't imagine (and probably couldn't have imagined) that this mixed state was temporary, a transition into fossil fuel dependence and massively greater industrial output.

The point is not to indict him (or anyone) for an inability to predict the future; the point is to recognize that this is hard to do, which is why energy transitions require care and vigilance, given what's at stake, another reason to identify the flaws in the distant climate-change fire drill that Franklin's stove project represents. After Franklin's death, organic power, including some of the most coercive forms of labor, gave way to a highly capitalized use of fossil energy and unqualified admiration of its new possibilities. That overvaluing of nonrenewable energies, despite their damaging effects, is the part of the industrial transition that must now be reversed. Franklin's generation went forward with their own set of adaptations without knowing where they would lead, but we can now see where this has put us.

Within this metamorphosis, the creation and fortunes of the United States played a part, though not as industrialization's original cause. Too many foundations had already been laid; too much inertia already existed. The new nation was, at most, an efficient cause, the vehicle or delivery system, a chugging railroad that carried the coal. The American Revolution would have been truly revolutionary if it had somehow reversed trends that had been set in motion long before 1776, especially in Pennsylvania. Instead, settlers spread over the land and got to work. Their voracious consumer demands and robust industrial activities were the continuing reasons for the United States's eventual (and enormous) impact on the natural world, fostering modern material life and a damaged planet to match.

Franklin and the other delegates at the 1789 Constitutional Convention in Philadelphia signed the Constitution—with quill pens—while witnessing some of the earliest steam-fired locomotion. In 1782, James Watt had developed a rotative engine that, through gearwork, transferred a piston's vertical motion to turn wheels. This adjustment would make railroads and paddle-driven steamboats possible, the biggest mobility breakthrough before invention of the internal combustion engine. Two American inventors, James Rumsey and John Fitch, began testing rival steamboats: smackdown. As white US citizens threatened to take over more of the continent, including the Mississippi River, steam power had national consequence in its ability to conquer distance. The prototype boats had enormous potential power, if somebody's design for them worked.

Franklin was skeptical. He said so most frankly to J. Hector St. John de Crèvecœur, Franco-American farmer and man of letters. (The two men met in France in 1781.) In 1788, Crèvecœur wrote: "Having accidentally become acquainted with a Mr. Fitch from your City . . . I ask'd him a great many questions concerning his new method of applying the power of Steam, to which he answer'd with much diffidence and modesty," but also confidence in his eventual success. Franklin was unsurprised—he'd had an early peek at steam engine design back in 1766, after all, when consulted about the Boulton and Watt separate condenser model. "I never doubted," he told Crèvecœur, "that the Force of Steam properly apply'd might be sufficient to move a Boat against the Current in most Rivers." His qualms were over the engine's cost. "But the Opinion you have sent me of Mr. Rittenhouse," he told Crèvecœur, "who is an excellent Judge, gives me more favourable sentiments of it."[2]

Franklin and Crèvecœur are famous for their pronouncements on American identity and politics—their thoughts on energy consumption are less famous, even though energy definitely affects identity and politics. This is apparent in Franklin's exasperation that steam entrepreneurs continued to want to patent their designs, linking technology to state-sponsored profits and to personal fame.

Both Fitch and Rumsey requested support from state governments; Fitch also angled for federal help. He made a point of demonstrating his boat on the Delaware River while the Constitutional Convention met, using the convocation as a pitch session for his startup. Both he and Rumsey appealed to the savants at the American Philosophical Society and to the society's president, Dr. Benjamin Franklin. Franklin's main response, as in his letter to Crèvecœur, was optimistic skepticism. He worried that any steamboat, whatever its force in resisting "Wind and Tide," was, for the moment, far too expensive. The APS did, however, offer to subsidize Rumsey's travel to England so he could buy a Boulton and Watt steam engine for his experiments. Were these successful, he might then apply for a British patent. Franklin's letter of introduction for Rumsey explained that "another Mechanician of this Country is endeavouring to deprive him of such Advantage, by pretending a prior Right to the Invention."[3]

In Britain, patents were indeed entering an era of roaring growth. More than six hundred of them can be identified for the 1790s, most of them related to industrial activity, principally new machinery and new modes of energy consumption. A majority of the patentees were at Rumsey's social rank—skilled artisans, not wealthy investors—and even many of the manufacturers who sought patents had once been apprenticed to trades, not unlike Franklin. Patented inventions designed to save fuel had steadily risen over the eighteenth century, constituting 16.4 percent of patents in the 1790s, second only to the 20.1 percent that promised to reduce running costs (some of which might have included fuel costs).[4]

Subsequent news from Rumsey indicated, however, a widening gap between those who'd arrived early to the industrial arts and those elbowing into a more crowded market at century's end. Armed with Franklin's letter, Rumsey connected with Boulton and Watt in

Birmingham. But, he complained, they "made me proposals that would have been Dishonourable of me to Except [accept]." "A large Sum of money might be made in this Country," Rumsey said, "if I had the prevelige of Erecting Boats to go by Steam without Interfereing with Messrs. Bolton & Watt's patent previliges, Or could form a conection with them on reasonable terms." Rumsey next sought French support, but would die in 1792 before releasing a successful design. Boulton and Watt won, yet again. When the first commercial steamboat in the United States, Robert Fulton's SS *Clermont*, made its historic run up the Hudson River in 1807, it was fired by a Boulton and Watt engine.[5]

Right after the steamboat drama, Franklin described his stove in his famous autobiography. He'd written the penultimate part of the memoirs in France in 1784, then decided to spend his time at sea, homeward bound, to finish some scientific essays, including his thoughts on smoky chimneys and on his vase stove. He wrote the third and final—and shortest—part of the never-finished memoirs during the brief interval between finalizing the text of the Constitution and his death: "I am now about to write at home, August 1788." Only after he'd watched the ruthless posturing that went into the making of steam power—the Fitch-and-Rumsey rivalry echoing certain refrains from the Boulton and Watt story—did Franklin relate the history of his first stove and assert his "having no Desire of profiting by Patents myself, and hating Disputes." It's a nice claim. And it was almost true.[6]

Franklin certainly didn't mind getting credit for his inventions. When he'd dashed out an outline for his memoirs in 1771, he made sure to include his heating devices, sandwiching a quick word, "Boxes," between "Project of Hospital" and "Made a Commissioner of the Treasury." The entry might cover both his stove models, firebox and firebox-plus-airbox, slightly out of chronological order, devices of the 1740s placed amid events from the 1750s, though, crucially and accurately, before his electrical experiments. The outline then leaps forward, into events Franklin never wrote out fully before his death, including two London projects, "Stoves and Chimney plates. Armonica." These entries listed the musical glass "armonica" he invented, the sliding chimney plate he'd described to James Bowdoin, and stoves—plural. Maybe this meant different models of his vase stove, or another design (or designs) tested and discarded along the way.[7]

When Franklin filled out the story of the "Boxes" in his memoirs, he gave a final reminiscence about his original fireplaces. He said they were "in 1742 invented" as "an open Stove, for the better warming of Rooms and at the same time saving Fuel." He'd made "a Present of the Model to Mr. Robert Grace, one of my early Friends," and gave Grace the "profitable" right to cast the plates of the stove, which "were growing in Demand." And Franklin admitted it was "to promote that Demand I wrote and published a Pamphlet," which "had a good Effect." He claimed that Pennsylvania's deputy governor, George Thomas, had offered him a patent for the design, which he declined on principle: "*That as we enjoy great Advantages from the Inventions of Others, we should be glad of an Opportunity to serve others by any Invention of ours, and this we should do freely and generously.*"[8]

An "Ironmonger in London" lacked this generosity, Franklin added, and that person (probably James Durno) had made some "small Changes in the Machine," also "assuming a good deal of my Pamphlet and working it up into his own." Here Franklin may have confused Durno with James Sharp; only the latter seems to have patented his heaters, while the former had definitely plagiarized Franklin's pamphlet. But the imitations were a tiny irritation. Franklin must have known that Sharp name-checked him, that Durno praised his work in Britain during the American Revolution—and that he was famous as a heating innovator throughout the former colonies and from

Edinburgh to Venice, for crying out loud. Above all, Franklin rejoiced that his original design "has been and is a great Saving of Wood" in Pennsylvania and neighboring "Colonies," a slip he would surely have corrected to "States" had he lived to revise and publish his memoirs.[9]

Unless, that is, he thought his fireplace was a thing of the colonial past, quaintly irrelevant as patented steam engines proliferated. But then Franklin's corresponding statement against patents seems mismatched, insisting on the virtues of a simpler era of innovation. It also doesn't match another statement, very much in favor of patents (and copyright), enshrined in Article I, Section 8, Clause 8, of the US Constitution—which Franklin accepted and signed, with a quill pen, as one of the world's first steamboats cruised the Delaware River at a speed that topped out at eight miles per hour.[10]

Beyond the words of his memoirs, or of the Constitution, Franklin made a far bolder objection to the old organic economy by attacking enslavement, though without any sense that steam power might substitute for human labor. Again, this noncausal sequence of events questions any assumption that fossil fuels have some intrinsically liberationist force. Lenapes and Black people had never needed the steam engine to demand freedom for themselves or for others. Nor were the white abolitionists who followed their lead necessarily inspired by the new power source.

In Franklin's case, his and his wife's decision to stop enslaving people, and his blossoming efforts as a public abolitionist, never referred to alternative energy. Remarkably, Franklin worked with the abolitionist brother of London iron merchant James Sharp, the much more famous Granville Sharp, who sent the American his "Tracts against Slavery." But Franklin never connected the two Sharps' concerns, burning coal and freeing people. (The only collision of the two—a pretty glancing one—was when Granville Sharp asked about the *Book of Common Prayer* Franklin had, with Francis Dashwood, abridged to protect invalids from long services in a cold church.) The point was to eliminate bound service, whether that was materially convenient or not. Moreover, steamboats and stationary steam engines would be smoothly adapted to service the cotton-producing Southern states. That terrible subsequent history shows how a technical device cannot by itself effect broader societal change; people have to do that.[11]

And, in Pennsylvania, people did. Quaker critics, joined by secular antislavery advocates, Black and white, had, before the war, lobbied colonial legislators to free all enslaved people. The Pennsylvania Abolition Society (formed in 1775, led by Franklin after 1785, and incorporated in 1789), achieved a partial victory in 1780 when the state passed a Gradual Abolition Act (the first such statute in the world), prohibiting new imports of enslaved people, declaring anyone born in the state henceforth to be free, and setting deadlines for freeing everyone else, depending on their age in 1780. Thereafter, members of the Abolition Society sued anyone (including ironmasters) who tried to evade the law.[12]

Gradually, the number of enslaved people in Pennsylvania diminished, from around 6,000 in 1765 to 795 by 1810. Many people had to free themselves, however, as had long been the case. After 1780, if anyone suspected their enslavers had not registered them and their years of birth (establishing a deadline for their emancipation), they might simply decamp, knowing that any organized pursuit would expose the lack of compliance. (While George Washington served his presidency in Philadelphia, he rotated his enslaved people in and out of Pennsylvania so they couldn't establish residency, thus requiring him to register them.) The gradualism of the legislation forced hundreds of young people to serve indentures until they were freed; a labor market in these contracts perpetuated conditions resembling enslavement.[13]

During the transition, it's notable that Black people used other lingering elements of an organic fuel economy to free themselves. This was a temporary escape route, a closing "opportunity" during the final decades of dependence on photosynthetically derived energy.

Richard Allen, Philadelphia's most prominent Black leader in the early Republic, did exactly this—with wood. Born in bondage in Maryland, Allen obtained in 1780 a written agreement from his enslaver: freedom for $2,000 paid over five years. Allen began to cut firewood to sell, chopping so much on the first day that his hands were too torn to continue. Once recovered, he cut one and a half to two cords per working day. Three or four thousand cords later, he paid his final installment early, in August 1783, just weeks before Franklin signed the Treaty of Paris. Allen proceeded into a life of civic engagement, becoming a Methodist minister, head of Philadelphia's Mother Bethel Church (the oldest African Methodist Episcopal church in the United States), and an abolitionist. He used print to lament the sin of enslaving human beings; he and Absalom Jones became, in 1794, the first copyrighted Black authors in US history, four years after the US Constitution had guaranteed copyrights and patents.[14]

While Black citizens forged a hard-won freedom within Pennsylvania, Indigenous people lived under constant pressure to clear out. In the same part of his memoirs where he discusses his heating "Boxes," Franklin recalls his military work leading into the Seven Years' War. He assumes violence between settlers and Natives is inevitable, and the latter people prone to self-annihilation. Recalling negotiation of the 1753 treaty at Carlisle, Pennsylvania, with Lenape and other Indigenous delegates from the Ohio Valley, he portrays the Natives as heatedly intemperate. He claims that at night, they made "a great Bonfire" and drunkenly quarreled with each other. The scene, lit "only by the gloomy Light of the Bonfire," resembled "our ideas of Hell."[15]

Here, Franklin overlooks any unifying legal significance, for Natives, of a shared fire. Still, he hadn't exactly forgotten council fires. While the Constitutional Convention was still meeting in June of 1787, Franklin wrote three letters to Indigenous correspondents (two Cherokees; one Choctaw-Chickasaw delegation) to explain there could be no firm agreement with any of them until the nation's political foundation was secure. Each time, he put it the way he thought they'd understand the transition: "I am sorry that the Great Council Fire of our Nation is not now burning . . . In a few Months the Coals will be rak'd out of the Ashes and the Fire will be again kindled." But when he wasn't addressing Natives directly, he reverted to an assumption that any outdoor fire must be a bonfire, a site of barely contained mayhem, as on Bonfire Night (Pope Day) in his native Boston.[16]

And as he recalled the Carlisle negotiations, he invoked his fantasy that North America's Indigenous population would go extinct. If it was "the Desire of Providence to extirpate these Savages in order to make room for Cultivators of the Earth," he wrote, "it seems not improbable that Rum may be the appointed Means. It has already annihilated all the Tribes who formerly inhabited the Sea-coast," a falsehood, concealing centuries of settler violence against Natives. Meanwhile, Franklin held eleven grants to western land, with title finalized in two sets of treaty negotiations with the Haudenosaunee (1784) and with Lenapes and Shawnees (1785). Profits from later sales of this land would enrich his daughter's family.[17]

Neolin's exhortation that Natives gather strength by returning to a precolonized mode of life had nonetheless persisted. A post-Pontiac generation of leaders took up the work of defending land and its resources. Shawnee prophet Tenskwatawa and his brother, military leader Tecumseh, would be the primary warriors in this continued battle, once the American Revolution left Native nations to face the emboldened citizens of the United States with fewer or weakened allies from rival European empires. "It is better to die as men at once than die a lingering death," the Shawnee leaders told the US military. When in the end they were removed from Ohio to Oklahoma in 1832, they refused to go by steamboat, having no desire "to move by fire" or risk being scalded "like the white man cleans his hog." Their preference registered the real dangers of early rotative steam engines and repudiated industrialized energy, as did other Indigenous people who continued to use sun-generated power and nourishment, whose processes they knew were natural.[18]

The long early modern effort to conserve wood—what did it achieve? It makes sense that people reconsidered their use of fuel during the Little Ice Age, living in chilly fear that trees were vanishing. Their emergency efforts prove that people can be aware of climate change, see the need to adapt, and take action. In the longer term, though, which of their solutions had consequences, especially positive ones?

There was one big payoff: modern climate science. Franklin and his contemporaries had redefined climate as complex and changing, altering not only over seasons but decades, centuries, even millennia, and possibly because of human action. A science of the atmosphere was one important result, interpreting how material forces generated long-term and short-term conditions. This transformation had been the work of many hands, including non-Western people, though they were rarely acknowledged. Contributors include the Lenapes, who described long-term patterns in winter weather, and the Shawnees, who led colonizers to fossilized ancient animals who resembled living creatures from contrasting climates, indicating past climatic change.

Within this larger history, Franklin's work is notable both for its content and its original inspiration. From his investigations of heat and circulation within a room, Franklin acquired a powerful way to explain the atmosphere, as when he interpreted the Atlantic coast's wind patterns, the Gulf Stream, and the dispersal of volcanic effluvia, which brought him back full circle to the question of indoor heating during an emergency in France.

Franklin's writings on these topics themselves circulated, and had an impact. In 1783, for example, an Italian edition of his scientific writings, *Opere filosofiche [Philosophical Works] di Beniamino Franklin*, included his work on atmospheric electricity, the aurora borealis, and ocean currents and temperatures. Richard Kirwan, in his influential 1802 summary of atmospheric science, cited Franklin's work on the movement of Atlantic storms and on the Gulf Stream.[19]

Alexander von Humboldt (1769–1859), one of the most important interpreters of ocean currents and climate as dynamic, nonlinear systems, would credit Franklin (and others of his generation) for establishing these atmospheric phenomena as scientific facts. In his account of his travels in the Americas from 1799 to 1804, and in his much later *Cosmos: A Sketch of a Physical Description of the Universe* (1845–1862), Humboldt cited Franklin's observations of varying temperatures in the Atlantic Ocean, and in the former work he credited Franklin and Franklin's British acolyte Charles Blagden for their thermal analysis of the Gulf Stream. Franklin's work is evidence that modern climate science reaches back at least into the eighteenth century and has roots in the colonizing cultures of Europe and its American settlements. His influence absolutely depended on his strategic decisions to publish his scientific observations and to link them to improved heating technology, making the scientific principles tactile, associated with consumer convenience, and easy to understand.[20]

The specific inventions themselves had an only fleeting positive impact, given their coal-bound trajectory. In Pennsylvania, inventors continued to produce devices that burned wood. But for consumers in the longest-colonized parts of Pennsylvania, this was becoming expensive; their wood had to come from New Jersey and Delaware. Wartime inflation had cooled, but the cost of living rose steadily from the late 1780s onward. Wages were, conversely, on a downward trend until the outbreak of a new war in 1793 (courtesy of Napoleon Bonaparte), though the uptick didn't benefit everyone. That year, after decades of giving Philadelphia's poor people firewood at public expense, a private fund did that, ending a rare government intervention into energy use.[21]

Private and public efforts thereafter supported fuel efficiency, as indirect guidance for consumers. Between 1790 and 1845, the US Patent Office registered more than eight hundred patents for improved stoves or grates. The APS offered a premium in early 1795 for "the best construction of stoves or fireplaces." Their aim was "the benefit of the poorer class of people, especially . . . where fuel is dear." Any successful heater should use "the least expence of fuel possible; & should be capable of being employed both for the purposes of warming the room & cooking." The APS also reported the work of French architect François Cointeraux, about brick being more effective than metal for constructing not just open fireplaces but closed stoves.[22]

It may never be clear which innovations worked and which merely burnished their inventors' egos. One of the learned Philadelphia Rittenhouse brothers, either Benjamin or David, modified Franklin's fireplace insert and had Cornwall Furnace manufacture the result. These were evidently popular; one Philadelphia merchant bought 195 of them in January 1795. But a rival polymath, Charles Willson Peale, said the device failed to keep indoor temperatures above freezing on the coldest days; the potted plants in his parlor died and the water dishes in his birdcages froze. He did his own upgrade, altering his chimney and adding a sliding mantel to better distribute the heat. And he installed brickwork stoves (inspired by Cointeraux) at his celebrated natural history museum inside the State House, later Independence Hall. On a January day when it was three degrees Fahrenheit, Peale claimed his exhibit space was a relatively balmy fifty-four degrees.[23]

The museum's stoves—"smoke eaters"—were Peale's most ambitious, and even better showcased in a space designed for public viewing than Franklin's originals had been at the Philadelphia post office. As described in 1798, they were shaped like classical columns and painted to look like marble. An open-ended metal pipe, widening as it rose, was suspended inside the column, which was closed but for sliding registers at its bottom. Firewood was placed inside the base and lit. At the top of the column, the smoke had nowhere to go except down the pipe and back into the flames. The registers adjusted the speed of combustion and showed the smoke being consumed, demonstrating how emissions might be eliminated. As visitors wandered the museum, moreover, Peale gave them a survey of Earth's climates, including climatic variation of animals and plants. And the museum was itself an exhibit of a building-sized artificial atmosphere, available by admission. On a bad winter's day, maybe Philadelphians considered the cost of their firewood and opted to buy a ticket to the warm museum, instead.[24]

At this point, many private Philadelphia houses and public buildings had at least one kind of stove, either freestanding or set into a fireplace. That was true of the State House, the Library Company, the War Office, the Old City Hall, and the Treasury Office, plus multiple churches and schools. A print of the State House by Philadelphia artist William Birch, done in 1800, shows stovepipes poking from two windows flanking a chimney, with a handy woodpile at the foot of the steps just outside. The trees around the State House and the presence of four Native men are also reminders of Pennsylvania's original nature, preserved or tolerated by the dominant white population. But the trees and the Natives were both (and relatedly) receding. Burning wood was becoming expensive. A luxurious fireplace insert, modeled on Franklin's design and dating from circa 1795, shows this in its fuel-gobbling big mouth and its fancy decoration, with profiles of Franklin and George Washington.[25]

That fireplaces could be marketed as luxury goods is also obvious in the case of Franklin's clearest intellectual heir, British scientist and inventor Benjamin Thompson, later Count Rumford. Thompson was a Massachusetts native (like Franklin) who remained loyal to Great Britain (unlike Franklin) and fled America in 1776. In service to the Duchy of Bavaria, he gained his title, Reichsgraf von Rumford,

IN THE FIRST CAPITAL OF THE UNITED STATES, THERE WERE
LIVING TREES AND FIREWOOD, NATIVES AND SETTLERS.
(William Russell Birch, Back of the State House, *Philadelphia [1799],
Library Company of Philadelphia)*

choosing the name of a small New Hampshire town to honor his American roots. He would later marry the widowed Marie-Anne Paulze Lavoisier (her second chemical bond).

Rumford's multiple scientific and technological interests included heat. In the 1790s, he developed an improved fireplace based on experiments he'd done with the heat conductivity of different materials and with air at varying levels of humidity. Like many other designs (not least Franklin's), Rumford's fireplace made the area that held the fire smaller, with tilted sides to absorb and radiate the heat, and a narrowed throat to constrict where smoke might try to sneak out. Rumford recommended his fireplace to maintain health and reduce fuel costs—at least by half, if not up to two-thirds. Its design was equally applicable, he said, to coal, peat, or wood.[26]

As well as describing his fireplaces and explaining how to install them, Rumford announced he would never patent them or let anyone else do so: "All persons are at full liberty to imitate them." Which they did—Rumford's became a standard design. An image by the contemporary British cartoonist James Gillray shows the beaming inventor basking in front of his invention in a parlor. The modish decor, with fashionable green paint or wallpaper instead of thick insulating paneling, signals that the design was costly; a Rumford fireplace was best built as a new house was going up or new room added on, unless someone was willing to retrofit an existing chimney and hearth.[27]

But the trend for smaller fireplaces continued. This was partly due to awareness of waste, but also to the uptake of coal, a mere hatful of which generates greater heat than many big logs of wood. In Jane Austen's novel *Northanger Abbey* (1817), the gothic-besotted main character finds that Northanger's largest fireplace, "where she had expected the ample width and ponderous carving of former times, was contracted to a Rumford." Such was their popularity that "Rumford" fireplace inserts were imported into the United States, with a sense of continuity from one American inventor to another; one Philadelphia reader bound together their copies of Franklin's work on smoky chimneys and Rumford's on fireplaces.[28]

The rage for Rumfords coincided with a fashion for light women's clothing, only possible in winter with sufficient indoor heat. Europeans' fascination with clothing they thought exotic but attractive—the flowing garments of ancient Greece, the white cotton dresses worn in India and the Caribbean—and a revolutionary era's desire for greater simplicity and freedom of movement set off a rage, in the 1790s, for simple chemises made from white cotton muslin, originally imported from India. Both fans and critics described the muslin-draped women as nude, with exposed arms and necklines and the body's contours quite obvious. Even with stockings and chemise worn beneath and a shawl above, the ensemble was flimsier than the everyday wear of just a decade earlier. Another of Gillray's satiric cartoons, "Advantages of wearing muslin dresses!" (1802), shows that female consumers always get burned. One of the women taking tea beside a coal fire has accidentally ignited her neoclassical gown; the picture of an erupting Vesuvius above the mantel should have warned her. And thus a modern fireplace could keep quite toasty someone wearing the rough equivalent of her grandmother's underwear. (The combination of light fabrics and open flame would remain deadly through the nineteenth century.)[29]

Even when it didn't set anyone's dress alight, a coal fire was unprecedentedly damaging, indoors and out. Franklin had remarked on London's polluted haze when he developed his third fireplace design and when he told Boulton and Watt their steam engine ought to minimize smoke.

It took almost twenty years, but Watt took the hint. In 1785, he applied for a British patent on an industrial furnace that reburned its smoke. That patent, Number 1485, awarded on July 9, specified that the "said new Invention consists only in the method of consuming the smoak and increasing the heat by causing the smoak and flame of the fresh fuel to pass through very hot funnels or pipes." This detail was meant to prevent infringement on any other Watt design, including the Boulton and Watt separate-condenser engine, whose patent had another fifteen years to go.[30]

It's significant that an American had prompted Watt to consider emissions. Both Franklin and Rumford were horrified by old-world energy use. "Nothing surely was ever more dirty, inelegant, and disgusting than a common coal fire," Rumford scolded, knowing, like

Franklin, of new-world places where coal was rarely burned and fresh air abounded. And like Franklin, Rumford deplored the squandered coal that drifted off as smoke. "The enormous waste of fuel in London may be estimated by the vast dark cloud which continually hangs over this great metropolis," he said, "and frequently overshadows the whole country," murk on a *national* scale. The British credited that American unease about air pollution in the *Encyclopaedia Britannica*'s third edition of 1797, whose entry on "smoke" was, more or less, Franklin's censorious essay on smoky chimneys. In reproducing the piece, the encyclopedia also broadcast Franklin's definition of a "tight room," an American judgment on the possibility of overengineering a problem.[31]

To have a "Rumford" or a "Franklin" in any room seemed a sign of progress, consumerism informed by science. But Franklin and Rumford were validating, among other things, coal's efficiency. If that encouraged more consumption of coal, it wasn't quite the improvement they claimed; calls for fuel efficiency, using some kind of a special grate or hearth, were, intentionally or not, an excellent foundation for dependence on coal long before factory industrialization.

There's a good case, for example, that improvements in indoor heating triggered the Jevons Paradox, named for William Stanley Jevons, who in 1865 would warn that energy efficiency invites greater consumption, canceling any gain. Although Jevons was talking about coal in a period of high industrialization, his paradox may have begun in Franklin's era, after some downward trends that, had they continued, might have offered a different history. Certainly, within Europe, lower rates of coal consumption briefly indicated that some efficiencies may have worked. Consumers bought new stoves, fireplace inserts, grates, sash windows, and so on, to cultivate warm yet ventilated interior spaces. But maybe too many such spaces, eventually negating the conservation.[32]

In Europe, Franklin had seen a new desire to heat every part of a house, a trend that could easily nullify the efficiency of improved fireplaces and stoves. This was happening in Pennsylvania as well.

Visual images and property records show that wealthier houses were, by the 1790s, using combinations of stoves and fireplaces to create artificial atmospheres throughout. And so, devices designed for efficiency may have only temporarily slowed fuel consumption, with an interval in which old and new heating systems coexisted. As late as 1803, the "stove room" in Warwick Furnace's main dwelling, built by Rebecca Grace's first mother-in-law, Anna Nutt, had a wood-burning Warwick Furnace stove. This was the second Franklin design, firebox-plus-airbox (though not a Grace version with the smiling sun). It was, sometime around that year, transferred or sold to another property, where a visitor saw it in 1868, noting it had *Warwick Furnace* cast on its front plate. Then it vanished. No matter, as Franklin didn't think his inventions, or his science of atmospheres—or the United States of America—were his only legacy.[33]

Franklin's status as a fire wrangler made him a transitional if not ambiguous figure, with one foot in the old wood-burning past and another in the coal-fired future. So, too, his final recommendations about which energy sources Americans should make use of. In the last four years of his life, 1787 to 1790, he made his semi-endorsement of steamboats, decided not to challenge enslavement in the Constitution (which in fact enshrined it), then petitioned Congress (in his competing role as president of the Pennsylvania Abolition Society) for the abolition of the slave trade and slavery. Put together, his actions represent the way energy transitions are never perfectly linear yet they move forward, with chronologies that extend well past any individual's lifetime, including Franklin's.

As his health declined and he retreated into private life, Franklin adjusted his circumstances, and his conscience, to what he regarded as optimal material choices. His wife had built his improvements for heating and ventilation into their Philadelphia home, which he continued to renovate. As he explained to his sister, the house's recent expansion supplied a room where he and his daughter's family could "dine a Company of 24 Persons, it being 16 Feet wide and 30 ½ long; and it has 2 Windows at each End, the North and South, which will make it an airy Summer Room; and for Winter there is a good Chimney in the Middle made handsome with marble Slabs." The renovations also yielded "a large Cellar for Wood." In parallel, Franklin reworked his final will and testament. He stipulated that, to inherit anything, his son-in-law had to free a man named "Bob." This guaranteed that Franklin's grandchildren from his daughter—his only surviving line of descendants—would not inherit any enslaved people through him. That was Franklin's final personal step away from the energy source that had earlier marked his status, assisted his family, and—through his stoves—given him his first spark of fame, the reason so many Philadelphians followed the news that their famous fellow citizen was, by 1789, in decline, then, by 1790, dying.[34]

Nine days after his death on April 17, 1790, Franklin's estate was inventoried, giving us a last look at his material life, its treasures and comforts. His moveable goods included a "stove" in his bedchamber, the room where he died, where iron entrepreneur Rebecca Grace, Robert's widow, was one of the last people outside Franklin's family to see him. The heater in his bedroom may have been the vase stove he'd invented in London, the one intended to consume smoke. If so, it was the "Chamberstove" that, in 1776, Quaker poet Hannah Griffitts had thought fit to decorate the inventor's grave. (John Adams, Franklin's long-ago, fresh-air-fearing roommate, sniped that the old man must have insisted on keeping a window open and caught cold from the draft.)[35]

So much for the modern Prometheus. In early 1792, Franklin's printer grandson, Benjamin Franklin Bache, advertised the sale of his grandfather's household goods in his newspaper, listing, among other things, some "circular and other Coal Grates; Franklin Stoves," the great man's invention becoming, with his death, increasingly eponymous. In 1802, Franklin's daughter and son-in-law would turn the well-heated home Deborah and Benjamin had built into a boardinghouse, later a private academy; the grandchildren would pull it down in 1812 and divide the land into lots for sale. Many consumer goods survive from Franklin's estate—from books to shoe buckles—but not a single heating device.[36]

Still, Franklin told us which projects he hoped might outlive him. In 1788, he inserted a codicil into his will, giving bequests to the cities of Boston and Philadelphia for societal improvements. He designated precise goals and time spans for the projects, in essence predictions about what the two major American cities might need over the next two centuries. (Why think small?) These desired outcomes turned out not to anticipate what was actually unfolding, as industrialization obliterated the organic economy, with problems Franklin didn't foresee, and which didn't proceed according to the two hundred-year periods of improvement he set into motion.

For the first hundred years after his death, Franklin wanted to invest in people, using his bequests to offer small loans to "young married artificers" who were, he thought, "most likely to make good citizens." Each city received £1,000 sterling (with an estimated minimum value of £140,700 but possibly representing as much as £14,060,000 in 2023). Franklin specified the funds' caretakers, how they should disburse the loans, and how the repayments (with fixed interest) would return to the original stock. Beginning the year of his death, this pioneering scheme of microfinance—a boon to the working poor in an era when (still) they had little or no capital—was to run until 1890. The increased funds were then to be invested in public works, to make Boston and Philadelphia "more convenient" and "agreeable," an effort that would run for a second century to 1990.[37]

By requiring the original loan recipients to be young and married, Franklin was underwriting his prediction that rapid population growth would supply a stock of industrious labor. This was his final—and positive—judgment about a trend he'd first worried about in his stove pamphlet, then applauded in his analysis of population. Within an organic economy, human energy was an essential source of wealth, individually and societally. Under some circumstances, however, it costs more than it yields, and high levels of consumption make the trade-off especially dubious as a threat to natural resources. The cumulative impact of US consumers would prove this, given the nation's comparatively high standard of living. But that hazard seemed distant in the 1790s when economic growth was a high priority, as represented not only in Franklin's will but also in Secretary of the Treasury Alexander Hamilton's "Report on the Subject of Manufactures."

Submitted to Congress in 1791, the report argued that the nation needed manufacturing to crown its agricultural base—specifically to grow the economy. Hamilton proceeded from Adam Smith's tenet that wealth resulted from labor, the number of workers and the productive division of their efforts. This followed from Franklin's analysis of population growth, which Smith had defined as essential to a thriving economy—with British North America the peak example. Hamilton agreed: an "extensive domestic market for the surplus produce of the soil is of the first consequence," and "the multiplication of manufactories not only furnishes a Market for those articles . . . it likewise creates a demand for such as were either unknown or produced in inconsiderable quantities. The bowels as well as the surface of the earth are ransacked for articles which were before neglected."[38]

"Ransack" was a good word for it. One year after Franklin's death, Pennsylvanians discovered the large coal deposits he'd hypothesized with a hopeful "if." Anthracite from the Lehigh Valley was the first coal to be commercially mined in North America on a large scale. During the embargos of the Napoleonic Wars, the fuel was a boon, initially for blacksmiths and public works. Wyoming Valley's bituminous and sub-bituminous coal beds thereafter serviced both domestic and industrial demand, with an accelerating extraction that itself used the fuel, as with railroads, which both consumed and carried coal. By the 1830s, stationary steam engines, railroads, steamboats, and industrial furnaces all burned coal, also generating new centers of coal mining and industrial production, as with Allentown in the Lehigh Valley, part of the Walking Purchase fraud, and Scranton in the Wyoming Valley, not far from where Teedyuscung was assassinated. Weissport, location of the Fort Allen that Franklin helped build, is now in Carbon County, formed in 1843 from parts of two earlier counties that turned out to abound in coal.[39]

A wealth of coal in a land still forested was historically distinctive, Appalachia's John Hancock. One visitor to Pittsburgh (the former Fort Pitt) compared its smoke—marking the city's location under an otherwise clear sky—to London's notorious haze, though here surrounding a town circled with the kind of woods Londoners hadn't seen nearby in centuries. And US citizens continued to consume quantities of wood alongside coal: in 1799, the average American used 58 board feet versus 10 for the average citizen of the United Kingdom; in 1809, 57 in the US, 10 in the UK; and in 1819, 59 in the US, 17 in the UK (with the restoration of Baltic timber imports after the Napoleonic Wars). Industrialization and a high standard of living are often associated with steel and coal, but in America, wood helped build them.[40]

Wood competed with American coal, initially anthracite, in part because that kind of coal is hard to ignite. A new generation of stove tinkerers sprang into action, as with New York clergyman and academic Eliphalet Nott, "the philosopher of caloric," who in the 1820s invented (and patented) stoves specifically for Pennsylvania anthracite. This kind of technical fix linked the impending fossil fuel dependence back to changes within an organic economy, when consumers had been prompted to adopt new technologies to lower fuel costs. But as was common at the time, Americans assumed coal must be inexhaustible, a vast and possibly renewing source within the earth, in contrast to the more easily depleted organic energy sources on the surface. They began to burn coal prodigiously. During his 1842 visit to the United States, Charles Dickens complained of the nation's constant companion, "the red-hot demon of a stove," whose "hard anthracite" made even hotel corridors and train cars unbearable.[41]

And it was in the first century of Franklin's posthumous project that Pennsylvania—slightly in advance of the rest of the United States—became an energy omnivore, burning everything available, wood, coal, and coke (coal from which impurities have been vaporized at high heat). In this way, the state was first to catch up with Britain. There, the decisive uptake of steam power occurred in the 1830s, when coal became the nation's main fuel, though not until the 1850s would more British coal be burned for industrial purposes than domestic ones. The comparable American takeoff was later, with coal accounting for 50 percent of energy use only by around 1885. Americans knew what was coming—even before wood and fireplaces vanished, they shed nostalgic tears for them. In his story "Fire-Worship" (1846), Nathaniel Hawthorne laments "this almost universal exchange of the open fire-place for the cheerless and ungenial stove," which masked with iron "that brilliant guest—that quick and subtle spirit whom Prometheus lured from Heaven." Hawthorne reckons that "sixty cords" once supplied his house's three fireplaces, before they ceded their place to stoves, with no concern for the trees that had provided the cords. (He mentions coal just once.)[42]

Within this sentimental vein, right as Daniel Boone was achieving his iconic status, for example, forests and frontier activities were becoming national centerpieces, a reinterpretation of American resources as primal possessions of rough, tough settler men, and forested American environments as their vital sustenance. That scenario only worked by somehow deleting America's Natives, with Pennsylvania yet again in the dismal vanguard for this integration of a perpetually dying organic economy into the nation's history.

Fictionalized Lenapes were the unfortunate stars of this story. Since the late colonial period, and reaching an apex in the early Republic, "Tammany" societies, holidays, and dinners were popular political activities for white men. These referred to the Lenape leader, Tamanend, who, in a mythic meeting with William Penn under an elm tree, had granted Pennsylvania's first English settlers a foothold. By appropriating Indigenous identities, including "Tammany's," white men designated themselves as successors to Natives they had displaced and expected to vanish entirely—"playing Indian" as if putting on colorful costumes (sometimes by wearing actual costumes), at last confident in their essential bodily difference from Native people. (New York's corrupt political machine, Tammany Hall, is the most famous abuse of the name.) The extinction narrative appeared, most notoriously, in James Fenimore Cooper's *The Last of the Mohicans: A Narrative of 1757* (1826). More overtly than the other four of Cooper's Leatherstocking Tales, the novel uses the "last man" narrative to prophesy the extinction of the Delawares or Lenapes; the Mohicans were a related Algonquian group whose name Cooper decided to use for his Lenape characters Uncas and Chingachgook, supposedly the last of their line—in 1757—an obvious falsehood. Lenapes still existed but were being rendered invisible through coerced removal, their lands turned into farms and timber lots, then mines and furnaces.[43]

The United States' all-devouring appetite for energy proceeded to grow even as labor narrowed into one standard option—waged work—and as levels of capitalization for industry rose along with the advent of business corporations. Civic corporations (including the Pennsylvania Hospital, which Franklin helped to establish) and joint-stock companies (through which colonial ironmakers had raised capital) have long histories in early America. But imperial law had banned business corporations. US independence made them possible—the very first was created in Philadelphia with the Bank of North America (1781). Thereafter, white Americans went on a corporatizing spree.

From 1790 to 1860, state governments chartered 22,419 businesses, with capital of at least $4.58 billion—more corporations and more authorized corporate capital than in the United Kingdom, France, or Prussia during the same period. In Pennsylvania, corporations organized banks, public works, canal projects, and, most momentously for US industrialization, steelworks and railroads—the colossal Pennsylvania Railroad would, by the turn of the century, be the world's largest private business corporation, able to plunk its name down where it pleased, as with Pennsylvania Station in Manhattan.[44]

These trends would reach an apogee when Andrew Carnegie formed US Steel in 1901, capitalized at just over $1.4 billion, making it the largest corporation in the world, one that for a time produced most of the steel made in the United States. This was a financial universe far removed from what colonial ironmasters had achieved with their quaint pooling of capital and strategic intermarriage. It's too apt that the Morgan Library, established by J. Pierpont Morgan, US Steel's major financer, owns a copy of Franklin's stove pamphlet that belonged to Franklin's brother. Indeed, industrial wealth made it possible to consign the colonial era to museums: Henry Mercer acquired Franklin's one surviving stove in 1910, and from the 1920s onward, an heir to the DuPont chemical fortune, Henry Francis du Pont, amassed the antiques that were the core of Winterthur Museum, where the *BE LIBERTY THINE* fireback made by enslaved workers is now displayed. Corporations on the scale of US Steel could, moreover, distance themselves from conditions of extraction and labor exploitation; corporate liability or culpability are designed to be harder to pin down than damages done on smaller and more local scales.[45]

It was under these conditions that the economic growth which had been percolating over the course of the eighteenth century surged into its modern dimensions. From the nineteenth century onward, highly capitalized industrialization delivered rising rates of national and annual output, with a material surplus that generated vast wealth at the corporate top, created a comfortable life for the bourgeois middle classes, and lifted millions from poverty. It wouldn't take a doctrinaire Marxist to note, however, that the new surplus of capital depended on the exploitation of labor; few of capitalism's defenders deny the truth of that charge.[46]

Marx had admired "the famous Franklin" for arguing that labor was the origin of all economic "value." Yet, by Marx's era, the "artificers" Franklin had prized—the people who made the items now behind glass in museums—had vanished. Their sons and grandsons (and granddaughters) had become manufacturing or industrial workers with lower social status. Before unionization and government protection, such workers were economic pawns, their capacity to produce and consume downgraded in financial and ethical terms. By the early 1800s, the trustees of Franklin's bequest saw no reason to help such people. In Boston, the fund's board steadily welcomed bankers, textile factory investors, insurance brokers, railroad magnates. They abandoned the microfinance plan in 1852 and turned the money into a private pool of capital. By 1872, Philadelphia's managers used nearly all the Franklin Fund to purchase long-term, interest-bearing loans to invest in urban infrastructure. In essence, the funds financed intra-class relationships, if not favors. And, instead of underwriting human energy, the funds sometimes endorsed carbon power, as when in 1841 the Philadelphia trustees bought stock in the Philadelphia Gas Works.[47]

The fact that industrial development—especially coal mining and factory production—proceeded with free labor reinforced a sense of functional connection between the two. Even before emancipation was complete, Pennsylvania iron became an icon of a slave-free modernity; when Thomas Paine offered a design for an unprecedented iron bridge over the Schuylkill River, he ignored the fact that the raw material for the span would rely on enslaved labor, implying a not-yet-accurate similarity between iron manufacturing in Europe and in the United States. Before the Civil War, Pennsylvanians began to present their state as an antithesis of the slaveholding South, a selective reading of the colonial past and a very loose (if not self-congratulatory) association of Philadelphia's contributions to political independence with its role in generating national wealth.[48]

Some of this was true. Though once similar, Pennsylvania and Virginia were diverging. Both states had long-standing iron industries to which they added coal mining, though Pennsylvania was replacing enslavement with waged work. And Pennsylvania outstripped all other states in developing heavy industry; in the 1830s and 1840s, Philadelphia operations were producing a quarter of US iron and steel, and Pennsylvania's population growth, urbanization, and railroad network were related measures of its economic modernization. But profits from enslavement went into this. Pennsylvania's industrialization built upon the colonial iron industry that exploited enslaved Black workers, some of whom were taken west by settlers who wanted to make iron there in the interval before the workers' legal emancipation, which came only slowly.[49]

Consider the twelve-year-old Black boy a visiting Englishman saw in Philadelphia on March 1, 1806. Given the provisions of the 1780 emancipation act, the child was indentured and would remain so until he turned twenty-eight—or managed to escape. In that late winter, he wore only "a light linsey jacket and trowsers, without hat, shoes, or stockings." And he bore an iron collar, from either side of which "an iron bow passed over his head," to make him conspicuous should he try to run away, as he'd already tried to do. In an age and place famous for claims of liberty, people were still being treated as mere units of economic production, denied the goods and comforts available to free white people, degraded with the cruelest artifacts of early American industrial production.[50]

Indeed, continuities from enslavement into a modern energy regime were obvious in Philadelphia. To prevent fire, city laws required chimneys to be swept, which became especially important once coal was a common household fuel. Because sweeps risked accidents, cancer, and pulmonary disease, white people avoided the work. Black men (and a few Black women) for a time dominated the trade in northern cities. Richard Allen, for example, who'd cut wood to buy his freedom, maintained his autonomy through chimney cleaning. He swept chimneys himself, including the US president's. Then he became an entrepreneur who employed Black boys, a common tactic; some of these apprentices were indentured, as part of the Pennsylvania's gradual emancipation, and they were unhappy—through *Pennsylvania Gazette* ads, Allen pursued two runaway boys who wanted out.[51]

Barely removed from slavery, Black sweeps purged the grime of a new energy regime, not always willingly. Multiple depictions of Philadelphia's urban life show the child workers, barefoot, dressed in rags. Cleaning chimneys was a grim option, and for Black workers a temporary opportunity. By introducing special tools too expensive for most Black people to afford, white entrepreneurs took over the business, once it no longer required such direct exposure to filth. This is but one example of how Black US citizens were, after their emancipation, and as industrialization proceeded, shunted into the worst options, trapped within dead-end intersections of organic and fossil fuel energy regimes. The position of Pennsylvania's free Black people was all the more precarious because of their proximity to the South's enslaving states; kidnappers could too easily remove them by force and sell them into bondage.[52]

Nevertheless, the idea of Pennsylvania as a free-labor bulwark against slavery was embellished in, among other things, Abraham Lincoln's Gettysburg Address. Commemorating the Civil War battle in Pennsylvania that repelled Confederate invasion, Lincoln lauded the Union dead buried in "consecrated" ground, marking "a new birth of freedom." That imagery had obvious merits in that place and by that time. But it was also, and not unlike Franklin's presentation of colonial ironworks as free of racist exploitation, a whitewashing of the nation's shared history of enslavement and Indian removal, as the United States, starting in Pennsylvania, moved into a fully modern economic life.[53]

Pennsylvania remained the heart of the American energy economy well into the nineteenth century. The Centennial Exposition of 1876, held in Philadelphia, had as its centerpiece a seven-hundred-ton Corliss steam engine. Through a maze of shafts and belts, its thirty-foot flywheel sent power to about eight hundred other machines scattered around a thirteen-acre exhibition hall. Visitors marveled at the range of manufactures thus produced: shoes cobbled, reading material printed, wallpaper lithographed, fabric woven. All done under the supervision of one man who sat, absorbed in a newspaper—with the sweating stokers heaving coal into the engine well out of sight. The monstrous machine was the exposition's most memorable exhibit, one example of an industrial-era fascination with steam technology, as if it powered itself without human assistance or visible fuel.[54]

Pennsylvania's industrial might was by this point world-famous. The state's coal yielded most of the nation's industrial energy and set a series of US firsts: the production of energy-rich coke, the use of coke to smelt iron, the large-scale production of steel, the mass manufacture of prefabricated construction items like I beams and H beams. With these milestones came pinnacles of machine-age and art deco design. Pennsylvania steel built nearly every iconic US bridge, from the steampunk elegance of the Brooklyn Bridge to the fog-defying blaze of the Golden Gate. Midway between those two spans rose the original Ferris wheel, erected for the 1893 World's Fair in Chicago, intended to rival Paris's Eiffel Tower, and made—naturally—of steel from the Keystone State. So, too, were the rails for the nation's railroads, tying the continent together from sea to sea. Alternatively, the oceans were connected via the Panama Canal, whose locks were also cast from Pennsylvania steel. Flying was at this point a rare option, but, yes, key parts of the engine for Charles Lindbergh's *Spirit of St. Louis*, first to make an aerial crossing of the Atlantic (1927), were forged by Bethlehem Steel.[55]

Bridges, especially those for railroads, were the first complex structures made of steel beams (as with the Benjamin Franklin Bridge across the Delaware River, opened in 1926 for the nation's sesquicentennial anniversary) and that design, tipped on end, became the frames for the first skyscrapers. Up went the Flatiron Building, the Chrysler Building, and the Empire State Building (New York), the Stock Exchange (Chicago), and the Crocker Building (San Francisco). Somewhat ironically, no comparable towers rose in Philadelphia. There, until 1986, nothing was permitted to rise above the broad-brimmed Quaker hat on William Penn, whose bronze effigy surveys the scene atop City Hall (1901)—itself built of masonry. Meanwhile, Pennsylvania steel mills made armored ships for both world wars, metal mustering forth to relieve the trees that for so long had served in battle. And thus the United States had its first true military-industrial complex, source not only of ships but tanks, jeeps, aircraft, and steel-cased munitions.[56]

By the 1950s, Pittsburgh was the largest steel-producing city in the world. Its rise opposite Philadelphia, on the far side of the state, effectively mapped the transformation of Indigenous territory into an arsenal for modern capitalism, source of gleaming symbols of beauty, mobility, wealth, power, and freedom, with a map unrolling well beyond anything Penn might have imagined. From the Haudenosaunee in Canada, migrant Mohawk construction workers—"skywalkers"—ascended fantastically high above the Manaháhtaan of Lënapehòkink to assemble New York City's skyline, using iron and steel mined from Lenape, Susquehannock, and Shawnee ancestral lands in Pennsylvania.[57]

One price of these accomplishments was Pennsylvania's appalling record in labor history: some of the most dangerous working conditions (in coal mines and steel factories), one of the most vicious efforts at breaking a strike and thwarting unionization (at Andrew Carnegie's Homestead steelworks in 1892), and the largest coal strike in US history in 1902. And, all along, Black and immigrant working communities were consigned to the worst labor options, shunted into the most polluted landscapes in which to live, and excluded from union membership or leadership. In his Afrofuturist science-fiction story "The Princess Steel" (c. 1908–1910), W.E.B. Du Bois represents US industrialization as a primordially racist extraction of wealth. From a Manhattan skyscraper, via a "megascope" that somehow bridges time and space, viewers can see, in distant Pennsylvania, two medieval European knights battle over an African princess, long buried in "the Pit of Pittsburg." Princess Steel's tough, silvery hair is the prize—even as it grows from her head, it is spun into a mat or net to tie the continent together.[58]

Now that large sections of the coal, iron, and steel industries have drastically shrunk if not collapsed, however, there seems to be little lasting benefit, except inherited wealth for a few. Pressured by unionized labor, Pennsylvania and US law began to safeguard workers in the early twentieth century, just in time for the state's key industries to begin a steep decline. The names of the state's two NFL franchises summarize a mixed legacy: the owners of the Philadelphia Eagles endorsed the New Deal in 1933 by naming their team for the Blue Eagle symbol of Franklin Delano Roosevelt's National Recovery Administration (1933), which, among other things, protected collective bargaining; more simply, the Pittsburgh Steelers' owners renamed their team (the former Pirates) in 1940 to claim the strength of their city's unbending metal. A hopeful sentiment. But Bethlehem Steel, the Pennsylvania Railroad, and other Keystone icons are no more; on the eve of the 2024 presidential election, US Steel's possible acquisition by Nippon Steel of Japan sparked state-of-the-nation anxiety.[59]

Many parts of the United States are now postindustrial zones—"rust belts"—but Pennsylvania is the historic original, the rustiest, its coal-rich west an early frontier in a now suffering Appalachia. Industrialization's human cost is exorbitant. So, too, is its toll on the natural world. This was the case even before Pennsylvania's truly distinctive contribution to global energy consumption: liquid petroleum.

The first oil boom in history took place not in Texas, the Middle East, or the Gulf of Mexico, but in the Allegheny Valley. From 1859 to 1873, Pennsylvania was the largest oil producer in the world. The extraction revived the fantasy, as with coal, that the mineral fuel must be inexhaustible: an early shareholder certificate for the Pennsylvania Rock Oil Company shows an allegorical figure holding a cornucopia, symbol of everlasting abundance. (The name of a later company, Seneca Oil, made clear whose ancestral land was being drilled for petroleum.) The race to extract oil despoiled entire landscapes—veterans who'd survived the Civil War's torn and scored battlefields were aghast at this new devastation. But the oil industrialists restated faith that technology could and must improve human life. New drilling and pumping technologies, and the commercialization of petroleum to generate light and power, became showpieces of American ingenuity—*Pennsylvanian* ingenuity, as the corporate names of Quaker State and Pennzoil proudly announced.[60]

In this, Pennsylvania was no longer imitating industrial practice that might have begun elsewhere but laying out a future for everyone else. With liquid fossil fuel, energy consumption would become even less effortful. Where people had once stocked or at least seen the woodpile beside their house or the heap of wood or coal in their cellar, and had tended their fires themselves (ladies and gentlemen alike, and maybe even some kings and queens), they no longer needed to wrangle or even see what kept them warm—or, eventually, cool. Power plants had begun to send coal gas into houses through pipes, and then coal-fired power plants generated electricity sent to consumers through wires, with any pollution kept at a distance, at least for a while. Kerosene would be the first mass-marketed liquid petroleum, initially to replace lighting done with whale oil, then for heating, and eventually for jet fuel. Supplied to consumers with no hint of its origin, petroleum carried new prestige for those whose energy needs it made invisible, effortless, wiped of grime.[61]

With the invention of the internal combustion engine in the second half of the nineteenth century, liquid fuel would become the means for unprecedented mobility, both for people and for commercial goods. Soaring demand would generate further uses for petroleum. Unless consumers lived near oil fields, they didn't see what the extraction process looked like, for the environment or for workers.[62]

In 1890, the trustees of the Franklin funds in both Philadelphia and Boston faced an embarrassment of lawsuits. The plaintiffs (including Franklin's daughter's descendants) raised many valid objections, given the abandonment of microloans in both cities (on the grounds that suitable candidates had become too hard to find) and the failure of either city's stock to achieve the financial results Franklin had expected, based on a 5 percent interest rate and steady stream of loans. The funds survived the legal objections, though the caretakers realized they were under greater scrutiny while pursuing Franklin's goals for a second century, with the bulk of the funds laid out for "public works" but with a reserve to continue financing young workers.[63]

After considering many options, each fund's managers designated a civic institution to underwrite. Boston's caretakers set up the Franklin Trade School (1893) as an institutional successor to the apprenticeships artisans had once had, and as an alternative to high school. The school would have, in 1904, an unexpected connection to industrial Pennsylvania when coal and steel magnate Andrew Carnegie promised it a substantial gift, boasting, "I'll match Ben Franklin." In Philadelphia, the trustees decided to support an existing institution, the Franklin Institute, founded in 1824 as the Franklin Institute of the State of Pennsylvania for the Promotion of the Mechanic Arts. The institute funded science education through public lectures, a library, and a high school. Underwriting from the Franklin Fund assisted these activities after 1906. Then, more than halfway through the twentieth century, the managers in both cities reaffirmed the task of lending money to young people. Starting in 1962, loans were available to medical students at three universities: Tufts, Boston University, and Harvard; in Philadelphia, the Franklin Institute underwrote student loans in 1977. And so, almost two hundred years later, Franklin's bequest was at last useful in the way he'd intended, and for populations beyond the young married white men he'd been thinking about in 1790.[64]

It's puzzling he didn't authorize both of his goals—microloans and civic projects—from the very start, especially given his history of founding public institutions, including the American Philosophical Society and the University of Pennsylvania. These organizations were doing work that, to a certain extent, directly or indirectly confronted what the transition out of an organic economy represented in both its human and environmental costs.

Of course, the human and the material worlds can't be separated, though, to some extent, that's what Franklin attempted in his last will, by investing in young skilled working men for one century and then in institutional projects for another. In this, there was a temporal and ethical mismatch, because projects offsetting industrialization's damage had started well before Franklin's transition year of 1890, and might have achieved more with an overt connection to human well-being, including workers' well-being, had Franklin and others argued for their intrinsic inseparability. This was especially the case for the hastening extraction of mineral resources (a key Pennsylvania story) despite the theorization of extinction (also a Pennsylvania story) that intersected with just that extraction. And yet the threat of species annihilation always seemed remote, buried in extreme antiquity, with no modern resonance, even as many discoveries made in Pennsylvania reaffirmed it as a fact of the past that might recur in the future.

Big Bone Lick's mastodon fossils had already indicated that nature might change dramatically enough to obliterate entire species. Some investigators used theories of climate change to explain these extinctions. Buffon, the French naturalist who theorized a seventh epoch in which humans would perfect earthly conditions, suggested that North America's greater cold and damp (compared to Western Europe) might have gradually shrunk new-world animals; big American creatures, like the huge, tusked *incognitum* from Big Bone Lick, had flourished in a warmer era, then died and been reduced to fossils and replaced by smaller species. Buffon likewise thought Natives were smaller and weaker than Northern Europeans—might white settlers suffer a similar diminishment? Colonists retorted that, while the climate might challenge tribal peoples, they had culture and technology to resist its effects. Buffon took that position himself, accepting Franklin's claim in his *Increase of Mankind* that colonists, unlike Natives, would flourish in North America—and that anthropogenic climate change was one of colonization's benefits.[65]

Based on these discussions, another Frenchman, comparative anatomist Georges Cuvier, defined the theory of extinction central to science today. In papers published between 1796 and 1806, Cuvier had concluded that fossils represented animals that no longer existed. His work cited Pennsylvania men of science, including Charles Willson Peale, who organized the excavation of an entire mastodon skeleton (displayed in his smoke-eater-heated museum) and University of Pennsylvania naturalist and physician Benjamin Smith Barton, who declared in 1807 that such fossils were of animals that were "*extinct*."[66]

If extinction were possible, was it possible to prevent it? How much of nature might need to be protected to do that? At last, some

Pennsylvanians decided on a serious plan of conserving their forests, a sharp reversal from previous settler practice.

In fact, Pennsylvania set the trend for American forest conservation, in which private groups took initial steps, inspiring government policies and agencies to proceed from their efforts. Although the protection of the Adirondacks, starting in 1872, is regarded as the origin of this history, the earliest instance actually comes in 1855, when French-born botanist François André Michaux left $14,000 to the American Philosophical Society to promote forestry along with agriculture. Michaux wrote an authoritative *North American Sylva*, published first in French (1812–1813), then in several English editions. He traveled extensively in the Allegheny Mountains and considered the APS a benefactor, hence his gift. From his funds, in 1877 the APS appointed Joseph T. Rothrock, professor of botany at the University of Pennsylvania, as Michaux Lecturer. Rothrock became the state Department of Agriculture's first forestry commissioner in 1895, leading the new Division of Forestry (created 1893). He would designate state forests in 1897, starting with three reserves of about forty thousand acres each.[67]

The protections helped—somewhat. Before 1492, 99 percent of Pennsylvania's 28.8 million acres had been wooded if not forested; by 1945, only 52 percent of them were. This means the state's territory can absorb only about half the carbon dioxide it once did and generate half the oxygen, even as industrial and domestic carbon emissions have risen. But this is better than the outcome in other northeastern states. Pennsylvania retains an estimated 2,400 to 3,200 hectares of old-growth forest; Massachusetts has only 200 hectares, and Connecticut, 20 to 40. It's a terrible irony that Pennsylvania's coal saved its trees by keeping them from becoming fuel. But also, and at best (as if running to stay in place), by offsetting the damage of mining and forest destruction. Only after the devastating Johnstown Flood of 1889

(2,209 people killed) did Pennsylvania create its Division of Forestry, once it was clear that deforested land absorbs much less rainfall.[68]

Pennsylvania (and US) forest conservationists were distinctive, as well, in assuming their project required massive displacement of human populations—specifically, American Indians. In Europe, forestry had regulated people and their economic activities, but didn't force them out on a mass scale. In Pennsylvania, intensive development of western and northern zones—as for mining coal and smelting metal—guaranteed constant settler migration into areas where Natives had been forcibly cleared. Once some of those lands were recovered from settlement and economic development, however, Lenapes and others weren't invited back (or given any of industrialism's profits), nor were their earlier efforts to maintain forests honored in any serious way. Instead, their disappearance was, as Franklin had thought, still interpreted in naturalized terms, as if their extinction were inevitable anyway. Only that's a lie. In 2006, the Delaware Nation of Oklahoma sued the Commonwealth of Pennsylvania, contesting the Walking Purchase. They lost. The court's decision admitted that the deal was fraudulent but stated that the colony's proprietors had had legal authority to acquire land—by any means.[69]

Pennsylvania was a modern, industrial keystone in one final way. The scientific term "Pennsylvanian" embeds the state's distinctive energy history into North America's geological substrate and the planet's deep past. In 1822, two English geologists had proposed "Carboniferous" to name the era stretching from 358.9 million to 298.9 million years ago. During that time, warm temperatures fostered plant growth, everything from lofty trees to minuscule algae. When conditions shifted, the plant life died and was buried, slowly becoming the slate, limestone, and coal that modern geologists—living in a carbon-dependent economy—used to date a carbon-rich place and time. "Carboniferous" designates the geological deposits of this era throughout the world.[70]

But, for North America, US geologist Henry Shaler Williams in 1891 proposed subdividing the Carboniferous period into two periods, Mississippian and Pennsylvanian. They had different fossils, Williams pointed out. Pennsylvanian specimens from the more recent period, specifically its coal deposits, were hard anthracite rather than soft bituminous. Tellingly, the fossils Williams used as his evidence came from oil wells being drilled in western Pennsylvania. As a well-regarded geology textbook would put it in the early twentieth century, the Pennsylvanian "was probably the greatest coal-forming period the world has ever seen." Yes. And alas. The name now appears on those epochal chronologies of Earth's history, scrolling back through unfathomable millennia, long before the tiny temporal interval when humans kindled fire, invented needles, forged iron, worried about trees, fought over empires, dug for coal.[71]

Little could William Penn have imagined the epochal abyss yawning beneath the wooded land that bore his family's name, the sea of trees lapping his green country town of Philadelphia. Just as Franklin's fireplace—variously called Pennsylvanian, Philadelphian, American—had linked a colonized place to its trees, so the Pennsylvanian period tied that same territory to a carboniferous economy whose plentiful fuel ended the Little Ice Age's emergency exercise in conservation. Franklin's adopted home was foundational to the history of modern materiality, the connected histories of energy, industrialization, exploited labor, and consumerism, whose effects on the planet have outstripped the impact of another of his inventions, the United States of America.

Since the late nineteenth century and the naming of the Pennsylvanian period—the second century of Franklin's final project—Earth's temperature has risen about two degrees Fahrenheit. The year 2023 was the hottest in recorded history and, as of September, 2024 was predicted to be even hotter—June of 2024 marked a full twelve months (since July 2023) that had average temperatures more than

THE PENNSYLVANIAN PERIOD
(SEEN ON THE LEFT EDGE OF
THE SECOND SPIRAL) NAMES
PART OF EARTH'S MOST
CARBONIFEROUS PERIOD, WHEN
ABUNDANT PLANT LIFE CAPTURED
MASSIVE CARBON FROM THE SUN,
NOW REDUCED TO COAL.
*(Joseph Graham, William Newman, John Stacy,
The Geologic Time Spiral—A Path to the Past
[2008], US Geological Survey, Department of
the Interior/USGS)*

1.5 degrees Celsius (2.7 degrees Fahrenheit) above preindustrial levels. The heat has been rising since the start of the industrial era. As American experimenter Eunice Foote demonstrated in 1856, fossil fuel was not only emitting the visible smoke that so annoyed Franklin, but invisible carbon dioxide, poisoner of Joseph Priestley's captive mice, disastrous re-forger of the planet's atmosphere, part of the Industrial Revolution, which, intentionally or not, has generated a very wicked problem, with solutions long promised, but long deferred.[72]

CODA

God helps them who help themselves.
—POOR RICHARD

"We are as gods and might as well get good at it," Stewart Brand declared in the first issue of his countercultural *Whole Earth Catalog* (1968). Bold new vision? No. It's the old Promethean claim: turning material things into tools gives humans superhuman power. And so Brand showcased a splendor of disruptive technologies in his catalog—a wood-burning stove, for instance, but also the Apple home computer. Far *out*: together, these two devices evoke a recent past when our now-ubiquitous consumer high tech was just emerging—but alongside nostalgia for ye olde appliances. A home computer was groovy, and wood-burning heat, cool, a nonconformist revival of organic energy. (The *Whole Earth Catalog* also championed solar power.) But as Immanuel Kant had warned, fantasies of possessing or stealing divine power have a history of leading to disaster, advice that Franklin endorsed in a proverb he put into his *Poor Richard's Almanack* for 1736:

"God helps them who help themselves." The adage works whether people think assuming godlike power is sinful or whether they don't believe in any god at all—either way, humans can't count on any mystical assistance. They must act.[1]

And that was the message of the third part of Franklin's codicil to his will: we're on our own now. He left directions for the two centuries after his death, investing first in his fellow citizens, and then in public works or institutions. But after that, as of 1990, he simply "presumed that there will always be found . . . virtuous and benevolent citizens" willing to work for public benefit, though he would no longer be "presuming [on them] to carry my views farther." Micromanager he may have been, but he realized that, at some point, times would change, and he might no longer know best. The question is: Do we?[2]

Twenty-first-century humans have many more technical fixes than the people who read the *Whole Earth Catalog* in 1968 or *Poor Richard* in 1736. But using technology wisely is hard at any time, especially in tackling a wicked problem like the climate crisis. Techno-escape fantasies, in which we (some of us) flee Earth's compromised condition, payloaded into artificial atmospheres that orbit our planet or light out for another, derive from eighteenth-century conceptions of outer space and aerial travel. But the technologies that generate the necessary air and heat for these voyages don't also protect passengers from the dangers of microgravity and solar radiation. A life beyond Earth is itself beyond speculative. Protecting human lives down here remains the imperative. This will require reassessment of industrialization's gifts and costs. Capitalism has lifted many boats—consider the wealth now extended, via income, capital, or credit, to millions of people today—but not everyone's fortunes rose, and the wealth that's there (especially at society's highest and narrowest tip) exists by inflicting damage, historic and ongoing, on many populations and all of the natural world.

Using indoor climate control to give comfort and health to a growing number of people, beyond the rich, is historically momentous. By now, billions have been able to claim those benefits, which increasingly include staying warm in cold weather and cool in hot, plus enjoying those conditions in smaller and moving spaces, such as cars, buses, trains, or airplanes, billions of mini-atmospheres roaming the planet. This kind of comfort is still more of a privilege than a right, however. In 2021, an estimated 3 billion of the world's 7.9 billion people lacked consistent access to modern energy (electricity or piped gas) in their homes. More than a third of humanity must rely on or revert to energy sources Franklin would have known—burning wood, dung, charcoal, peat, or coal to cook or keep warm—with significant health hazards from inhaling smoke indoors, causing millions of deaths each year, as well as the danger of burns from open fires. And, as temperatures rise, these people lack reliable ways to cool down, as does anyone if their energy source fails. Above all, the carbon footprint of domestic climate control is a gigantic cost—residential and commercial energy consumption, including electricity end use for heating and air-conditioning, accounted for 30 percent of US greenhouse gas emissions in 2021.[3]

We need to become a post–fossil fuel society, but we don't exactly know what that will look like. Certain dimensions are coming into focus; certain predictions remain futuristic, if not suspiciously Promethean. Franklin's era, which underwent the transition into a fossil fuel society we must reverse, won't tell us everything, but it conveys some useful things, definitely in its adaptation to climate change, the whole point of the Little Ice Age fire drill.

First, the mixed news: there are promising solutions to our emergency, but for anyone hoping we can get away with one, two, or just a few of them—I'm very sorry. The material adaptations to the Little Ice Age, overlapping with and co-constituting the Industrial Revolution, were multiple and entangled, with connected knock-on effects that (just to keep things lively and super-complicated) changed all the time. Today's discussions of climate solutions are unhelpful if they become shouting matches among proponents of different projects. No one has a silver bullet. True, there will need to be sequencing of steps, some ranking of remedies, and rejection of outright nonsense. But multivariable complexity is baked into the problem.

It's especially revealing that, in the case of the Franklin stove, consumer choice was necessary but not sufficient to guide an optimal energy transition during climate change. Government regulation was the other obvious remedy, as with the schemes to subsidize firewood for poorer colonists or the attempts to slow deforestation, but Franklin was not unique in rejecting economic regulation, most overtly when he criticized the Iron Act in his essay on population, proposing that colonial economies would grow themselves without imperial interference.

In fact, his entire stove project depended on colonial law's power to designate land as property, nullifying Indigenous people's sovereignty, and on laws that regulated labor to the point of making some people into property. Pennsylvanian fireplaces, made on Lenape land and with enslaved labor, express perfectly how American nature was being legally redefined for modern consumer-producers, as a revolution merely industrious was morphing into something semi-industrial. Franklin's never-reconciled contradiction—accepting governmental power (including the post office where he displayed and sold his devices) when that benefited him but resenting it when it favored people he considered unworthy—is a continuing problem. Unregulated consumer behavior had enormous power, but not necessarily in a positive direction, given that it led to the adoption of coal.

Franklin's example shows, too, that basic science, and respect for it, matter equally. That Franklin's status as a man of science survived the divisions of the American Revolution, with his work republished in London and stoves based explicitly on his design sold there, is somewhat astonishing. It's hard to think of anything comparable happening today. Franklin's international fame depended on his contemporaries' belief that science and technology should cross borders, a conviction that yielded some striking acts of international unity even during the Seven Years' War and the American Revolution, as with the US passport Franklin brokered to get a British steam engine into France in 1779. Such amity survived into later eras, most recently during (and despite) the Cold War, as with the activities of the International Geophysical Year (1957–1958). This was when ham-radio enthusiasts around the world shared a no-need-to-translate wordless geek-out over the signals oscillating from the Soviet Union's Sputnik 1, harbinger of the many satellite-assisted global communications to come.[4]

Certain parts of the earlier history are not so much advice as they are warnings. It's instructive that Franklin and his contemporaries acknowledged the possible need for both climate adaptation and mitigation. But it's dismaying that, in placing their bets on mitigation, many colonists set off to chop down forests to make the climate warmer. They turned out to be right about the ultimate need for mitigation, though not in the way they expected, and while sacrificing woodlands that, in many places, have never recovered. Their climate crisis, with its own distinctive multivariable complex of complexities, may have required adaptation instead of mitigation. Until the end of the twentieth century, our climate crisis was probably the reverse. But despite our diminishing prospects for mitigation, it's important not to give up on it, even as adaptation becomes more critical. One of the best features of Franklin's era was its faith in improvement and in projects. Some of the projectors' goals were misplaced (their failure to consider everyone's benefit most notably), but that's a reason to *improve* their ethic of improvement, not reject it.

One crucial project links all climate solutions: reconciling ecology and economics. If modern economics, and growth economics in particular, distanced human happiness from the state of the natural world, that division of assets is now bankrupt. Since the 1970s and

1980s, several social science variants—ecological economics, cultural ecology, political ecology—have been trying to suture the two back together. Doing this cosmic mending job will require, as well, appreciating people, including America's Indigenous peoples, who never separated humans from the rest of nature. Franklin attended and recorded many Indigenous council fires, with their noneconomic ethics—a philosophy of intercultural and interspecies mutuality that offered not just an admirable and alternative idea, but a conceptualization of power that, through treaties, continues to be a living presence within US law. But the council fires, though always an invitation to build bonds beyond nations and even species, never influenced Franklin's signal contributions to modern economic theory—a tragic missed opportunity, perhaps his worst failure of imagination.[5]

It's not just Franklin. Of the thirteen original states, Pennsylvania is one of three with no recognized Native group (the other two are Rhode Island and New Hampshire), and it doesn't have any federally recognized tribes, either. This even as Sweden—thousands of miles away—maintains diplomatic ties to the Delaware Nation (in Oklahoma) based on historic acknowledgment of those people when they were still in Lenapehoking, their ancestral land: if the Delaware Nation of Oklahoma were to petition the United Nations for membership, an existing member state would endorse them with a firm *ja*. Recognizing Native peoples as legal sovereigns is the necessary first step in correcting past wrongs against them and, relatedly, in understanding how to protect a natural world they first conserved. The Indigenous ethic of a life concurrent with nature is now recognized as essential to solving the climate crisis, though this is hardly the only reason to acknowledge Natives' sovereign statuses within what is now the United States.[6]

As Franklin's republic approaches its 250th anniversary, it's worth considering the climate crisis against that kind of time span, shorter than a geological epoch, longer than an electoral cycle, rather like Franklin's multicentury posthumous project. In his will, he shows a marked and remarkable patience with what might be called the medium-long term. But his willingness to back away, after the first two centuries, is equally impressive, and possibly his most radical statement about his life's works. These include the US Constitution, now subjected to scrutiny of its authors' original intent or historical context. Let the record show: one of those authors believed that, as of the end of the last century, his intentions and his era were irrelevant. We're on our own, but maybe with a lingering exhortation paraphrased from Franklin's codicil, suited to national or international communities alike: *Are there no longer any virtuous and benevolent citizens among you?*[27]

Two earlier US anniversaries used energy consumption to say something about the state of the nation: the coal-fired Corliss engine that was a centerpiece of the 1876 Centennial Exhibition in Philadelphia, and the wind-powered tall ships that, at a more countercultural, *Whole Earth*, and oil-crisis-inflected point in American history, marked the nation's bicentennial in 1976. These technical objects are reminders, as well, that 1776 may be overpromoted as a historic watershed compared to the bigger background transition from organic to fossil energy. So, what might a comparable symbol be for 2026—and what does a span of 250 years represent, anyway? The damage of the climate crisis might last much longer than that, perhaps thousands of years, yet another way in which the consequences of the Industrial Revolution will almost certainly trump those of the American Revolution. But how long might it take to undo the damage?

Leaving aside the industrious revolution, the succeeding Industrial Revolution—main driver of anthropogenic climate change—has lasted slightly more than two hundred years, roughly the duration of the United States (so far). Might that length of time be long enough to reverse the damage and set a new course? Another logical "start" date for such a reversal would be just after World War II, during the Great Acceleration of everything Franklin's era set into motion (consumerism, industrialization, commercial extraction of resources, generation of pollution and waste), when multiple scientific authorities began to sound the alarm about anthropogenic damage to the planet. This periodization would signify a revolution that might yield results by 2145, far too late to stave off outright disaster. And yet using only half the time to undo what took two hundred years to accomplish would suggest a target year of 2045, likely too soon to expect sufficient decarbonization of the global economy.[8]

Alternatively, what about a century starting from the first energy crises of the 1970s? We've already used up half that time, trying to decide what to do, starting to do some of it, though not enough. The net-zero target of ending greenhouse emissions by 2050, to keep global warming to 1.5 degrees Celsius (the Paris Agreement's original goal, as stated at the 2015 UN Climate Change Conference), is now impossible. Holding to no more than 2.0 degrees Celsius of warming by the 2070s is the next best goal. Industrialization's pioneers, including the United States, have the greatest responsibility to hit that target. Millions of people are already dying each year from fossil fuel pollution and extreme weather events, and some very sober authorities predict that more than a billion people might become climate refugees by the end of the century; these estimates will likely be worse by the time you read this page. And yet most people in a position to affect energy consumption and climate policy (even just by voting) may be looking at such statements while thinking: *But that won't be me.* That presumption is wrong ethically, and may very well be wrong factually. The statistics already indicate that *a quarter* of the world's population (now roughly 8 billion) will, in this century, very likely be severely affected by climate change and dependence on fossil fuel energy. Most tellingly, climate change's costs are now calculated as severe threats to GDP.[9]

That being the case, there might be little to celebrate by the time of the United States's three hundredth birthday in 2076. Although my book questions whether technology can tackle the climate crisis by itself, I can only hope *all* the relevant technologies—electric cars, solar panels, heat pumps, carbon capture, geoengineering, gigantic orbiting parasols, other wondrous devices not yet even thought of—are going to work. They'd better.

But I have to remind you that making energy-transition technologies dependent on consumer preference is, historically speaking, a bad idea. It was the part of Franklin's climate-change fire drill that never quite worked. Government action was and is essential. We'll need it to stop carbon emissions—the only way to slow the crisis and delay its worst effects—which will require decarbonizing the world's energy systems and switching to renewable power, reversing the transition that Franklin witnessed. For this reason, a good marker for the 2026 US celebrations might be not a physical device, but, societally, something else humans can devise: energy programs with teeth and legal challenges to fossil fuel interests. These public policies, on every level, city to nation, would have to recognize the scale of the emergency that everyone, in the United States and in the rest of the world, is facing. We have a choice: radical change now, while we can exert some control over it, or even more radical change later, as control becomes less and less possible.

Can we act fast enough, making good on the mere half century we may have left? One hundred years is much longer than a generation, or even the wartime or moon-shot chronologies that are sometimes discussed in relation to the climate crisis. And yet it's half the duration of Franklin's last project, a speed at which mistakes will need to be identified and corrected at a pace that will challenge careful thought. But the care and the speed are both necessary. Dear reader, please, we must make haste slowly.

September 23, 2024

NOTES

Abbreviations

PBF   Leonard W. Labaree et al., eds., *The Papers of Benjamin Franklin*, 44 vols. to date (New Haven: Yale University Press, 1959–)
WMQ   *William and Mary Quarterly*, 3rd Series
HSP   Historical Society of Pennsylvania
APS   American Philosophical Society
BF   Benjamin Franklin

Introduction: Make Haste Slowly

1. BF, *An Account of the New Invented Pennsylvanian Fire-Places* . . . (Philadelphia, 1744), in *PBF*, 2: 438; Asa Briggs, *The Age of Improvement* (London: Longman, 1959).
2. Bill Joy, "Why the Future Doesn't Need Us," *Wired*, April 1, 2000; Steven Connor, *Dream Machines* (London: Open Humanities Press, 2017).
3. John M. Staudenmaier, *Technology's Storytellers: Reweaving the Human Fabric* (Cambridge, MA: MIT Press, 1985); Michael Adas, *Machines as the Measure of Men: Science, Technology, and Ideologies of Western Dominance* (Ithaca, NY: Cornell University Press, 1989); Leo Marx, "Technology: The Emergence of a Hazardous Concept," *Social Research* 64, no. 3 (Fall 1997): 965–88; David Edgerton, "From Innovation to Use: Ten Eclectic Theses on the Historiography of Technology," *History and Technology* 16, no. 2 (1999): 111–36; Ian Inkster, "Technological and Industrial Change: A Comparative Essay," in *The Cambridge History of Science: The Eighteenth Century*, ed. Roy Porter (Cambridge: Cambridge University Press, 2003), 845–58; Leandro Prados de la Escosura, ed., *Exceptionalism and Industrialization: Britain and its European Rivals, 1688–1815* (Cambridge: Cambridge University Press, 2004); Thomas J. Misa, *Leonardo to the Internet: Technology and Culture from the Renaissance to the*

Present (Baltimore: Johns Hopkins University Press, 2004); David Edgerton, *The Shock of the Old: Technology and Global History since 1900* (London: Profile Books, 2006); Lissa Roberts, Simon Schaffer, and Peter Dear, *The Mindful Hand: Inquiry and Invention from the Late Renaissance to Early Industrialisation* (Amsterdam: Edita KNAW, 2007).

4. Cyclical variations in solar radiation or ocean circulation are possible causes of the Little Ice Age, as is unusual volcanic activity. Independently, disastrous declines of human population, as with the Black Death in Europe and among Indigenous populations after the Spanish invasion of the Americas, let cleared areas revert to forest, temporarily increasing the absorption of atmospheric carbon, with consequent cooling. See H. H. Lamb, *Climate, History, and the Modern World*, 2nd ed. (London: Routledge, 1995); Brian Fagan, *The Little Ice Age: How Climate Made History, 1300–1850* (New York: Basic Books, 2000); National Research Council of the National Academies, *Surface Temperature Reconstructions for the Last 2,000 Years: Consensus Study Report* (Washington, DC: National Academies Press, 2006), https://www.nap.edu/catalog/11676/surface-temperature-reconstructions-for-the-last-2000-years; M. Sigl et al., "Timing and Climate Forcing of Volcanic Eruptions for the Past 2,500 Years," *Nature* 523 (2015): 543–49, esp. fig. 3; Simon L. Lewis and Mark A. Maslin, "Defining the Anthropocene," *Nature* 519 (2015): 171–80; *Earth Observatory Glossary*, NASA Goddard Space Flight Center, s.v. "Little Ice Age," accessed September 22, 2024, https://earthobservatory.nasa.gov/glossary/l/n.

5. Geoffrey Parker, *Global Crisis: War, Climate Change and Catastrophe in the Seventeenth Century* (New Haven: Yale University Press, 2013); William M. Cavert, "Winter and Discontent in Early Modern England," in *Governing the Environment in the Early Modern World: Theory and Practice*, ed. Sara Miglietti and John Morgan (New York: Routledge, 2017), 114–33; Dagomar Degroot, *The Frigid Golden Age: Climate Change, the Little Ice Age, and the Dutch Republic, 1560–1720* (Cambridge: Cambridge University Press, 2018).

6. The oldest scholarship mostly considered whether the first stove worked as Franklin said it did, not why he invented it—and therefore wanted it to work the way he said it should: Henry C. Mercer, *The Bible in Iron . . .* , ed. Joseph E. Sandford and Horace M. Mann, 3rd ed. (Doylestown, PA: Bucks County Historical Society, 1914), 128–31; Samuel Y. Edgerton Jr., "Heat and Style: Eighteenth-Century House Warming by Stoves," *Journal of the Society of Architectural Historians* 20, no. 1 (March 1961): 20–26; John F. Fitchen III, "The Problem of Ventilation Through the Ages," *Technology and Culture* 22, no. 3 (July 1981): 485–511; Samuel Y. Edgerton Jr., "Supplement: The Franklin Stove," in I. Bernard Cohen, *Benjamin Franklin's Science* (Cambridge, MA: Harvard University Press, 1990), 199–211 (which, in declaring the stove a failure, compared the only surviving example of a Franklin stove to illustrations of the second model, when in fact the surviving device is a modification of that design, one that Franklin himself had deemed inadequate, given that he subsequently produced three more). More recently, historians of design have analyzed Franklin's heating technology as an innovative architectural element, proof that domestic design was intended, by the eighteenth century, to give greater comfort to more people: Priscilla J. Brewer, *From Fireplace to Cookstove: Technology and the Domestic Ideal in America* (Syracuse, NY: Syracuse University Press, 2000); John E. Crowley, *The Invention of Comfort: Sensibilities and Design in Early Modern Britain and Early America* (Baltimore: Johns Hopkins University Press, 2001), 175–85; William B. Meyer, "Why Indoor Climates Change: A Case Study," *Climatic Change* 55 (2002): 395–407; Paul G. E. Clemens, "The Consumer Culture of the Middle Atlantic, 1760–1820," *WMQ* 62, no. 4 (October 2005): 577–624; Howell John Harris, "Conquering Winter: US Consumers and the Cast-Iron Stove," *Building Research and Information* 36, no. 4 (July/August 2008): 337–50; Harris, "'The Stove Trade Needs Change Continually': Designing the First Mass-Market Consumer Durable, ca. 1810–1930," *Winterthur Portfolio* 43, no. 4 (Winter 2009): 365–406.

7. Joyce E. Chaplin, "The Other Revolution," *Early American Studies* 13, no. 2 (Spring 2015): 297–301.
8. Christine MacLeod, *Inventing the Industrial Revolution: The English Patent System, 1660–1800* (Cambridge: Cambridge University Press, 1988), 101–102; Marie Thébaud-Sorger, "Changing Scale to Master Nature: Promoting Small-Scale Inventions in Eighteenth-Century France and Britain," *Technology and Culture* 61, no. 4 (October 2020): 1076–1107.
9. Henry Home, Lord Kames, to BF, February 18, 1768, in *PBF*, 15: 50–51; Vladimir Janković, *Confronting the Climate: British Airs and the Making of Environmental Medicine* (New York: Palgrave Macmillan, 2010), esp. 8–12, 15–40.
10. On energy, see J. U. Nef, *The Rise of the British Coal Industry*, 2 vols. (London: G. Routledge, 1932), 1: 165–89, 224–61; 2: 3–134; Kenneth Pomeranz, *The Great Divergence: China, Europe, and the Making of the Modern World Economy* (Princeton, NJ: Princeton University Press, 2000); E. A. Wrigley, *Energy and the English Industrial Revolution* (Cambridge: Cambridge University Press, 2010); Andreas Malm, *Fossil Capital: The Rise of Steam Power and the Roots of Global Warming* (London: Verso, 2016). On consumerism, see Neil McKendrick, John Brewer, and J. H. Plumb, *The Birth of a Consumer Society: The Commercialization of Eighteenth-Century England* (London: Europa Publications, 1982); Jan de Vries and Ad van der Woude, *The First Modern Economy: Success, Failure, and Perseverance of the Dutch Economy, 1500–1815* (Cambridge: Cambridge University Press, 1997), 693–722; Jan de Vries, *The Industrious Revolution: Consumer Behavior and the Household Economy, 1650 to the Present* (Cambridge: Cambridge University Press, 2008), esp. 10–72, 122–85.
11. Eric Williams, *Capitalism and Slavery* (Chapel Hill: University of North Carolina Press, 1944); Sidney W. Mintz, *Sweetness and Power: The Place of Sugar in Modern History* (New York: Viking, 1985); Joyce E. Chaplin, *An Anxious Pursuit: Agricultural Innovation and Modernity in the Lower South, 1730–1815* (Williamsburg, VA, and Chapel Hill, NC: Institute of Early American History and Culture, 1993); John Bellamy Foster, Hannah Holleman, and Brett Clark, "Imperialism in the Anthropocene," *Monthly Review* 71 (2019): 70–88; Gurminder K. Bhambra and John Holmwood, *Colonialism and Modern Social Theory* (Cambridge: Cambridge University Press, 2021); Amitav Ghosh, *The Nutmeg's Curse: Parables for a Planet in Crisis* (Chicago: University of Chicago Press, 2021); Maxine Berg and Pat Hudson, *Slavery, Capitalism and the Industrial Revolution* (Cambridge: Polity Press, 2023).
12. Edmund S. Morgan, *American Slavery, American Freedom: The Ordeal of Colonial Virginia* (New York: W. W. Norton, 1975), esp. 3–24; Jack P. Greene, *Pursuits of Happiness: The Social Development of Early Modern British Colonies and the Formation of American Culture* (Chapel Hill: University of North Carolina Press, 1988); William Cronon, *Nature's Metropolis: Chicago and the Great West* (New York: W. W. Norton, 1991); Nikole Hannah-Jones et al., eds., *The 1619 Project: A New Origin Story* (New York: One World, 2021).
13. On how "the middle states quietly eloped with the nation," see John M. Murrin, "A Roof without Walls: The Dilemma of American National Identity," in *Beyond Confederation: Origins of the Constitution and American National Identity*, ed. Richard Beeman, Stephen Botein, and Edward C. Carter II (Williamsburg, VA, and Chapel Hill, NC: Institute of Early American History and Culture, 1987), 333–48 (quotation 347); Gary B. Nash, *First City: Philadelphia and the Forging of Historical Memory* (Philadelphia: University of Pennsylvania Press, 2002), 1–13.
14. *OED Online*, s.v. "economics," "political economy," "ecology"; BF to Peter Collinson, May 9, 1753, in *PBF*, 4: 482; Donella H. Meadows et al., *The Limits to Growth* (Falls Church, VA: Potomac Associates, 1972); Walter Eltis, *The Classical Theory of Economic Growth* (London: Macmillan, 1984), esp. 68–105; World Commission on Environment and Development, *Our Common Future* (New York: Oxford University Press, 1987); Joseph E.

Stiglitz, Amartya Sen, and Jean-Paul Fitoussi, *Mismeasuring Our Lives: Why GDP Doesn't Add Up* (New York: New Press, 2010); Robert Costanza et al., "Development: Time to Leave GDP behind," *Nature* 505 (2014): 283–85; Iris Borowy and Matthias Schmelzer, "Introduction: The End of Economic Growth in Long-Term Perspective," in *History of the Future of Economic Growth: Historical Roots of Current Debates on Sustainable Degrowth*, ed. Iris Borowy and Matthias Schmelzer (London: Routledge, 2017), 1–26; Giorgos Kallis et al., *The Case for Degrowth* (Cambridge: Polity Press, 2020); Tim Jackson, *Post Growth: Life after Capitalism* (Cambridge: Polity Press, 2021).

15. On Indigenous belief in the interconnectedness of all lives, and reverence for them, see Vine Deloria Jr., *God Is Red: A Native View of Religion* (New York: Grosset and Dunlap, 1973). On treaty relationships and Indigenous sovereignty, see Vine Deloria Jr., *Behind the Trail of Broken Treaties: An Indian Declaration of Independence* (New York: Delacorte Press, 1974); Russel Lawrence Barsh and James Youngblood Henderson, *The Road: Indian Tribes and Political Liberty* (Berkeley: University of California Press, 1980), 31–49, 270–82; David Murray, *Forked Tongues: Speech, Writing and Representation in North American Indian Texts* (Bloomington: University of Indiana Press, 1991); Francis Paul Prucha, *American Indian Treaties: The History of a Political Anomaly* (Berkeley: University of California Press, 1994); Joanne Barker (Lenape), "For Whom Sovereignty Matters," in *Sovereignty Matters: Locations of Contestation and Possibility in Indigenous Struggles for Self-Determination*, ed. Joanne Barker (Lincoln: University of Nebraska Press, 2006), 4–17; Taiaiake Alfred (Mohawk), "Sovereignty," in *Sovereignty Matters*, 34, 39–40, 45–46; Colin G. Calloway, *Pen and Ink Witchcraft: Treaties and Treaty Making in American Indian History* (New York: Oxford University Press, 2013); Jeffrey Glover, *Paper Sovereigns: Anglo-Native Treaties and the Law of Nations, 1604–1664* (Philadelphia: University of Pennsylvania Press, 2014); Hayley Negrin, "Return to the Yeokanta/River: Powhatan Women and Environmental Treaty Making in Early America," *Environmental History* 28, no. 3 (July 2023): 522–53. On the centrality of Native people to all American history, Ned Blackhawk, *The Rediscovery of America: Native Peoples and the Unmaking of U.S. History* (New Haven: Yale University Press, 2023).

16. A. R. Ubbelohde, *Man and Energy* (London: Hutchinson's Scientific and Technical Publications, 1954); Jean-François Mouhot, "Past Connections and Present Similarities in Slave Ownership and Fossil Fuel Usage," *Climatic Change* 105 (2011), 329–55; Chris Hayes, "The New Abolitionism," *The Nation*, April 22, 2014, https://www.thenation.com/article/archive/new-abolitionism/; Bob Johnson, "Energy Slaves: Carbon Technologies, Climate Change, and the Stratified History of the Fossil Economy," *American Quarterly* 68, no. 4 (December 2016): 955–79. On the historical contingency of nineteenth-century abolitionism—its lack of inevitability—see also Christopher Leslie Brown, *Moral Capital: Foundations of British Abolitionism* (Williamsburg, VA, and Chapel Hill, NC: Omohundro Institute of Early American History and Culture, 2006).

17. Leo Marx, *The Machine in the Garden: Technology and the Pastoral Ideal in America* (New York: Oxford University Press, 1964); Andrew Kirk, "Appropriating Technology: The *Whole Earth Catalog* and Counterculture Environmental Politics," *Environmental History* 6, no. 3 (July 2001): 374–94. See also Hayden White, *Tropics of Discourse: Essays in Cultural Criticism* (Baltimore: Johns Hopkins University Press, 1985), introduction.

18. Alondra Nelson, "Future Texts," *Social Text* 20, no. 2 (2002): 1–15; International Labour Organization, "Guidelines for a Just Transition Towards Environmentally Sustainable Economies and Societies for All," February 2, 2016, https://www.ilo.org/publications/guidelines-just-transition-towards-environmentally-sustainable-economies.

19. BF, *Pennsylvanian Fire-Places*, 441.

20. *Poor Richard, 1744. An Almanack...*, in *PBF*, 2: 396; *OED Online*, s.v. "device," esp. 7a. and 9.

1. Promethean

1. Immanuel Kant, *Gesammelte Schriften* (Berlin: G. Reimer, 1902 [1997]), 1: 472.
2. Ann Gibbons, "Who Were the Denisovans?" *Science* 333 (August 2011): 1084–87; Alan J. Osborn, "Eye of the Needle: Cold Stress, Clothing, and Sewing Technology During the Younger Dryas Cold Event in North America," *American Antiquity* 79, no. 1 (2014): 45–68.
3. Roger C. Echo-Hawk, "Ancient History in the New World: Integrating Oral Traditions and the Archaeological Record in Deep Time," *American Antiquity* 65, no. 2 (2000), 267–90; Robin Wall Kimmerer, *Braiding Sweetgrass: Indigenous Wisdom, Scientific Knowledge and the Teachings of Plants* (Minneapolis: Milkweed Editions, 2013), 3–9; Paulette F. C. Steeves, *The Indigenous Paleolithic of the Western Hemisphere* (Lincoln: University of Nebraska Press, 2021); Bonnie L. Pitblado, "On Rehumanizing Pleistocene People of the Western Hemisphere," *American Antiquity* 87, no. 2 (2022): 217–35.
4. Mary Byrd Davis, ed., *Eastern Old-Growth Forests: Prospects for Rediscovery and Recovery*, foreword by John Davis (Washington, DC: Island Press, 1996), 20, 78, 127, 181; Rolf Peter Sieferle, *The Subterranean Forest: Energy Systems and the Industrial Revolution* (Cambridge: White Horse Press, 2001), 44; Michael C. Stambaugh et al., "Wave of Fire: An Anthropogenic Signal in Historical Fire Regimes across Central Pennsylvania, USA," *Ecosphere* 9, no. 5 (2018), article e02222; Scott H. Markwith and Asha Paudel, "Beyond Pre-Columbian Burning: The Impact of Firewood Collection on Forest Fuel Loads," *Canadian Journal of Forest Research* 52, no. 3 (2022): 365–71. But see also W. Wyatt Oswald et al., "Conservation Implications of Limited Native American Impacts in Pre-Contact New England," *Nature Sustainability* 3 (2020): 241–46.
5. A. G. Brown et al., "Roman Vineyards in Britain: Stratigraphic and Palynological Data from Wollaston in the Nene Valley, England," *Antiquity* 75 (2001): 745–57; Raphael Neukom et al., "No Evidence for Globally Coherent Warm and Cold Periods Over the Preindustrial Common Era," *Nature* 571 (2019): 550–54.
6. Jed O. Kaplan et al., "Large Scale Anthropogenic Reduction of Forest Cover in Last Glacial Maximum Europe," *PLoS One* 11 (2016).
7. Astrid Kander, Paolo Malanima, and Paul Warde, *Power to the People: Energy in Europe over the Last Five Centuries* (Princeton, NJ: Princeton University Press, 2013), 37–80, 60–70.
8. Charles F. Carroll, *The Timber Economy of Puritan New England* (Providence: Brown University Press, 1973), 3–21; Oliver Rackham, *Trees and Woodlands in the British Landscape* (London: J. M. Dent, 1976); John F. Richards, *The Unending Frontier: An Environmental History of the Early Modern World* (Berkeley: University of California Press, 2003); Michael Williams, *Deforesting the Earth: From Prehistory to Global Crisis* (Chicago: University of Chicago Press, 2003); Kander et al., *Power to the People*, 56–60, 96–100. But see Paul Warde, "Fear of Wood Shortage and the Reality of the Woodland in Europe, c. 1450–1850," *History Workshop Journal* 62, no. 1 (Autumn 2006): 28–57; Keith Pluymers, *No Wood, No Kingdom: Political Ecology in the English Atlantic* (Philadelphia: University of Pennsylvania Press, 2021), on management of wood scarcities.
9. Fernand Braudel, *The Structures of Everyday Life: The Limits of the Possible*, trans. Siân Reynolds (London: Collins, 1981), 362–67; Paul Warde, *Ecology, Economy, and State Formation in Early Modern Germany* (Cambridge: Cambridge University Press, 2006); Karl Appuhn, *A Forest on the Sea: Environmental Expertise in Renaissance Venice* (Baltimore: Johns Hopkins University Press, 2009); Hamish Graham, "Fleurs-de-Lis in the Forest: 'Absolute' Monarchy and Attempts at Resource Management in Eighteenth-Century France," *French History* 23, no. 3 (September 2009): 311–35; Paul Warde, *The Invention of Sustainability: Nature and Destiny, c. 1500–1870* (Cambridge: Cambridge University Press, 2018), esp. 58–101.

10. Robert Albion, *Forests and Sea Power: The Timber Problem of the Royal Navy, 1652–1862* (Cambridge, MA: Harvard University Press, 1926); Paul W. Bamford, *Forests and French Sea Power, 1660–1789* (Toronto: University of Toronto Press, 1956); Williams, *Deforesting the Earth*, 193–206; Appuhn, *Forest on the Sea*, 30–33, 54–57; N.A.M. Rodger, *The Command of the Ocean: A Naval History of Britain, 1649–1815* (New York: W. W. Norton, 2005), 482, 607–609; John T. Wing, "Keeping Spain Afloat: State Forestry and Imperial Defense in the Sixteenth Century," *Environmental History* 17, no. 1 (January 2012): 116–45; Pluymers, *No Wood, No Kingdom*, 14–58. Cf. Baihui Duan and Rebekah Clements, "Fighting for Forests: Protection and Exploitation of Kŏje Island Timber during the East Asian War of 1592–1598," *Environmental History* 27, no. 3 (July 2022): 415–40.
11. Pluymers, *No Wood, No Kingdom*, 59–107.
12. Jan de Vries and Ad van der Woude, *The First Modern Economy: Success, Failure, and Perseverance of the Dutch Economy, 1500–1815* (Cambridge: Cambridge University Press, 1997), 50–57; Joyce E. Chaplin, *Subject Matter: Technology, the Body, and Science on the Anglo-American Frontier, 1500–1676* (Cambridge, MA: Harvard University Press, 2001), 124–30; Kander et al., *Power to the People*, 82–84.
13. John Evelyn, *Sylva, or, A Discourse of Forest-Trees, and the Propagation of Timber in His Majesties Dominions*, 2nd ed. (London, 1670), 213; Williams, *Deforesting the Earth*, 190–209.
14. J. U. Nef, *The Rise of the British Coal Industry* (London: G. Routledge, 1932), 1: 156–64, 190–223; John Hatcher, *The History of the British Coal Industry*: vol. 1: *Before 1700: Towards the Age of Coal* (Oxford: Clarendon Press, 1993), 409–18; Peter Brimblecombe, *The Big Smoke: A History of Air Pollution in London since Medieval Times* (London: Methuen, 1987), 22–38; Sieferle, *Subterranean Forest*, 78–137; Richards, *Unending Frontier*, 193–241; Kander et al., *Power to the People*, 60–62, 107–16; William M. Cavert, *The Smoke of London: Energy and Environment in the Early Modern City* (Cambridge: Cambridge University Press, 2016), esp. 17–31; Ruth Goodman, *The Domestic Revolution: How the Introduction of Coal into Victorian Homes Changed Everything* (New York: W. W. Norton, 2020).
15. J[ohn] E[velyn], *Fumifugium, or, The Inconveniencie of the Aer and Smoak of London . . .* (London, 1661), 5, 15, 16; Hatcher, *British Coal Industry*, 1: 409–18; Brimblecombe, *Big Smoke*, 39–62; Cavert, *Smoke of London*, 32–99, 173–94.
16. Brimblecombe, *Big Smoke*, 26–29.
17. Brian Fagan, *The Little Ice Age: How Climate Made History, 1300–1850* (New York: Basic Books, 2000); Geoffrey Parker, *Global Crisis: War, Climate Change and Catastrophe in the Seventeenth Century* (New Haven: Yale University Press, 2013), esp. xv–xx, 660–67; Wolfgang Behringer, *A Cultural History of Climate* (Cambridge: Polity Press, 2010), 121–67; Joyce E. Chaplin, "Ogres and Omnivores: Early American Historians and Climate History," *WMQ* 72, no. 1 (January 2015): 25–32; Anya Zilberstein, *A Temperate Empire: Making Climate Change in Early America* (New York: Oxford University Press, 2016).
18. Wolfgang Behringer, "Weather, Hunger and Fear: Origins of the European Witch-Hunts in Climate, Society and Mentality," *German History* 13 (1995): 1–27; Kander et al., *Power to the People*, 72–73; Parker, *Global Crisis*, esp. 3–109.
19. Antonello Gerbi, *The Dispute of the New World: The History of a Polemic, 1750–1900*, trans. Jeremy Moyle (Pittsburgh: University of Pittsburgh Press, 1973); Karen Ordahl Kupperman, "Fear of Hot Climates in the Anglo-American Colonial Experience," *WMQ* 41, no. 2 (April 1984): 213–40; Antonello Gerbi, *Nature in the New World: From Christopher Columbus to Gonzalo Fernández de Oviedo*, trans. Jeremy Moyle (Pittsburgh: University of Pittsburgh Press, 1985); Nicolás Wey-Gómez, *The Tropics of Empire: Why Columbus Sailed South to the Indies* (Cambridge, MA: MIT Press, 2008).
20. Karen Ordahl Kupperman, "The Puzzle of the American Climate in the Early Colonial Period," *American Historical Review* 87, no. 5 (December 1982): 1262–89 (which focuses on the climate/latitude conundrum and stops before the eighteenth century); Sam White, "Unpuz-

zling American Climate: New World Experience and the Foundations of a New Science," *Isis* 106, no. 3 (September 2015): 544–66 (with a similar emphasis on pre-eighteenth-century comprehension of climate and latitude); Dagomar Degroot et al., "Towards a Rigorous Understanding of Societal Responses to Climate Change," *Nature* 591 (2021): 539–50.
21. Carroll, *Timber Economy of Puritan New England*, 75–97; Richards, *Unending Frontier*, 412–62; William Beinart and Lotte Hughes, *Environment and Empire* (New York: Oxford University Press, 2007), 22–39; Alexander Kocha et al., "Earth System Impacts of the European Arrival and Great Dying in the Americas after 1492," *Quaternary Science Reviews* 207 (2019): 13–36.
22. Richard H. Grove, *Green Imperialism: Colonial Expansion, Tropical Island Edens and the Origins of Environmentalism, 1600–1860* (Cambridge: Cambridge University Press, 1996); Ian Baucom, *History 4° Celsius: Search for a Method in the Age of the Anthropocene* (Durham, NC: Duke University Press, 2020); Pluymers, *No Wood, No Kingdom*, 108–66, 193–238.
23. Carroll, *Timber Economy of Puritan New England*, 101–19; Michael Williams, *Americans and Their Forests: A Historical Geography* (Cambridge: Cambridge University Press, 1989), 82–110; Priscilla J. Brewer, *From Fireplace to Cookstove: Technology and the Domestic Ideal in America* (Syracuse, NY: Syracuse University Press, 2000), 28.
24. Neil McKendrick, John Brewer, and J. H. Plumb, *The Birth of a Consumer Society: The Commercialization of Eighteenth-Century England* (London: Europa Publications, 1982); de Vries and van der Woude, *First Modern Economy*, 693–710, 711–22; Jan de Vries, *The Industrious Revolution: Consumer Behavior and the Household Economy, 1650 to the Present* (Cambridge: Cambridge University Press, 2008), esp. 10–72, 122–85; Thomas Piketty, *Capital in the Twenty-First Century*, trans. Arthur Goldhammer (Cambridge, MA: Harvard University Press, 2014), esp. 93–99.
25. Daniel Defoe, *Every-Body's Business is No-Body's Business* (London, 1725), 2; Alan Hunt, *Governance of the Consuming Passions: A History of Sumptuary Law* (Houndmills, UK: Macmillan, 1996); John Styles and Amanda Vickery, eds., *Gender, Taste, and Material Culture in Britain and North America, 1700–1830* (New Haven: Yale University Press, 2006); de Vries, *Industrious Revolution*, 154–85.
26. Joel Mokyr, *The Lever of Riches: Technological Creativity and Economic Progress* (New York: Oxford University Press, 1990), 57–80.
27. Braudel, *Structures of Everyday Life*, 298–302; Klaus Bingenheimer, *Die Luftheizungen des Mittelalters: Zur Typologie und Entwicklung eines technikgeschichtlichen Phänomens* (Hamburg: Kovač, 1998)—I am indebted to Jin-Woo Choi for this reference; Daniel Roche, *A History of Everyday Things: The Birth of Consumption in France, 1600–1800*, trans. Brian Pearce (New York: Cambridge University Press, 2000), 123–26; John E. Crowley, *The Invention of Comfort: Sensibilities and Design in Early Modern Britain and Early America* (Baltimore: Johns Hopkins University Press, 2001), 7–22. On the longer history of indoor atmospheres (and their environmental cost), see Reyner Banham, *The Architecture of the Well-Tempered Environment* (London and Chicago: Architectural Press and University of Chicago Press, 1969), esp. 18–70.
28. Merry Wiesner-Hanks, "Having Her Own Smoke: Employment and Independence for Singlewomen in Germany, 1400–1750," in *Singlewomen in the European Past, 1250–1800*, ed. Judith M. Bennett and Amy M. Froide (Philadelphia: University of Pennsylvania Press, 1999), 192–216; Crowley, *Invention of Comfort*, 8–23, 56–57.
29. George L. Phillips, *England's Climbing Boys: A History of the Long Struggle to Abolish Child Labor in Chimney Sweeping*, pref. Arthur H. Cole (Cambridge, MA: Harvard University Press, 1949), 1–3; Braudel, *Structures of Everyday Life*, 298–302; Roche, *History of Everyday Things*, 127–30; Crowley, *Invention of Comfort*, 22–36, 49–62; de Vries, *Industrious Revolution*, 40–58.
30. William Shakespeare, *Cymbeline* (1611?), 4.2.262–63.

31. René Descartes, *Discourse on the Method*, in *The Philosophical Writings of Descartes*, vol. 1, trans. John Cottingham, Robert Stoothoff, and Dugald Murdoch (Cambridge: Cambridge University Press, 1985), 116, 127; Joan DeJean, *The Age of Comfort: When Paris Discovered Casual—and the Modern House Began* (New York: Bloomsbury, 2009), 93–101; Robert M. Hazen and Margaret Hindle Hazen, *Keepers of the Flame: The Role of Fire in American Culture, 1775–1925* (Princeton, NJ: Princeton University Press, 2014); Jason Nguyen, "Fire, Décor, and Heating Machines," *Oxford Art Journal* 40, no. 3 (2017): 371–96.
32. De Vries, *Industrious Revolution*, 31–37, 133–44; Crowley, *Invention of Comfort*, 36–44, 62–69, 71–72; Sasha Handley, *Sleep in Early Modern England* (New Haven: Yale University Press, 2016), 39–61.
33. Crowley, *Invention of Comfort*, 69–78.
34. *PBF*, 1: lvi; Thomas M. Wickman, *Snowshoe Country: An Environmental and Cultural History of Winter in the Early American Northeast* (New York: Cambridge University Press, 2018), 56–90.
35. Wickman, *Snowshoe Country*, 233–35, 241–50.
36. Cotton Mather, *Winter Meditations: Directions How to Employ the Liesure of the Winter for the Glory of God* (Boston, 1693), 25, 29, 63.
37. Mather, *Winter Meditations*, introduction (n.p.), 44–45.
38. Jim Egan, *Authorizing Experience: Refigurations of the Body Politic in Seventeenth-Century New England Writing* (Princeton, NJ: Princeton University Press, 1999); Chaplin, *Subject Matter*, esp. 157–98.
39. Brooke Hindle, ed., *America's Wooden Age: Aspects of Its Early Technology* (Tarrytown, NY: Sleepy Hollow Restorations, 1975); Williams, *Americans and Their Forests*, 5–9, 53–81; Crowley, *Invention of Comfort*, 79–103; Strother E. Roberts, *Colonial Ecology, Atlantic Economy: Transforming Nature in Early New England* (Philadelphia: University of Pennsylvania Press, 2019), 97–126.
40. Roger Williams, *A Key into the Language of America* (1643), ed. John J. Teunissen and Evelyn J. Hinz (Detroit: Wayne State University Press, 1973), 107, 138; R. V. Reynolds and Albert H. Pierson, *Fuel Wood Used in the United States, 1630–1930* (Washington, DC: United States Department of Agriculture, 1942), 1, 7; Carroll, *Timber Economy of Puritan New England*, 57–71; William Cronon, *Changes in the Land: Indians, Colonists, and the Ecology of New England* (New York: Hill and Wang, 1983), 119–21; Carolyn Merchant, *Ecological Revolutions: Nature, Gender, and Science in New England* (Chapel Hill: University of North Carolina Press, 1989), 157.
41. Wickman, *Snowshoe Country*, 20–55, 87–90.
42. William N. Fenton, "Structure, Continuity, and Change in the Process of Iroquois Treaty Making," in *The History and Culture of Iroquois Diplomacy: An Interdisciplinary Guide to the Treaties of the Six Nations and Their League*, ed. Francis Jennings et al. (Syracuse, NY: Syracuse University Press, 1985), 15, 118; Daniel K. Richter, *The Ordeal of the Longhouse: The Peoples of the Iroquois League in the Era of European Colonization* (Williamsburg, VA, and Chapel Hill, NC: Institute of Early American History and Culture, 1992), 9–10, 19, 21, 42–43, 225.
43. Heidi Bohaker, *Doodem and Council Fire: Anishinaabe Governance Through Alliance* (Toronto: University of Toronto Press, 2020), xii–xxi, 147–66, 180–81; Hayley Negrin, "Return to the Yeokanta/River: Powhatan Women and Environmental Treaty Making in Early America," *Environmental History* 28, no. 3 (July 2023): 522–53. See also David Graeber and David Wengrow, *The Dawn of Everything: A New History of Humanity* (New York: Farrar, Straus and Giroux, 2021), on Indigenous liberty versus European hierarchy, 27–77.
44. Fenton, "Iroquois Treaty Making," 5–7; Heidi Kiiwetinepinesiik Stark, "Marked by Fire: Anishinaabe Articulations of Nationhood in Treaty Making with the United States and Canada," *American Indian Quarterly* 36, no. 2 (Spring 2012): 119–49.

45. *OED Online*, s.v. "bonfire"; *The New-England Courant*, December 24–31, 1722.
46. David Cressy, *Bonfires and Bells: National Memory and the Protestant Calendar in Elizabethan and Stuart England* (Berkeley: University of California Press, 1989), 67–87; Francis D. Cogliano, "Deliverance from Luxury: Pope's Day, Conflict and Consensus in Colonial Boston, 1745–1765," *Studies in Popular Culture* 15 (1993): 15–28; James Sharpe, *Remember, Remember: A Cultural History of Guy Fawkes Day* (Cambridge, MA: Harvard University Press, 2005), 141–45.
47. [Daniel Defoe], *An Essay upon Projects* (London, 1697), 1, 19, 24–25; Asa Briggs, *The Age of Improvement, 1783–1867* (London: Longmans, Green, 1959); Paul Slack, *The Invention of Improvement: Information and Material Progress in Seventeenth-Century England* (Oxford: Oxford University Press, 2014), esp. 1–3.
48. Robert Boyle, *New Experiments and Observations Touching Cold, or, An Experimental History of Cold Begun: to Which Are Added an Examen of Antiperistasis and an Examen of Mr. Hobs's Doctrine about Cold* (London, [1665]), quotations from preface (n. p.).
49. Boyle, *Touching Cold*, 3, 18, 32; Boyle, *The Philosophical Works of the Honourable Robert Boyle Esq . . .* , ed. Peter Shaw, 2nd ed., vol. 1 (London, 1738); 579.
50. On atmosphere, see Milton Kerker, "Herman Boerhaave and the Development of Pneumatic Chemistry," *Isis* 46, no. 1 (March 1955): 36–49; Viktor Harsch, "Otto von Guericke (1602–1686) and His Pioneering Vacuum Experiments," *Aviation, Space, and Environmental Medicine* 78, no. 11 (2007): 1075–77; Craig Martin, "The Invention of Atmosphere," *Studies in History and Philosophy of Science* 52 (August 2015): 44–54. On instruments, see Lissa Roberts, "The Death of the Sensuous Chemist: The 'New' Chemistry and the Transformation of Sensuous Technology," *Studies in History and Philosophy of Science* 26, no. 4 (1995): 503–20; Jan Golinski, "'Fit Instruments': Thermometers in Eighteenth-Century Chemistry," in *Instruments and Experimentation in the History of Chemistry*, ed. Frederic L. Holmes and Trevor H. Levere (Cambridge, MA: MIT Press, 2000), 185–210; John C. Powers, "Measuring Fire: Herman Boerhaave and the Introduction of Thermometry into Chemistry," *Osiris* 29 (2014): 158–77.
51. Everett Mendelsohn, *Heat and Life: The Development of the Theory of Animal Heat* (Cambridge, MA: Harvard University Press, 1964), 8–107; Tristram Stuart, *The Bloodless Revolution: A Cultural History of Vegetarianism from 1600 to Modern Times*, 1st US ed. (New York: W. W. Norton, 2006), 60–77; Joyce E. Chaplin, "Why Drink Water? Diet, Materialisms, and British Imperialism," *Osiris* 35 (2020): 99–122.
52. H. W. Dickinson, *Matthew Boulton* (Cambridge: Cambridge University Press, 1937), ix, 64–74, 75–87, 113–14, 108; Mokyr, *Lever of Riches* 10–16; Marie Thébaud-Sorger, "Capturing the Invisible: Heat, Steam and Gases in France and Great Britain, 1750–1800," in *Compound Histories: Materials, Governance and Production, 1760–1840*, ed. Lissa Roberts and Simon Werrett (Leiden: Brill, 2017), 85–105; Steven Shapin, *The Scientific Revolution*, 2nd ed. (Chicago: University of Chicago Press, 2018), part 3.
53. Joyce O. Appleby, *Economic Thought and Ideology in Seventeenth-Century England* (Princeton, NJ: Princeton University Press, 1978); Slack, *Invention of Improvement*, 113–214; Gareth Dale, "Seventeenth-Century Origins of the Growth Paradigm," *History of the Future of Economic Growth: Historical Roots of Current Debates on Sustainable Degrowth*, ed. Iris Borowy and Matthias Schmelzer (London: Routledge, 2017), 27–51.
54. René Antoine Ferchault de Réaumur cited in J.-F. Belhoste, "Une sylviculture pour les forges, XVIe–XIXe siècles," in *Forges et forêts: Recherches sur la consommation proto-industrielle de bois*, ed. Denis Woronoff (Paris: École des hautes études en sciences sociales, 1990), 242.
55. James D. Drake, *King Philip's War: Civil War in New England, 1675–1676* (Amherst: University of Massachusetts Press, 1999), 4.
56. *The New-England Courant*, March 1, 1725; Wickman, *Snowshoe Country*, esp. chs. 3, 4, 5; Lisa Brooks, *Our Beloved Kin: A New History of King Philip's War* (New Haven: Yale University Press, 2019); Christine M. DeLucia, *Memory Lands: King Philip's War and the Place of*

Violence in the Northeast (New Haven: Yale University Press, 2019). See also Jean M. O'Brien, *Dispossession by Degrees: Indian Land and Identity in Natick, Massachusetts, 1650–1790* (New York: Cambridge University Press, 1997); Chaplin, *Subject Matter*, 82–108.

57. *The New-England Courant*, November 20, 1721.
58. Lydia Barnett, *After the Flood: Imagining the Global Environment in Early Modern Europe* (Baltimore: Johns Hopkins University Press, 2019).
59. Karen Ordahl Kupperman, "Climate and Mastery of the Wilderness in Seventeenth-Century New England," in *Seventeenth-Century New England*, ed. David D. Hall, David Grayson Allen, and Philip Chadwick Foster Smith (Boston: Colonial Society of Massachusetts, 1984), 3–37; Jan Golinski, *British Weather and the Climate of Enlightenment* (Chicago: University of Chicago Press, 2007), 192–202; Zilberstein, *Temperate Empire*, 164–73.
60. Frank G. Speck, "The Eastern Algonkian Wabanaki Confederacy," *American Anthropologist* 17, no. 3 (1915): 492–508; Willard Walker, Robert Conkling, and Gregory Buesing, "A Chronological Account of the Wabanaki Confederacy," in *Political Organization of Native North Americans*, ed. Ernest L. Schusky (Washington, DC: University Press of America, 1980), 41–50; Jean-Pierre Sawaya, *La fédération des sept feux de la vallée du Saint-Laurent, xviie au xixe siècle* (Sillery, QC: Septentrion, 1998), esp. 23–40; Wickman, *Snowshoe Country*, 187–88.
61. Wickman, *Snowshoe Country*, 190–92; Sam White, *A Cold Welcome: The Little Ice Age and Europe's Encounter with North America* (Cambridge, MA: Harvard University Press, 2017), 204, 221–23.
62. *The New-England Courant*, August 21, 1721.

2. Forging

1. *Benjamin Franklin's Autobiography*, ed. Joyce E. Chaplin (New York: W. W. Norton, 2012), 27.
2. Amy C. Schutt, *Peoples of the River Valleys: The Odyssey of the Delaware Indians* (Philadelphia: University of Pennsylvania Press, 2007), 7–30.
3. Gunlög Fur, *A Nation of Women: Gender and Colonial Encounters Among the Delaware Indians* (Philadelphia: University of Pennsylvania Press, 2009); Jean R. Soderlund, *Lenape Country: Delaware Valley Society Before William Penn* (Philadelphia: University of Pennsylvania Press, 2015), 12–34; Michael Goode, "'Rancontyn Marenit': Lenape Peacemaking Before William Penn," in *The Worlds of William Penn*, ed. Andrew R. Murphy and John Smolenski (New Brunswick, NJ: Rutgers University Press, 2018), 217–31.
4. Soderlund, *Lenape Country*, 35–54.
5. James H. Merrell, *Into the American Woods: Negotiators on the Pennsylvania Frontier* (New York: W. W. Norton, 1999), 27–31, 38, 190; Michael Dean Mackintosh, "Ink and Paper, Clamshells and Leather: Power, Environmental Perception, and Materiality in the Lenape-European Encounter at Philadelphia," in *A Greene Country Towne: Philadelphia's Ecology in the Cultural Imagination*, ed. Alan C. Braddock and Laura Turner Igoe (University Park, PA: Penn State University Press, 2016), 19–33.
6. Soderlund, *Lenape Country*, 149–76; calculations for converting the sum of £1,200 from 1683 (an average; payments were made from 1682–1684) to pounds in 2023 (and pounds converted to dollars for 2023) done with https://www.measuringworth.com, September 22, 2024.
7. Schutt, *Peoples of the River Valleys*, 31–40, 63–81.
8. Charles W. Eliot, ed., *French and English Philosophers: Descartes, Rousseau, Voltaire, Hobbes: With Introductions and Notes* (New York: P. F. Collier, 1910), 77.
9. William Petty, *The Petty Papers: Some Unpublished Writings of Sir William Petty*, ed. Marquis of Lansdowne, 2 vols. (London: Constable, 1927), 2: 109–20; Nicholas Canny, "The Irish

Background to Penn's Experiment," in *The World of William Penn*, ed. Richard S. Dunn and Mary Maples Dunn (Philadelphia: University of Pennsylvania Press, 1986), 139–56; Joyce E. Chaplin, *Subject Matter: Technology, the Body, and Science on the Anglo-American Frontier, 1500–1676* (Cambridge, MA: Harvard University Press, 2001), 318–20; Ted McCormick, *William Petty and the Ambitions of Political Arithmetic* (Oxford: Oxford University Press, 2009); Shuichi Wanibuchi, "William Penn's Imperial Landscape: Improvement, Political Economy, and Colonial Agriculture in the Pennsylvania Project," in *The Worlds of William Penn*, 378–402.
10. Petty, *Petty Papers*, 2: 125–29.
11. William Penn, *A Letter from William Penn...* (London, 1683), 1, 2.
12. Act of November 27, 1700, *The Statutes at Large of Pennsylvania from 1682 to 1801*, comp. James T. Mitchell and Henry Flanders ([Harrisburg]: State Printer of Pennsylvania, 1896–), 2: 66–67; Henry Clepper, "Rise of the Forest Conservation Movement in Pennsylvania," *Pennsylvania History* 12, no. 3 (1945): 200–16, esp. 201, 202; Shuichi Wanibuchi, "Political Ecologies and the Transformation of Landscape in the Early Modern Delaware Valley" (PhD diss., Harvard University, 2023), ch. 5.
13. William Penn, "Instructions given by me ...," [September 30, 1681], *The Papers of William Penn*, vol. 2, *1680–1684*, ed. Richard S. Dunn and Mary Maples Dunn (Philadelphia: University of Pennsylvania Press, 1982), 121; Robert I. Alotta, *Street Names of Philadelphia* (Philadelphia: Temple University Press, 1975), 106–108, 132–33, 139, 142; Adam Levine, "The Grid versus Nature: The History and Legacy of Topographical Change in Philadelphia," in *Nature's Entrepôt: Philadelphia's Urban Sphere and Its Environmental Thresholds*, ed. Brian C. Black and Michael J. Chiarappa (Pittsburgh: University of Pittsburgh Press, 2012), 17–34; Craig Zabel, "William Penn's Philadelphia: The Land and the Plan," *Nature's Entrepôt*, 139–44.
14. Chaplin, *Benjamin Franklin's Autobiography*, 25–29.
15. Thomas Tryon, *The Way to Health, Long Life, and Happiness: or, a Discourse of Temperance...* (London, 1683), quotations p. 50; Everett Mendelsohn, *Heat and Life: The Development of the Theory of Animal Heat* (Cambridge, MA: Harvard University Press, 1964), 8–26, 67–107; Tristram Stuart, *The Bloodless Revolution: A Cultural History of Vegetarianism from 1600 to Modern Times*, 1st US ed. (New York: W. W. Norton, 2006); Joyce E. Chaplin, "Why Drink Water?: Diet, Materialisms, and British Imperialism," *Osiris* 35 (2020): 99–122.
16. Thomas Tryon, *Tryon's Letters upon Several Occasions* (London: 1700), 49–50; Chaplin, *Benjamin Franklin's Autobiography*, 37.
17. Chaplin, *Benjamin Franklin's Autobiography*, 45–46.
18. Chaplin, *Benjamin Franklin's Autobiography*, 78–80; Silence Dogood, no. 12, *The New-England Courant*, September 10, 1722, in *PBF*, 1: 40–41.
19. BF, *A Modest Enquiry into the Nature and Necessity of a Paper-Currency* (Philadelphia, 1729), in *PBF*, 1: 149, 150; *The Pennsylvania Gazette*, January 5, 1731, in *PBF*, 1: 213–14.
20. John J. McCusker and Russell R. Menard, *The Economy of British America, 1607–1789* (Williamsburg, VA, and Chapel Hill, NC: Institute of Early American History and Culture, 1985), 228–29; Marianne S. Wokeck, *Trade in Strangers: The Beginnings of Mass Migration to North America* (University Park, PA: Penn State University Press, 1999).
21. McCusker and Menard, *Economy of British America*, 52–57, 198–208; Wanibuchi, "Transformation of Landscape in the Early Modern Delaware Valley."
22. Alan Gallay, *The Indian Slave Trade: The Rise of the English Empire in the American South, 1670–1717* (New Haven: Yale University Press, 2002); Margaret Ellen Newell, *Brethren by Nature: New England Indians, Colonists, and the Origins of American Slavery* (Ithaca, NY: Cornell University Press, 2015); Andrés Reséndez, *The Other Slavery: The Uncovered Story of Indian Enslavement in America* (Boston: Houghton Mifflin Harcourt, 2016);

Wendy Warren, *New England Bound: Slavery and Colonization in Early America* (New York: W. W. Norton, 2016).

23. Gary B. Nash and Jean R. Soderlund, *Freedom by Degrees: Emancipation in Pennsylvania and Its Aftermath* (New York: Oxford University Press, 1991), 41–73; Soderlund, *Lenape Country*, 196–204.

24. James T. Lemon, *The Best Poor Man's Country: A Geographical Study of Early Southeastern Pennsylvania* (Baltimore: Johns Hopkins University Press, 1972); Arthur Cecil Bining, *Pennsylvania Iron Manufacture in the Eighteenth Century*, 2nd ed. (Harrisburg: Pennsylvania Historical and Museum Commission, 1973), 6–18, 39–50; Ronald L. Lewis, *Coal, Iron, and Slaves: Industrial Slavery in Maryland and Virginia, 1715–1865* (Westport, CT: Greenwood Press, 1979), 10–13.

25. Bining, *Pennsylvania Iron Manufacture*, 151–58; Priscilla J. Brewer, *From Fireplace to Cookstove: Technology and the Domestic Ideal in America* (Syracuse, NY: Syracuse University Press, 2000), 32.

26. Bining, *Pennsylvania Iron Manufacture*, 55–81 (the blowing tub would replace bellows in the late 1700s; hot blast furnaces, in which the air was preheated, saved fuel and became a new industry standard in the early 1800s); Lewis, *Coal, Iron, and Slaves*, 17–18; John Bezís-Selfa, *Forging America: Ironworkers, Adventurers, and the Industrious Revolution* (Ithaca, NY: Cornell University Press, 2004), 13–16.

27. The National Museum of Scotland's Newcomen and Boulton and Watt engines stand 9.5 meters tall (31.2 ft.), https://www.nms.ac.uk/explore-our-collections/stories/science-and-technology/newcomen-engine.

28. "To Be Let," *The Pennsylvania Gazette* (Philadelphia), March 21, 1744, [3]; Bining, *Pennsylvania Iron Manufacture*, 19–38 (an outdated, pastoral vision of iron plantations); Bezís-Selfa, *Forging America*, 16–21.

29. Joseph E. Walker, "Negro Labor in the Charcoal Iron Industry of Southeastern Pennsylvania," *Pennsylvania Magazine of History and Biography* 93, no. 4 (1969): 466–86; Gary B. Nash, "Slaves and Slaveholders in Colonial Philadelphia," *WMQ* 30, no. 4 (October 1973): 223–56, esp. 243, 237, 244; Lewis, *Coal, Iron, and Slaves*, 21–28, 35. For antebellum enslavement and iron production, see Charles B. Dew, *Bond of Iron: Master and Slave at Buffalo Forge* (New York: W. W. Norton, 1994).

30. Bezís-Selfa, *Forging America*, 6–8, 105–12.

31. *The Pennsylvania Gazette* (Philadelphia), June 7, 1775; S. M. Epstein, "A Coffin Nail from the Slave Cemetery at Catoctin, Maryland," *MASCA Journal* 1 (1981): 208–10; Jean Libby, "Technological and Cultural Transfer of African Ironmaking into the Americas and the Relationship to Slave Resistance," in *Rediscovering America, 1492–1992: National, Cultural and Disciplinary Boundaries Re-Examined*, ed. Leslie Bary et al. (Baton Rouge: Louisiana State University Press, 1992), 57–76, esp. 65–67; Gary B. Nash, *First City: Philadelphia and the Forging of Historical Memory* (Philadelphia: University of Pennsylvania Press, 2002), 41–42. On enslaved people as discerning consumers of iron goods, see Chris Evans, "The Plantation Hoe: The Rise and Fall of an Atlantic Commodity, 1650–1850," *WMQ* 69, no. 1 (January 2012): 71–100; on enslaved ironworkers as skilled producers, Jenny Bulstrode, "Black Metallurgists and the Making of the Industrial Revolution," *History and Technology* 39, no. 1 (2023): 1–41.

32. Bining, *Pennsylvania Iron Manufacture*, 40–50, 171.

33. [Isabella Batchelder] James, *Memorial of Thomas Potts, Junior* (Cambridge [MA]: private press, 1874), 29–30, 42–43, 52–54, 69–70; Thomas Maxwell Potts, ed., *Historical Collections Relating to the Potts Family in Great Britain and America . . .* (Canonsburg, PA: The Compiler, 1901); Henry C. Mercer, *The Bible in Iron . . .*, ed. Joseph E. Sandford and Horace M. Mann, 3rd ed. (Doylestown, PA: Bucks County Historical Society, 1961), 104–108; Bining, *Pennsylvania Iron Manufacture*, 120–26; Bezís-Selfa, *Forging America*, 125–26.

34. James, *Memorial of Thomas Potts*, 32–33; Mercer, *Bible in Iron*, 104, 108.

35. Bining, *Pennsylvania Iron Manufacture*, 59–65, 119–21; Michael Williams, *Americans and Their Forests: A Historical Geography* (Cambridge: Cambridge University Press, 1989), 104–10; Rolf Peter Sieferle, *The Subterranean Forest: Energy Systems and the Industrial Revolution* (Cambridge: White Horse Press, 2001), 63–64; Bezís-Selfa, *Forging America*, 27–28.
36. McCusker and Menard, *Economy of British America*, 326. For the acreage estimated from cubic feet of fuel per ton of wrought iron, multiplied by annual output, converted into cords of wood (thirty cords roughly equivalent to one acre of forest), see William Cronon, *Changes in the Land: Indians, Colonists, and the Ecology of New England* (New York: Hill and Wang, 1983), 120–21. The use of wrought iron in this calculation overestimates fuel use; the use of one acre (rather than two) underestimates it.
37. Entries for 1735 (Captain Palmer, firewood), "Scates" for John Read, p. 132, Ledger B, Miscellaneous Benjamin Franklin Collections (Misc. BF Collections), APS; fountain pen, p. 61, wood, pp. 115, 255, 256, 382, billiard balls and "Galloshes," p. 207, Terence, p. 218, Ledger D, Misc. BF Collections, APS; June 9, 1737 (telescope), March 26, 1739 (books for woodman), Shop Book, Misc. BF Collections, APS.
38. Entry for February 13, 1738, Shop Book, Misc. BF Collections, APS (quotation); *Poor Richard, 1733. An Almanack...*, in *PBF*, 1: 310; November 1, 1737, entry under William Dewees Jr., November 1, 1737 (paper molds), Colebrook Dale Furnace, daybook, 1734–1741, vol. 216, collection (coll.) 212: Forges and Furnaces Collection, HSP; Mercer, *Bible in Iron*, 109, 117.
39. Entries on the Nutt-Grace family: December 1, 1730, August 4, 1731, Ledger A, Misc. BF Collections, APS; Ledger D, pp. 47, 369, Misc. BF Collections, APS.
40. Karen Ordahl Kupperman, "Climate and Mastery of the Wilderness in Seventeenth-Century New England," in *Seventeenth-Century New England*, ed. David D. Hall, David Grayson Allen, and Philip Chadwick Foster Smith (Boston: Colonial Society of Massachusetts, 1984), 3–37; Sam White, "Climate Change in Global Environmental History," in *A Companion to Global Environmental History*, ed. J. R. McNeill and Erin Stewart Mauldin (New York: Wiley Blackwell, 2012), 394–410. On the hope of warming North America: Jan Golinski, *British Weather and the Climate of Enlightenment* (Chicago: University of Chicago Press, 2007), 170–202; Golinski, "American Climate and the Civilization of Nature," in *Science and Empire in the Atlantic World*, ed. James Delbourgo and Nicholas Dew (New York: Routledge, 2008), 153–74; Anya Zilberstein, *A Temperate Empire: Making Climate Change in Early America* (New York: Oxford University Press, 2016).
41. Ian K. Steele, *The English Atlantic, 1675–1740: An Exploration of Communication and Community* (New York: Oxford University Press, 1986); Sara S. Gronim, *Everyday Nature: Knowledge of the Natural World in Colonial New York* (New Brunswick, NJ: Rutgers University Press, 2007); Thomas Wickman, "'Winters Embittered with Hardships': Severe Cold, Wabanaki Power, and English Adjustments, 1690–1710," *WMQ* 72, no. 1 (January 2015): 57–98; Early American Imprints, Series I, Evans (1639–1800), Readex/American Antiquarian Society, accessed on September 22, 2024.
42. *The Pennsylvania Gazette* (Philadelphia), February 19, 1730, December 3, 1730, January 4, 1732, November 23, 1732, in *PBF*, 1: 185, 188, 271, 277–78. Scholarship on winter and colonial communication has focused on New England: Katherine Grandjean, *American Passage: The Communications Frontier in Early New England* (Cambridge, MA: Harvard University Press, 2015), 15–44; Jordan E. Taylor, "Now Is the Winter of Our Dull Content: Seasonality and the Atlantic Communications Frontier in Eighteenth-Century New England," *New England Quarterly* 95, no. 1 (2022): 8–38.
43. *Poor Richard, 1737. An Almanack...*, in *PBF*, 2: 163.
44. "The Rape of Fewel. A Cold-Weather Poem," *The Pennsylvania Gazette*, October 19–26, 1732.
45. *Poor Richard, 1733*, in *PBF*, 1: 311, 312, 314; *Poor Richard, 1736. An Almanack...*, in *PBF*, 2: 142; *Poor Richard, 1737*, in *PBF*, 2: 169.

46. *The Pennsylvania Gazette*, November 30, 1732, in *PBF*, 1: 252–54.
47. Act of November 27, 1700, *Statutes at Large of Pennsylvania*, 2: 68; Cornel Zwierlein, *Prometheus Tamed: Fire, Security, and Modernities, 1400 to 1900* (Leiden and Boston: Brill, 2021).
48. *The Pennsylvania Gazette*, April 10, 1735, November 17, 1737, November 30, 1738, April 3, 1740, in *PBF*, 2: 132, 189 (and n. 6, pp. 188–89 on Shedid Allhazar), 216, 283.
49. Jacob F. Field, *London, Londoners and the Great Fire of 1666: Disaster and Recovery* (London: Routledge, 2018), 34–37, 43–44.
50. BF, "On Protection of Towns from Fire," *The Pennsylvania Gazette*, February 4, 1735, in *PBF*, 2: 12–15; Act of February 20, 1736, *Statutes at Large of Pennsylvania*, 2: 301—the act was repealed in 1762.
51. "Articles of the Union Fire Company," in *PBF*, 2: 150–53.
52. *The Pennsylvania Gazette*, March 21, 1734, in *PBF*, 1: 377; Lydia Barnett, *After the Flood: Imagining the Global Environment in Early Modern Europe* (Baltimore: Johns Hopkins University Press, 2019), 88–128.
53. Charles Tomlinson, *A Rudimentary Treatise on Warming and Ventilation . . .*, 3rd ed. (London: Virtue Brothers, 1864), 82–83; J. Pickering Putnam, *The Open Fireplace in All Ages* (Boston: Ticknor and Company, 1886), 31–34; Astrid Kander, Paolo Malanima, and Paul Warde, *Power to the People: Energy in Europe over the Last Five Centuries* (Princeton, NJ: Princeton University Press, 2013), 101–102.
54. J. T. Desaguliers, *A Course of Experimental Philosophy . . .*, vol. 2 (London, 1734–1744), 114–21, 141–43, 249, 291, 379–81; I. Bernard Cohen, *Franklin and Newton: An Inquiry into Speculative Newtonian Experimental Science and Franklin's Work in Electricity as an Example Thereof* (Philadelphia: American Philosophical Society, 1956), 243–45; Larry R. Stewart, *The Rise of Public Science: Rhetoric, Technology, and Natural Philosophy in Newtonian Britain, 1660–1750* (New York: Cambridge University Press, 1992), 26–27, 213–30; Jeffrey R. Wigelsworth, "Competing to Popularize Newtonian Philosophy: John Theophilus Desaguliers and the Preservation of Reputation," *Isis* 94, no. 3 (September 2003): 435–55.
55. [Nicolas] Gauger, *Fires Improv'd: or, A new Method of Building Chiminies, so as to Prevent their Smoaking . . .*, trans. J. T. Desaguliers, 2nd ed. (London, [1736]), i–vi (this second edition is most likely the one Franklin knew); Jason Nguyen, "Fire, Décor, and Heating Machines," *Oxford Art Journal* 40 (2017): 371–96, esp. 390–95.
56. Gauger, *Fires Improv'd*, i–iii, 15–24, 29, 63, 69, 71, 137–43. See also Tomlinson, *Warming and Ventilation*, 88–94.
57. Gauger, *Fires Improv'd*, 70–71, 155–58; Herman Boerhaave, *Elements of Chemistry . . .*, trans. Timothy Dallowe, vol. 1 (London, 1735), 175–81, and plate IV.
58. M[artin] Clare, *The Motion of Fluids, Natural and Artificial . . .* (London, 1735), 171, 221–22, 224–26.
59. Francis Jennings, *The Invasion of America: Indians, Colonialism and the Cant of Conquest* (Williamsburg, VA, and Chapel Hill, NC: Institute of Early American History and Culture, 1975), 201; Merrell, *Into the American Woods*, 130–32, 137, 150, 152; James O'Neil Spady, "Colonialism and the Discursive Antecedents of Penn's Treaty with the Indians," in *Friends and Enemies in Penn's Woods: Indians, Colonists, and the Racial Construction of Pennsylvania*, ed. William A. Pencak and Daniel K. Richter (University Park, PA: Penn State University Press, 2004), 18–40.
60. Carolyn Merchant, *Ecological Revolutions: Nature, Gender, and Science in New England* (Chapel Hill: University of North Carolina Press, 1989), 160; Williams, *Americans and Their Forests*, 77–78; James E. McWilliams, *A Revolution in Eating: How the Quest for Food Shaped America* (New York: Columbia University Press, 2005), 1–4, 65–73, 233–37.
61. Nicole Eustace, *Covered with Night: A Story of Murder and Indigenous Justice in Early America* (New York: W. W. Norton, 2021).

62. *The Pennsylvania Gazette*, October 14, 1736, September 22, 1737, in *PBF*, 2: 160–61, 188; Francis Jennings, "Iroquois Alliances in American History," in *The History and Culture of Iroquois Diplomacy: An Interdisciplinary Guide to the Treaties of the Six Nations and Their League*, ed. Francis Jennings et al. (Syracuse, NY: Syracuse University Press, 1985), 44–47; Francis Jennings, "Brother Miquon: Good Lord!" in *World of William Penn*, 196–205; David L. Preston, "Squatters, Indians, Proprietary Government, and Land in the Susquehanna Valley," in *Friends and Enemies in Penn's Woods*, 180–200; Susan Kalter, ed., *Benjamin Franklin, Pennsylvania, and the First Nations: The Treaties of 1736–62* (Urbana: University of Illinois Press, 2006). Recent analysis suggests Franklin's transcription of treaty negotiations, including Indians' statements, were generally accurate; see James H. Merrell, "'I Desire All That I Have Said . . . May Be Taken down Aright': Revisiting Teedyuscung's 1756 Treaty Council Speeches," *WMQ* 63, no. 4 (October 2006): 777–826.
63. William N. Fenton, "Structure, Continuity, and Change in the Process of Iroquois Treaty Making," *History and Culture of Iroquois Diplomacy*, 16–17; Jennings, "Glossary of Figures of Speech in Iroquois Political Rhetoric," *History and Culture of Iroquois Diplomacy*, 115–22; Alan Ojiig Corbiere, "'Their Own Forms of Which They Take the Most Notice': Diplomatic Metaphors and Symbolism on Wampum Belts," in *Anishinaabewin Niiwin: Four Rising Winds*, ed. Alan Ojiig Corbiere et al. (M'Chigeeng, ON: Ojibwe Cultural Foundation, 2014), 47–64.
64. Kalter, ed., *Treaty of Friendship* (1736), in *First Nations Treaties*, 51.
65. Kalter, ed., *Treaty of Friendship* (1736), in *First Nations Treaties*, 52, 53.
66. The Great Treaty of 1722, Albany, http://treatiesportal.unl.edu/earlytreaties/treaty.00001.html.
67. Francis Jennings, "The Scandalous Indian Policy of William Penn's Sons: Deeds and Documents of the Walking Purchase," *Pennsylvania History* 37, no. 1 (January 1970): 19–39; C. A. Weslager, *The Delaware Indians: A History* (New Brunswick, NJ: Rutgers University Press, 1989), 187–91; Jane T. Merritt, *At the Crossroads: Indians and Empires on a Mid-Atlantic Frontier, 1700–1763* (Williamsburg, VA, and Chapel Hill, NC: Omohundro Institute of Early American History and Culture, 2003), 46–49; Gregory Evans Dowd, *War Under Heaven: Pontiac, the Indian Nations, and the British Empire* (Baltimore: Johns Hopkins University Press, 2004), 35–37; Schutt, *Peoples of the River Valleys*, 81–89; Steven C. Harper, "Making History: Documenting the 1737 Walking Purchase," *Pennsylvania History: A Journal of Mid-Atlantic Studies* 77, no. 2 (Spring 2010): 217–33. Fraud inflected all colonial land acquisition; Colin G. Calloway, *Pen and Ink Witchcraft: Treaties and Treaty Making in American Indian History* (New York: 2013).
68. Jennings, "Scandalous Indian Policy," 19–39.
69. Schutt, *Peoples of the River Valleys*, 7–9.
70. BF, speech for Governor William Denny, November 12, 1756, in *PBF*, 7: 16–17; BF, speech for Denny, November 15, 1756, in *PBF*, 7: 21.
71. *The Pennsylvania Gazette*, April 17, 1740, July 18, 1745.
72. Schutt, *Peoples of the River Valleys*, 89–93; Dowd, *War Under Heaven*, 34–37.
73. William Penn, "Letter to the Free Society of Traders," August 16, 1683, in *William Penn and the Founding of Pennsylvania: A Documentary History*, ed. Jean R. Soderlund (Philadelphia: University of Pennsylvania Press, 1983), 314.
74. Alfred P. James, "Benjamin Franklin's Ohio Valley Lands," *Proceedings of the American Philosophical Society* 98, no. 4 (1954): 255–65.

3. The Pennsylvanian Fireplace

1. On the history of the home as a laboratory, see Deborah E. Harkness, "Managing an Experimental Household: The Dees of Mortlake and the Practice of Natural Philosophy," *Isis* 88,

no. 2 (June 1997): 247–62; Donald L. Opitz, Staffan Bergwik, and Brigitte Tiggelen, eds., *Domesticity in the Making of Modern Science* (Basingstoke, UK: Palgrave Macmillan, 2016).
2. "Philadelphia," *The Pennsylvania Gazette* (Philadelphia), March 5, 1741; "Philadelphia," *The Pennsylvania Gazette*, April 9, 1741; *Poor Richard, 1742. An Almanack . . .*, in *PBF*, 2: 333.
3. *The American Magazine; or, A Monthly View of the Political State of the British Colonies* (1741), with a bibliographical note by Lyon N. Richardson (New York: Columbia University Press, 1937), 115, 120; editorial note, *PBF*, 2: 301; *Poor Richard Improved, 1749*, in *PBF*, 3: 338.
4. *The Pennsylvania Gazette*, March 2, 1740; see also Sean Patrick Adams, "Warming the Poor and Growing Consumers: Fuel Philanthropy in the Early Republic's Urban North," *Journal of American History* 95 (June 2008): 69–94.
5. David Dickson, *Arctic Ireland: The Extraordinary Story of the Great Frost and Forgotten Famine of 1740–41* (Belfast: White Row Press, 1997); K. R. Briffa and P. D. Jones, "Unusual Climate in Northwest Europe During the Period 1730 to 1745 Based on Instrumental and Documentary Data," *Climatic Change* 79 (December 2006): 361–79; S. Engler et al., "The Irish Famine of 1740–1741: Famine Vulnerability and 'Climate Migration,'" *Climate of the Past* 9, no. 3 (2013): 1161–79. But see Michael Sigl et al., "Timing and Climate Forcing of Volcanic Eruptions for the Past 2,500 Years," *Nature* 523 (2015), fig. 3.
6. *Poor Richard Improved, 1748*, in *PBF*, 3: 245–48.
7. *The Pennsylvania Gazette*, February 11, 1752.
8. *The Pennsylvania Gazette*, January 3, 10, 17, 1765, February 7, 1765, March 28, 1765; James Parker to BF, January 14, 1765, in *PBF*, 12: 21.
9. Daniel George, *An Almanack . . .* (Boston, 1776); Nathanael Low, *An Astronomical Diary . . .* (Boston, 1776).
10. Edwin Wolf II, ed., *The Library of James Logan of Philadelphia, 1674–1751* (Philadelphia: Library Company of Philadelphia, 1974), 181; Mark Reinberger and Elizabeth McLean, *The Philadelphia Country House: Architecture and Landscape in Colonial America* (Baltimore: Johns Hopkins University Press, 2015), 217, 222–23.
11. Gary B. Nash, "Up from the Bottom in Franklin's Philadelphia," *Past and Present* 77 (1977): 57–83, esp. 75 n. 40; Gary B. Nash, *The Urban Crucible: Social Change, Political Consciousness, and the Origins of the American Revolution* (Cambridge, MA: Harvard University Press, 1979), 251; Tony C. Perry, "In Bondage When Cold Was King: The Frigid Terrain of Slavery in Antebellum Maryland," *Slavery and Abolition* 38, no. 1 (2017): 23–36; Office of State and Community Energy Programs, "Low-Income Energy Affordability Data (LEAD) Tool and Community Energy Solutions," https://www.energy.gov/scep/low-income-energy-affordability-data-lead-tool-and-community-energy-solutions.
12. Cotton Mather, *Winter Meditations: Directions How to Employ the Liesure of the Winter for the Glory of God* (Boston, 1693), 19.
13. Jared Sparks, ed., *The Works of Benjamin Franklin* (Boston: Hillard Gray, 1836), 2: 9n-10n; I. Bernard Cohen, *Franklin and Newton: An Inquiry into Speculative Newtonian Experimental Science and Franklin's Work in Electricity as an Example Thereof* (Philadelphia: American Philosophical Society, 1956), 216–22; Wolf, ed., *Library of James Logan*, 61–62; I. Bernard Cohen, *Benjamin Franklin's Science* (Cambridge, MA: Harvard University Press, 1990), 159–71.
14. Henry C. Mercer, *The Bible in Iron . . .*, ed. Joseph E. Sandford and Horace M. Mann, 3rd ed. (Doylestown, PA: Bucks County Historical Society, 1961), vii, 30–31, 82–88, 90, 95, 129–31, 255–56 (not all of the 409 illustrations are of stoves or stove plates).
15. February 6, 1737, Coventry Forge, ledger, 1736–1741, vol. 370, p. 102, coll. 212, HSP; account, BF with William Branson, 1737/38, LXVI, 73a, BF Papers, Part 13, APS.
16. *The Pennsylvania Gazette*, February 5, 1741, in *PBF*, 2: 316.
17. "To be Sold," *The Pennsylvania Gazette*, December 3, 1741, in *PBF*, 2: 331 ("Post-Office"); "Just came down . . . ," *The Pennsylvania Gazette*, January 20, 1742, in *PBF*, 2: 355 ("fresh Parcel"). On marketing and pricing, see editors' note, *PBF*, 2: 419–21.

18. *Benjamin Franklin's Autobiography*, ed. Joyce E. Chaplin (New York: W. W. Norton, 2012), 51, 58; "Collection 212: Forges and Furnaces Collection, 1727–1921), HSP, 18, 58, https://www.hsp.org/sites/default/files/legacy_files/migrated/findingaid212forgesandfurnaces.pdf; Thomas Maxwell Potts, ed., *Historical Collections Relating to the Potts Family in Great Britain and America* . . . (Canonsburg, PA: The Compiler, 1901), 282–86 (Potts family), 456–57 (Stephen Potts); Hannah Benner Roach, "Benjamin Franklin Slept Here," *Pennsylvania Magazine of History and Biography* 84 (1960): 127–74, esp. 141–42; entries for October 5, December 21, 1730, February 19, 1731, January 6, January 7, August 10–11, 1732, December 3, 1736, and reference to "Stephen's Book" for July 1731 (but under 1733), all in Ledger A, Miscellaneous Benjamin Franklin Collections (Misc. BF Collections), APS.
19. Coventry Forge, ledger, 1742–1748, vol. 371, p. 87 (quotation), coll. 212, HSP; entry for April 7, 1735 (lampblack for Thomas Potts), Ledger A, Misc. BF Collections, APS; Colebrook Dale Furnace, journal, 1735–1742, vol. 219, pp. 109, 168, ibid.; index list for "stove plates," (n. p.), Coventry Forge, ledger, 1727–1730, vol. 366, ibid.; Mercer, *Bible in Iron*, 109, 241–42. See also the "fireplace" shipped from Mt. Pleasant Forge, also a Potts property, on November 9, 1742, Mt. Pleasant Furnace, daybook, 1740–1745, vol. 634, p. 195, coll. 212, HSP.
20. Entries on "new fashioned" stove, Coventry Forge, ledger, 1742–1748, vol. 371, pp. 84, 85, 90, 112, coll. 212, HSP; Coventry Forge, ledger, 1742–1748, vol. 372, p. 90 ("Grace & Potts"), coll. 212, HSP; Mercer, *Bible in Iron*, 109, 241–42.
21. June 27, 1733 ("Paid Mr Grace 10 £ in part of his Bond"), Ledger A, Misc. BF Collections, APS; on May 19, 1739, the Franklins' business records include the entry "Landlord / Mr Grase," Shop Book, APS; see also Robert Grace to BF, December 30, 1745, in *PBF*, 3: 50–51; editorial note on Robert Grace, *PBF*, 3: 330n; Roach, "Benjamin Franklin Slept Here," 144–46; the Franklins moved out in 1748, though the printshop and post office remained on Market Street until January 1752. Grace may have helped sell the fireplaces from Coventry before the switch to Warwick; see entry for November 1, 1742, Coventry Forge, ledger, 1742–1748, vol. 371, p. 39, coll. 212, HSP, where a buyer returns a seven-plate "Fireplace" to Grace.
22. *The Pennsylvania Gazette*, December 6, 1739, in *PBF*, 2: 217 (almanac and newspaper prices); *The Pennsylvania Gazette*, November 15, 1744, in *PBF*, 2: 454–55 (pamphlet, soap); entry for December 17, 1744 (charge to Grace), Ledger D, Misc. BF Collections, APS; Claire Walsh, "The Advertising and Marketing of Consumer Goods in Eighteenth-Century London," *Advertising and the European City: Historical Perspectives*, ed. Clemens Wischermann and Elliott Shore (Burlington, VT: Ashgate, 2000), 79–95. On BF destroying documents, see Joyce E. Chaplin, *The First Scientific American: Benjamin Franklin and the Pursuit of Genius* (New York: Basic Books, 2006), 216.
23. Chaplin, *Benjamin Franklin's Autobiography*, 110 (quotation); BF, *An Account of the New Invented Pennsylvanian Fire-Places* . . . (Philadelphia, 1744), in *PBF*, 2: 419, 421; George Simpson Eddy, ed., *Account Books Kept by Benjamin Franklin, Ledger "D," 1739–1747* (New York: Columbia University Press, 1929), 61–63.
24. [Nicolas] Gauger, *Fires Improv'd: or, A new Method of Building Chiminies, so as to Prevent their Smoking* . . . , trans. J. T. Desaguliers, 2nd ed. (London, 1736), note on title page: "Given to the Library Company of Philadelphia by Rob.t Grace," rare books, Iu Gaug 399.O, Library Company of Philadelphia (LCP); [Nicolas] G[auger], *La Mechanique du feu, ou l'art d'en augmenter les effets, & d'en diminuer la dépense* (Amsterdam, 1714), Loganian Library, IU Gaug Log.259.D, LCP—Franklin would use the "Mechanique" spelling of this title when he cited Gauger. Logan owned a copy of another book Franklin consulted, [Robert Molesworth], *An Account of Denmark, as It Was in the Year 1692* (London, 1694), Loganian Library, Wing M 2382.A Log.403.O, LCP.
25. The 1741 catalog of the LCP includes a 1719 edition of Desaguliers, Grace's 1736 edition of *Fires Improv'd* (which, unlike that of 1715, mentioned the French title of the original, which was what Franklin cited), and a 1737 edition of Clare. See *A Catalogue of Books*

Belonging to the Library Company of Philadelphia (Philadelphia, 1741), 18–19, 37–38; Cohen, *Franklin and Newton*, 245–46, 261–66.
26. BF, *Pennsylvanian Fire-Places*, 421–22.
27. BF, *Pennsylvanian Fire-Places*, 422; A. William Hoglund, "Forest Conservation and Stove Inventors, 1789–1850," *Forest History* 5, no. 4 (Winter 1962): 2–8; William Cronon, *Changes in the Land: Indians, Colonists, and the Ecology of New England* (New York: Hill and Wang, 1983); Alan Taylor, "'Wasty Ways': Stories of American Settlement," *Environmental History* 3, no. 3 (July 1998): 291–310; Richard W. Judd, "A 'Wonderfull Order and Ballance': Natural History and the Beginnings of Forest Conservation in America, 1730–1830," *Environmental History* 11, no. 1 (January 2006): 8–36, which focuses mostly on the nineteenth century; John Lauritz Larson, *Laid Waste!: The Culture of Exploitation in Early America* (Philadelphia: University of Pennsylvania Press, 2020).
28. BF, *Pennsylvanian Fire-Places*, 428; [Nicolas Gauger], *La Mecanique du feu* . . . (Paris, 1713), vi.
29. Roach, "Benjamin Franklin Slept Here," 145, describes the building's footprint as seventeen feet wide by forty-five feet deep.
30. BF, *Pennsylvanian Fire-Places*, 422.
31. BF, *Pennsylvanian Fire-Places*, 422–23. On thermometers, cf. Samuel Y. Edgerton Jr., "Supplement: The Franklin Stove," in Cohen, *Benjamin Franklin's Science*, 200; Jan Golinski, *British Weather and the Climate of Enlightenment* (Chicago: University of Chicago Press, 2007), 114, 135.
32. BF, *Pennsylvanian Fire-Places*, 423.
33. BF, *Pennsylvanian Fire-Places*, 423.
34. BF, *Pennsylvanian Fire-Places*, 422; Steven Shapin and Simon Schaffer, *Leviathan and the Air-Pump: Hobbes, Boyle, and the Experimental Life* (Princeton, NJ: Princeton University Press, 1985), esp. 22–79.
35. BF, *Pennsylvanian Fire-Places*, 424–25.
36. BF, *Pennsylvanian Fire-Places*, 424, 425–27; Reinberger and McLean, *Philadelphia Country House*, 127.
37. BF, *Pennsylvanian Fire-Places*, 425–27. Franklin's "quotation" from Porzio paraphrases several longer pages; he wisely never claimed proficiency in Latin and might have had help from either Grace or James Logan. Luca Antonio Porzio, *De militis in castris sanitate tuenda* [*Concerning the Health of the Soldier in Camp*] (The Hague, 1739), with Franklin's "quotation" distilled from the preface, b^r–b2^r.
38. BF, *Pennsylvanian Fire-Places*, 425–27.
39. BF, *Pennsylvanian Fire-Places*, 428.
40. BF, *Pennsylvanian Fire-Places*, 428–29.
41. BF, *Pennsylvanian Fire-Places*, 429.
42. BF, *Pennsylvanian Fire-Places*, 429.
43. John Bunyan, *The Pilgrim's Progress*, ed. Roger Sharrock (London: Penguin, 1987), 75–76; Chaplin, *Benjamin Franklin's Autobiography*, 17; Mercer, *Bible in Iron*, 89–100; John Bezís-Selfa, *Forging America: Ironworkers, Adventurers, and the Industrious Revolution* (Ithaca, NY: Cornell University Press, 2004), 120–24; Rolf Peter Sieferle, *The Subterranean Forest: Energy Systems and the Industrial Revolution* (Cambridge: White Horse Press, 2001), 66.
44. Warwick Furnace, daybook, 1747–1748, vol. 920, pp. 432–33, coll. 212, HSP; Warwick Furnace, ledger, 1745–1747, vol. 929, p. 5, coll. 212, HSP; Warwick Furnace, fragments of ledger or daybook, entry for November 5, 1744, Pottstown Historical Society, Pottstown, PA.
45. BF, *Pennsylvanian Fire-Places*, 429.
46. BF, *Pennsylvanian Fire-Places*, 442–45 (the directions are on pp. 32–37 of the original pamphlet); entry for May 6, 1746, Thomas Hart paid £3, 7 shillings, and 6 pence in cash for Boerhaave, Ledger D, p. 138, Misc. BF Collections, APS.

47. BF, *Pennsylvanian Fire-Places*, 429–31, plate reproduced between 444 and 445.
48. BF, *Pennsylvanian Fire-Places*, 429–31. On the first model's lack of an airbox (the explanation supplied later), see Hugh Roberts to BF, November 27, 1765, in *PBF*, 12: 386–87.
49. BF, *Pennsylvanian Fire-Places*, 431.
50. BF, *Pennsylvanian Fire-Places*, 431–32 (quotation, 431), 442–43, 444.
51. BF, *Pennsylvanian Fire-Places*, 430, 432.
52. BF, *Pennsylvanian Fire-Places*, 431–32.
53. BF, *Pennsylvanian Fire-Places*, 432, 436–37, 445.
54. BF, *Pennsylvanian Fire-Places*, 432–33, 437; Sandra Cavallo, "Health, Air and Material Culture in the Early Modern Italian Domestic Environment," *Social History of Medicine* 29 (2016): 695–716.
55. BF, *Pennsylvanian Fire-Places*, 444.
56. BF, *Pennsylvanian Fire-Places*, 433, 442, 444.
57. BF, *Pennsylvanian Fire-Places*, 433–35; Adolph B. Benson, ed., *Kalm's Travels in North America*, vol. 2 (New York: Wilson-Erickson, 1937), 654 (firewood size).
58. BF to Jared Eliot, September 12, 1751, in *PBF*, 4: 194; entry for Lewis Evans, November 27, 1744, Ledger D, p. 241, Misc. BF Collections, APS.
59. BF, *Pennsylvanian Fire-Places*, 435–37, 438.
60. BF, *Pennsylvanian Fire-Places*, 438–39.
61. BF, *Pennsylvanian Fire-Places*, 438, 444.
62. BF, *Pennsylvanian Fire-Places*, 440–41.
63. BF, *Pennsylvanian Fire-Places*, 435, 437, 441.
64. BF, *Pennsylvanian Fire-Places*, 435.
65. BF, *Pennsylvanian Fire-Places*, 441.
66. BF, *Pennsylvanian Fire-Places*, 441.
67. Joyce E. Chaplin, *An Anxious Pursuit: Agricultural Innovation and Modernity in the Lower South, 1730–1815* (Williamsburg, VA, and Chapel Hill, NC: Institute of Early American History and Culture, 1993), 66–91; Conevery Bolton Valenčius, *The Health of the Country: How American Settlers Understood Themselves and Their Land* (New York: Basic Books, 2002); Katherine Johnston, *The Nature of Slavery: Environment and Plantation Labor in the Anglo-Atlantic World* (New York: Oxford University Press, 2022), esp. 101–23.
68. Coventry Forge, ledger, 1727–1730, vol. 366, list of Native people in index (n. p.), entries on pp. 79–83 (quotation, 83), coll. 212, HSP.
69. Colebrook Dale Furnace, journal, 1735–1742, vol. 219, pp. 115, 153 ("Indian Doctor"), coll. 212, HSP; Coventry Forge, ledger, 1736–1741, vol. 370, p. 89 ("Indian" paid 8 pence), ibid.; Colebrook Dale Furnace, daybook, 1734–1741, vol. 216, pp. 1, 17, 160, ibid.; Coventry Forge, daybook, 1746–1754, vol. 361, p. 234, ibid.; Warwick Furnace, daybook, 1750–1753, vol. 921, p. 440, ibid.
70. Coventry Forge, ledger, 1742–1748, vol. 372, pp. 105 ("pork man"), 74 ("Th.s Pott's Man"), 218 ("Servant"), 56, 58, 61, 79, 119, coll. 212, HSP; Coventry Forge, Robert Grace, ledger, 1744–1754, vol. 380, pp. 123 ("Labourer"), 84, ibid. For indentured servants and convicts at ironworks, see advertisements for runaway ironworkers: Edward McFarland, "Run away, on the 29th of November . . . ," *The Pennsylvania Gazette*, December 25, 1750, [3]; Amos Garret, "Run away from Cornwall iron-works . . . ," ibid., May 16, 1751, [2]; Richard Croxall, "Maryland, Patapsco, June 4, 1753," ibid., June 28, 1753, [3]; Croxall, "Eight Pounds Reward," ibid., Jan. 1, 1761, [3].
71. Coventry Forge, ledger 1742–1748, vol. 372, pp. 110 (the collier), 202 ("our Carter"), 247 (Meridith), coll. 212, HSP.
72. For Mills, see Coventry Forge, ledger 1742–1748, vol. 372, p. 80, coll. 212, HSP. See also BF account with Robert Grace, 1755, LXVI, 86, BF Papers, part 13, APS.
73. Coventry Forge, Robert Grace, ledger, 1744–1754, vol. 380, p. 64, coll. 212, HSP.

74. Anna Nutt, "To Be Sold or Let for a Term of Years," *The Pennsylvania Gazette*, July 19, 1744 [3] (quotation); [Isabella Batchelder] James, *Memorial of Thomas Potts, Junior* (Cambridge [MA]: private press, 1874), 44–46, 86–87, 112; Bezís-Selfa, *Forging America*, 6–8, 105–12; Gary B. Nash, "Franklin and Slavery," *Proceedings of the American Philosophical Society* 150, no. 1 (March 2006): 618–35, esp. 620. For women enslavers in the antebellum period, see Stephanie E. Jones-Rogers, *They Were Her Property: White Women as Slave Owners in the American South* (New Haven: Yale University Press, 2020).
75. For scattered references to "Negro" and "Mulatto" workers, see Colebrook Dale Furnace, journal, 1735–1742, vol. 219, coll. 212, HSP. For Ishmael, see Colebrook Dale Furnace, ledger, 1740–43, vol. 221, p. 3, ibid. See also 1748 account for "Negro Andrew," with expenses deducted against work payments, Warwick Furnace, ledger, 1745–1747, vol. 929, p. 15, ibid.; 1751 payments for "Negro York," Warwick Furnace, daybook, 1750–1752, vol. 921, p. 227, ibid., likely the same arrangement.
76. Coventry Forge, daybook, 1746–1754, vol. 361, p. 171 (list, including "Little George"), coll. 212, HSP; Coventry Forge, ledger, 1727–1730, vol. 366, p. 51 (Will), ibid.; Gary B. Nash and Jean R. Soderlund, *Freedom by Degrees: Emancipation in Pennsylvania and Its Aftermath* (New York: Oxford University Press, 1991), 28–29, 37, 39. On effacement of Black women, see Jennifer L. Morgan, *Reckoning with Slavery: Gender, Kinship, and Capitalism in the Early Black Atlantic* (Durham, NC: Duke University Press, 2021).
77. Ronald L. Lewis, *Coal, Iron, and Slaves: Industrial Slavery in Maryland and Virginia, 1715–1865* (Westport, CT: Greenwood Press, 1979), 25–26; Nash and Soderlund, *Freedom by Degrees*, 12–13; Guy Standing, *The Precariat: The New Dangerous Class* (London: Bloomsbury, 2014); Shaun Armstead et al., "'And I Poor Slave Yet': The Precarity of Black Life in New Brunswick, 1766–1835," in *Scarlet and Black: Slavery and Dispossession in Rutgers History*, ed. Marisa J. Fuentes and Deborah Gray White (New Brunswick: NJ: Rutgers University Press, 2011), 91–122; Sidney Chalhoub, "The Precariousness of Freedom in a Slave Society (Brazil in the Nineteenth Century)," *International Review of Social History* 56, no. 3 (2011): 405–39; Franco Barchiesi, "Precarity as Capture: A Conceptual Deconstruction of the Worker-Slave Analogy," in *On Marronage: Ethical Confrontations with Antiblackness*, ed. P. Khalil Saucier and Tryon P. Woods (Trenton, NJ: Africa World Press, 2015); Beatriz G. Mamigonian and Keila Grinberg, "The Crime of Illegal Enslavement and the Precariousness of Freedom in Nineteenth-Century Brazil," in *The Boundaries of Freedom: Slavery, Abolition, and the Making of Modern Brazil*, ed. Brodwyn Fischer and Keila Grinberg (New York: Cambridge University Press, 2022), 35–56.
78. Coventry Forge, Robert Grace, ledger, 1744–1754, vol. 380, pp. 88 ("Privateer"), 171 ("Driving"), 172 ("Washing"), coll. 212, HSP; Warwick Furnace, ledger, 1745–1747, vol. 929, p. 158 (woodcutter), ibid.; Coventry Forge, ledger, 1734–1740, vol. 369, p. 102 (Ben), ibid.
79. Coventry Forge, daybook, 1746–1754, vol. 361, p. 90 (cash), coll. 212, HSP; Coventry Forge, ledger, 1727–1730, vol. 366, lists for "Servants" and for "Negroes," n.p., ibid.; Warwick Furnace, daybook, 1747–1748, vol. 920, account lists, ibid. And see Tony C. Perry, "In Bondage When Cold Was King: The Frigid Terrain of Slavery in Antebellum Maryland," *Slavery and Abolition* 38, no. 1 (March 2017): 23–36.
80. For Cesar and Streaphon, see Coventry Forge, Robert Grace, ledger, 1744–1754, vol. 380, p. 61 (quotations), coll. 212, HSP. For the two Cesars, see also Coventry Forge, ledger, 1742–1748, vol. 372, p. 230, ibid. On the continuing history of manufactures (especially textiles) produced for enslaved people, see Seth Rockman, *Plantation Goods: A Material History of American Slavery* (Chicago: University of Chicago Press, 2024).
81. Coventry Forge, Robert Grace ledger, vol. 380, p. 61, coll. 212, HSP; Coventry Forge, ledger 1742–1748, vol. 371, p. 39, ibid. See Shane White and Graham White, "Slave Clothing and African-American Culture in the Eighteenth and Nineteenth Centuries,"

Past and Present 148 (1995): 149–86; David Waldstreicher, "Reading the Runaways: Self-Fashioning, Print Culture, and Confidence in Slavery in the Eighteenth-Century Mid-Atlantic," *WMQ* 56, no. 2 (April 1999): 243–72.

82. John Potts Ledger, 1739–1740, vol. 693, pp. 139, 253, coll. 212, HSP; Coventry Forge, ledger, 1742–1748, vol. 372, p. 230 ("Old Jenny"), ibid.; Coventry Forge, ledger, 1736–1741, vol. 370, p. 151 ("Betty's grave"), ibid. For another reference to Jenny, see Coventry Forge, ledger 1742–1748, vol. 371, p. 39, ibid.

83. Warwick Furnace, ledger, 1745–1747, vol. 929, pp. 118 (Streaphon's payment), 221 ("Wages"), coll. 212, HSP; on former enslavers as continuing employers, see Orlando Patterson, *Slavery and Social Death: A Comparative Study* (Cambridge, MA: Harvard University Press, 1982), 240–47.

84. McCusker and Menard, *Economy of British America*, 58–61.

85. Bezís-Selfa, *Forging America*, 107–20; the fireback/mantelpiece are in Winterthur Museum's Philadelphia Empire Bedroom.

86. BF, *Pennsylvanian Fire-Places*, 439.

87. Nash, "Franklin and Slavery," 618–35; David Waldstreicher, *Runaway America: Benjamin Franklin, Slavery, and the American Revolution* (New York: Farrar, Straus and Giroux, 2005), 145–74.

4. ATMOSPHERES

1. Henry C. Mercer, *The Bible in Iron* . . . , ed. Joseph E. Sandford and Horace M. Mann, 3rd ed. (Doylestown, PA: Bucks County Historical Society, 1961), 109–10, 242; Samuel Y. Edgerton Jr., "Heating Stoves in Eighteenth Century Philadelphia," *Bulletin of the Association for Preservation Technology* 3 no. 2/3 (1971): 15–104, esp. 42; October 24, 1751, Warwick Furnace, daybook, 1750–1752, vol. 921, p. 469, coll. 212, HSP. See also Warwick Furnace, ledger, 1745–1747, vol. 929, p. 34, coll. 212, HSP, for four fireplaces cast on order for specific buyers or merchants in 1745.

2. The Miscellaneous BF Collections (Misc. BF Collections) at the APS include records of Franklin's selling fireplaces or stoves to individuals (Ledger D, pp. 50, 61, 85, 155) or advancing them on account to or through his and Grace's distributors (Ledger D, pp. 205, 365, unpaginated leaf between 365–66, pp. 386, 395, 400), though not all of these may be his model. Mercer, *Bible in Iron*, 242.

3. Claire Walsh, "The Advertising and Marketing of Consumer Goods in Eighteenth-Century London," in *Advertising and the European City: Historical Perspectives*, ed. Clemens Wischermann and Elliott Shore (Burlington, VT: Ashgate, 2000), 80.

4. Priscilla J. Brewer, *From Fireplace to Cookstove: Technology and the Domestic Ideal in America* (Syracuse, NY: Syracuse University Press, 2000), 26, 35–36.

5. Mercer, *Bible in Iron*, 39–47, 82; Peter Ward-Jackson, *English Furniture Designs of the Eighteenth Century* (London: HM Stationery Office, 1958), 6–8, plate 30; Morrison H. Heckscher, *American Rococo, 1750–1775: Elegance in Ornament* (New York: Metropolitan Museum of Art; [Los Angeles]: Los Angeles County Museum of Art, 1992), 220.

6. BF, *An Account of the New Invented Pennsylvanian Fire-Places* . . . (Philadelphia, 1744), in *PBF*, 2: 445–46; January, *Poor Richard, 1748. An Almanack* . . . , in *PBF*, 3: 249.

7. On excluding Indians from anything modern, see Jean M. O'Brien, *Firsting and Lasting: Writing Indians out of Existence in New England* (Minneapolis: University of Minnesota Press, 2010).

8. *Boston Evening-Post*, September 8, 1746.

9. [John Bartram], "An Essay for the Improvement of Estates . . . ," *Poor Richard Improved, 1749*, in *PBF*, 3: 331.

10. [Bartram], "Essay for the Improvement of Estates," 331–32.
11. [Bartram], "Essay for the Improvement of Estates," 331–332, 334.
12. [Bartram], "Essay for the Improvement of Estates," 332–34.
13. BF, *Pennsylvanian Fire-Places*, 420–21; George S. Eddy, ed., *Account Books Kept by Benjamin Franklin, Ledger "D," 1739–1747* (New York: Columbia University Press, 1929), 61–63; Peter Franklin, account with Robert Grace, 1744–1745, in Peter Franklin Ledger, APS; *New-York Weekly Post-Boy*, September 11, 1746; *Boston News-Letter* February 7, 1745. On Colden, see John M. Dixon, *The Enlightenment of Cadwallader Colden: Empire, Science, and Intellectual Culture in British New York* (Ithaca, NY: Cornell University Press, 2016).
14. Cadwallader Colden to Johannes Fredericus Gronovius, December 1744, in *Letters and Papers of Cadwallader Colden*, vol. 3: *1743–1747* (New York: New-York Historical Society, 1919), 91; Gronovius to Bartram, June 2, 1746, and June 10, 1754, in *Memorials of John Bartram and Humphry Marshall: With Notices of Their Botanical Contemporaries*, comp. William Darlington (Philadelphia: Lindsay and Blakiston, 1849), 355, 363; [BF], *Beschreivinge van die Nieuwe Uitgevondene Pensilvanische Schoorsteenen . . .* (Leiden, 1746).
15. BF, *Nieuwe Uitgevondene Pensilvanische Schoorsteenen*, 5n (I am grateful to Hans Martin Pech for his help in translating this material); Jan de Vries and Ad van der Woude, *The First Modern Economy: Success, Failure, and Perseverance of the Dutch Economy, 1500–1815* (Cambridge: Cambridge University Press, 1997), 37–40; Astrid Kander, Paolo Malanima, and Paul Warde, *Power to the People: Energy in Europe over the Last Five Centuries* (Princeton, NJ: Princeton University Press, 2013), 112–14.
16. De Vries and van der Woude, *First Modern Economy*, 270–329, 578; E. A. Wrigley, *Energy and the English Industrial Revolution* (Cambridge: Cambridge University Press, 2010), 95, 174.
17. BF, *Nieuwe Uitgevondene Pensilvanische Schoorsteenen*, 8–11, 30; Wrigley, *Energy and the English Industrial Revolution*, 221–22.
18. Adolph B. Benson, ed., *Peter Kalm's Travels in North America*, 2 vols. (New York: Wilson-Erickson, 1937), 1: vii–xvi; 2: 235–36, 652–54. See also Lisbet Koerner, *Linnaeus: Nature and Nation* (Cambridge, MA: Harvard University Press, 1999), 113–14, 117–18; Kenneth Nyberg, "Linnaeus's Apostles and the Globalization of Knowledge, 1729–1756," in *Global Scientific Practice in an Age of Revolutions, 1750–1850: Discussing the Contingency/Inevitability Problem*, ed. Patrick Manning and Daniel Rood (Pittsburgh: University of Pittsburgh Press, 2016), 73–89; Hanna Hodacs, Kenneth Nyberg, and Stéphane Van Damme, eds., *Linnaeus, Natural History, and the Circulation of Knowledge* (Oxford: Voltaire Foundation, 2018).
19. Benson, ed., *Kalm's Travels*, 2: 235–36, 652–54.
20. Benson, ed., *Kalm's Travels*, 2: 653–54.
21. Pehr Kalm, *En resa til norra America: på Kongl. swenska wetenskaps academiens befallning, och publici kostnad, förrättad af Pehr Kalm . . .* , 3 vols. (Stockholm, 1753–1761), 2: dated entries for December 3, 5, 6, 7, 9, 1748, but missing December 8, where Kalm begins his description of the Pennsylvanian fireplace in his manuscript; cf. John Bezís-Selfa, *Forging America: Ironworkers, Adventurers, and the Industrious Revolution* (Ithaca, NY: Cornell University Press, 2004), 113.
22. Ulf Sundberg, "Ecological Economics of the Swedish Baltic Empire: An Essay on Energy and Power, 1560–1720," *Ecological Economics* 5, no. 1 (March 1992): 51–72; Per Eliasson and Sven G. Nilsson, "'You Should Hate Young Oaks and Young Noblemen': The Environmental History of Oaks in Eighteenth- and Nineteenth-Century Sweden," *Environmental History* 7, no. 4 (October 2002): 659–77; Michael Williams, *Deforesting the Earth: From Prehistory to Global Crisis* (Chicago: University of Chicago Press, 2003), 416–17; Matti Enbuske, "Lapland's Taxation as a Reflection of 'Otherness' in the Swedish Realm in the 17th and 18th Centuries: Colonialism, or a Priority Right of the Sami People?" in *Facing Otherness in Early Modern Sweden: Travel, Migration and Material Transformations, 1500–1800*, ed. Magdalena Naum and Fredrik Ekengren (Woodbridge, UK: Boydell and Brewer, 2018), 227–37.

23. Magnus Lindmark and Lars Fredrik Andersson, "Household Firewood Consumption in Sweden During the Nineteenth Century," *Journal of Northern Studies* 4, no. 2 (2010): 55–78; Magnus Lindmark and Fredrik Olsson Spjut, "From Organic to Fossil and In-between: New Estimates of Energy Consumption in the Swedish Manufacturing Industry During 1800–1913," *Scandinavian Economic History Review* 66, no. 1 (2018): 18–33.
24. Friedrich Solmsen, *Aristotle's System of the Physical World: A Comparison with His Predecessors* (Ithaca, NY: Cornell University Press, 1960), 397–420; Mary Louise Gill, "The Theory of the Elements in *De Caelo* 3 and 4," *New Perspectives on Aristotle's De Caelo*, ed. Alan C. Bowen and Christian Wildberg (Boston: Brill, 2009), 123–24; Mohan Matthen, "Why Does Earth Move to the Center? An Examination of Some Explanatory Strategies in Aristotle's Cosmology," *New Perspectives on Aristotle's De Caelo*, 142–46; Theokritos Kouremenos, *Heavenly Stuff: The Constitution of the Celestial Objects and the Theory of Homocentric Spheres in Aristotle's Cosmology* (Stuttgart: Franz Steiner Verlag, 2010), 18–19, 20–24, 62–66, 113–14; Craig Martin, "The Invention of Atmosphere," *Studies in History and Philosophy of Science* 52 (August 2015): 44–54.
25. BF to Cadwallader Colden, August 15, 1745, in *PBF*, 3: 33–34.
26. BF to John Perkins, February 4, 1753, in *PBF*, 4: 429–30, 433–34.
27. Note for July 20, 1749, Ledger D, p. 395, APS; BF to William Strahan, November 23, 1748, in *PBF*, 3: 327. The catalog for the library of the Royal Society of London, https://catalogues.royalsociety.org/calmview/librarysearch.aspx?src=CalmView.CatalogL (accessed September 22, 2024) records no copy of Franklin's *Account of the New Invented Pennsylvanian Fire-Places.*
28. BF to John Mitchell, April 29, 1749, in *PBF*, 3: 365–76 (quotation 372).
29. *Poor Richard Improved, 1753*, in *PBF*, 4: 403; BF to Cadwallader Colden, December 6, 1753, in *PBF*, 5: 146–47.
30. BF to Peter Collinson, May 25, 1747, in *PBF*, 3: 129; BF to Collinson, July 28, 1747, in *PBF*, 3: 157; BF, "Opinions and Conjectures," [July 29, 1750], in *PBF*, 4: 11; Joyce E. Chaplin, *The First Scientific American: Benjamin Franklin and the Pursuit of Genius* (New York: Basic Books, 2006), 103–32.
31. BF to Peter Collinson, April 29, 1749, in *PBF*, 3: 359–60; entries for 1748 and 1749, freight charges to lottery manager, Benjamin Eastburn, for dollars from Boston, New York, Rhode Island, Ledger D, 156, Misc. BF Collections, APS; W. G. Sumner, "The Spanish Dollar and the Colonial Shilling," *American Historical Review* 3 (July 1898): 607–19, esp. 613.
32. BF to Collinson, April 29, 1749, in *PBF*, 3: 360–61.
33. BF to Collinson, April 29, 1749, in *PBF*, 3: 361–62; Joel Mokyr, *The Lever of Riches: Technological Creativity and Economic Progress* (New York: Oxford University Press, 1990), 122–23.
34. "Extracts from the Gazette," August 12, 1736, April 29, 1742, August 25, 1743, *The Pennsylvania Gazette*, in *PBF*, 2: 160, 360, 391.
35. BF to Peter Collinson, June 29, 1751, in *PBF*, 4: 143–45; keyword searches for "atmosphere," using digitized copies, in *Eighteenth Century Collections Online* (Gale Cengage), of the *Philosophical Transactions*, vols. 46 (1752) and 47 (1753), consulted July 28, 2022.
36. Chaplin, *First Scientific American*, 124–28.
37. BF, "Physical and Meteorological Observations, Conjectures, and Suppositions," c. 1751, in *PBF*, 4: 235–43; BF to Cadwallader Colden, April 23, 1752, in *PBF*, 4: 298–99.
38. Jill Lepore, *New York Burning: Liberty, Slavery, and Conspiracy in Eighteenth-Century Manhattan* (New York: Penguin Random House, 2006); Hannah Benner Roach, "Benjamin Franklin Slept Here," *Pennsylvania Magazine of History and Biography* 84 (1960): 127–74, esp. 160.
39. *Poor Richard Improved, 1753*, in *PBF*, 4: 408–409.
40. Immanuel Kant, *Gesammelte Schriften . . .* , vol. 1 (Berlin: G. Reimer, 1902–[1997]), 472; Sergio Perosa, "Franklin to Frankenstein: A Note on Lightning and Novels," *Scienza e*

immaginazione della cultura Ingelese del settecento, ed. Sergio Rossi (Milan: Edizioni Aseq, 1987), 321–28.

41. J[ames] Durno, *A Description of a New-Invented Stove-Grate* . . . (London, 1753), esp. 3–24 ("Island," p. 3, "Stove-Grate," p. 19), 31; BF, *Pennsylvanian Fire-Places* in *PBF*, 2: 421 ("Northern Colonies"); John E. Crowley, *The Invention of Comfort: Sensibilities and Design in Early Modern Britain and Early America* (Baltimore: Johns Hopkins University Press, 2001), 182–84; Colin T. Ramsey, "Stealing Benjamin Franklin's Stove: A New Identification for the 'Ironmonger in London,'" *ANQ* 20, no. 2 (2007): 25–30, though this doesn't note Durno's stress on type of fuel.
42. Durno, *New-Invented Stove-Grate*, 10–19.
43. Durno, *New-Invented Stove-Grate*, 19–24, 31.

5. Grow or Die

1. BF, *Observations Concerning the Increase of Mankind* . . . (1755), in *PBF*, 4: 233.
2. Franklin's significance in embedding population into modern economic theory is well established: Dennis Hodgson, "Benjamin Franklin on Population: From Policy to Theory," *Population and Development Review* 17, no. 4 (1991): 639–61; Joyce E. Chaplin, *Benjamin Franklin's Political Arithmetic: A Materialist View of Humanity* (Washington, DC: Smithsonian Institution, 2009); Alison Bashford and Joyce E. Chaplin, *The New Worlds of Thomas Robert Malthus: Rereading the Principle of Population* (Princeton, NJ: Princeton University Press, 2016), 43–47, 51–53, 70–72; Ted McCormick, *Human Empire: Mobility and Demographic Thought in the British Atlantic World, 1500–1800* (New York: Cambridge University Press, 2022). But scholars still overlook Franklin's role and his colonial perspective on abundance and scarcity: Deborah Valenze, *The Invention of Scarcity: Malthus and the Margins of History* (New Haven: Yale University Press, 2023), 100, only considers Franklin via the scholarship of others, not by examining his own work; Fredrik Albritton Jonsson and Carol Wennerlind, *Scarcity: A History from the Origins of Capitalism to the Climate Crisis* (Cambridge, MA: Harvard University Press, 2023), doesn't consider Franklin at all. The emphasis in these works is on how European theorists were the architects of modern economics, eliding the experience and observations of anyone in European empires, including non-European people; neither of these last two books engages with Jennifer L. Morgan, *Reckoning with Slavery: Gender, Kinship, and Capitalism in the Early Black Atlantic* (Durham, NC: Duke University Press, 2021).
3. *A Collection of All the Statutes Now in Force: Relating to the Revenue and Officers of the Customs in Great Britain and the Plantations* (London: C. Eyre and W. Strahan, 1780), 2: 976–79; Arthur Cecil Bining, *Pennsylvania Iron Manufacture in the Eighteenth Century* (Harrisburg, PA: Pennsylvania Historical and Museum Commission, 1973), 139–44; statement of account, Benjamin Franklin and Robert Grace, July 14, 1755, LXVI, 86, BF Papers, Part 13, APS.
4. Printed in [William Clarke], *Observations on the Late and Present Conduct of the French* . . . *To which is added* . . . [BF], *Observations Concerning the Increase of Mankind* . . . (Boston, 1755), in *PBF*, 4: 227–29.
5. BF, *Increase of Mankind*, 230, 231, 233.
6. BF, *Increase of Mankind*, 228, 230.
7. Eric Hinderaker, *Elusive Empires: Constructing Colonialism in the Ohio Valley, 1673–1800* (New York: Cambridge University Press, 1997); James H. Merrell, "Shamokin, 'the Very Seat of the Prince of Darkness': Unsettling the Early American Frontier," in *Contact Points: American Frontiers from the Mohawk Valley to the Mississippi, 1750–1830*, ed. Andrew R. L. Cayton and Fredrika J. Teute (Williamsburg, VA, and Chapel Hill, NC: Omohundro

Institute of Early American History and Culture, 1998), 16–59, quotation p. 29; Amy C. Schutt, *Peoples of the River Valleys: The Odyssey of the Delaware Indians* (Philadelphia: University of Pennsylvania Press, 2007), 94–103.
8. Merrell, "Shamokin," 46, 54, 55, 58–59.
9. BF, *Increase of Mankind*, 230, 231.
10. BF, *Increase of Mankind*, 231, 233, 234; on the possibility of multiple inhabited planets, see BF, "Articles of Belief and Acts of Religion," November 20, 1728, in *PBF*, 1: 102.
11. BF, *Increase of Mankind*, 233; John Rieder, *Colonialism and the Emergence of Science Fiction* (Middletown, CT: Wesleyan University Press, 2008); Christopher F. Loar, *Political Magic: British Fictions of Savagery and Sovereignty, 1650–1750* (New York: Fordham University Press, 2014); Chris Pak, *Terraforming: Ecopolitical Transformations and Environmentalism in Science Fiction* (Liverpool: Liverpool University Press, 2016).
12. BF to Peter Collinson, May 9, 1753, in *PBF*, 4: 482.
13. Chaplin, *Franklin's Political Arithmetic*.
14. Chaplin, *Franklin's Political Arithmetic*.
15. Gary B. Nash, "Slaves and Slaveholders in Colonial Philadelphia," *WMQ* 30, no. 2 (April 1973) 223–56, esp. 243, 237, 244; Gary B. Nash, "Franklin and Slavery," *Proceedings of the American Philosophical Society* 150, no. 1 (March 2006): 618–35; "Heads of Families at the First Census, 1790," https://www2.census.gov/prod2/decennial/documents/1790m-02.pdf, p. 8.
16. John Bezís-Selfa, *Forging America: Ironworkers, Adventurers, and the Industrious Revolution* (Ithaca, NY: Cornell University Press, 2004), 6–8, 105–12.
17. Peter H. Lindert and Jeffrey G. Williamson, "American Colonial Incomes, 1650–1774," National Bureau of Economic Research (NBER) Working Paper 19861 (2014), http://www.nber.org/papers/w19861; Glenda Taylor and Evelyn Auer, "Heat Pump Cost Guide 2024: Installation and Replacement," *Bob Vila*, last updated April 18, 2024, https://www.bobvila.com/articles/heat-pump-cost/; "QuickFacts: Philadelphia County, Pennsylvania," United States Census Bureau, accessed September 22, 2024, https://www.census.gov/quickfacts/fact/table/philadelphiacountypennsylvania/AFN120217.
18. Coventry Forge, ledger, vol. 371, p. 71, coll. 212: Forges and Furnace, HSP; Henry C. Mercer, *The Bible in Iron* . . . (Doylestown, PA: Bucks County Historical Society, 1914), 241; wood-to-iron calculations based on Rolf Peter Sieferle, *The Subterranean Forest: Energy Systems and the Industrial Revolution* (Cambridge: White Horse Press, 2001), 63–64. See also Michael Williams, *Americans and Their Forests: A Historical Geography* (Cambridge: Cambridge University Press, 1989), 106; Strother E. Roberts, *Colonial Ecology, Atlantic Economy: Transforming Nature in Early New England* (Philadelphia: University of Pennsylvania Press, 2019), 104. I am grateful to Clint Flack, Exhibit Specialist and Preparator, Mercer Museum and Fonthill Castle, for data on the surviving stove.
19. Bining, *Pennsylvania Iron Manufacture*, 24; John J. McCusker and Russell R. Menard, *The Economy of British America, 1607–1789* (Williamsburg, VA, and Chapel Hill, NC: Institute of Early American History and Culture, 1985), 203.
20. R. V. Reynolds and Albert H. Pierson, *Fuel Wood Used in the United States, 1630–1930* (Washington, DC: United States Department of Agriculture, 1942), 3, 5–6.
21. John Evelyn, *Silva: Or, a Discourse of Forest-Trees*, 5th ed. (London, 1729), LCP *Il Evel Log.422.F, Library Company of Philadelphia; *The Pennsylvania Gazette*, August 18, 1748, August 25, 1748, August 20, 1752, October 4, 1753, November 6, 1760, November 20, 1760.
22. BF to Jared Eliot, February 13, 1750, in *PBF*, 3: 463–65.
23. Geo[rge] Hadley, "Concerning the Cause of the General Trade-Winds," *Philosophical Transactions* 39 (1735): 58–62; BF to Jared Eliot, February 13, 1750, 464; Anna Neill, "Buccaneer Ethnography: Nature, Culture, and Nation in the Journals of William Dampier," *Eighteenth-Century Studies* 33, no. 2 (Winter 2000): 165–80.

24. John Bartram, *Observations on the Inhabitants, Climate, Soil, Rivers, Productions, Animals, and Other Matters Worthy of Notice* . . . (London, 1751), 34, 72–74; Thomas P. Slaughter, *The Natures of John and William Bartram* (New York: Knopf, 1996), 71–72.
25. Bartram, *Observations*, 39–40, 45, 66, 71.
26. Slaughter, *Natures of John and William Bartram*, 47–48.
27. Lewis Evans, *Geographical, Historical, Political, Philosophical, and Mechanical Essays* . . . (Philadelphia, 1755), 5, 6–7; Walter Klinefelter, "Lewis Evans and His Maps," *Transactions of the American Philosophical Society*, n. s., 61 (1971): 3–65, esp. 14–16.
28. Evans, *Essays*, 11, 13, 14–15, 32; Schutt, *Peoples of the River Valleys*, 103–14.
29. Lewis Evans, "A Map of Pensilvania, New-Jersey, New-York, and the Three Delaware counties ([Philadelphia], 1749), https://www.loc.gov/item/74691938; BF to William Franklin, [September 27, 1766], in *PBF*, 13: 424–25; Klinefelter, "Lewis Evans and His Maps," 17–24.
30. Instructions to Lewis Evans (1750), in *The Statutes at Large of Pennsylvania from 1682 to 1801*, comp. James T. Mitchell and Henry Flanders ([Harrisburg]: State Printer of Pennsylvania, 1896–), 2: 48; Lewis Evans et al., *A General Map of the Middle British Colonies, in America . . . Wherein is Also Shewn the Antient and Present Seats of the Indian Nations* ([Philadelphia], 1755), https://www.loc.gov/item/gm71005449; *The Pennsylvania Gazette*, July 17, 1755; Klinefelter, "Lewis Evans and His Maps," 38–50.
31. Joyce E. Chaplin, *Subject Matter: Technology, the Body, and Science on the Anglo-American Frontier, 1500–1676* (Cambridge, MA: Harvard University Press, 2001), esp. 157–98.
32. *Poor Richard, 1744. An Almanack* . . . , in *PBF*, 2: 395.
33. Ezra Stiles to BF, March 30, 1758, in *PBF*, 7: 394–95; BF to John Lining, June 17, 1758, in *PBF*, 8: 111 ("Would not"); I. Bernard Cohen, *Franklin and Newton: An Inquiry into Speculative Newtonian Experimental Science and Franklin's Work in Electricity as an Example Thereof* (Philadelphia: American Philosophical Society, 1956), 266–79; D. G. C. Allan and R. E. Schofield, *Stephen Hales: Scientist and Philanthropist* (London: Scolar Press, 1980).
34. BF to Ezra Stiles, May 29, 1763, in *PBF*, 10: 264–65.
35. "Journal of the Proceedings of the Conference Held at Albany, in 1754," *Collections of the Massachusetts Historical Society*, ser. 3, vol. 5 (1836), esp. 37, 41, 42, 43, 61 (Native references to council fires), 45, 48, 61 (English references).
36. *Poor Richard Improved, 1757*, in *PBF*, 6: 74.
37. *The Pennsylvania Gazette* (Philadelphia), January 10, 1760; Gary B. Nash, *The Urban Crucible: Social Change, Political Consciousness, and the Origins of the American Revolution* (Cambridge, MA: Harvard University Press, 1979), 251.
38. Gottlieb Mittelberger, *Journey to Pennsylvania in the Year 1750* . . . (1756), trans. Carl Theodor Eben (Philadelphia: John Joseph McVey, 1898), 104.
39. Gary B. Nash, "Up From the Bottom in Franklin's Philadelphia," *Past and Present*, 77 (1977): 57–83, esp. 75 n. 40.
40. Gary B. Nash, "Poverty and Poor Relief in Pre-Revolutionary Philadelphia," *WMQ*, 1, no. 33 (January 1976): 3–30, esp. 12–13; Nash, *Urban Crucible*, 251; Nash, *Forging Freedom: The Formation of Philadelphia's Black Community, 1720–1840* (Cambridge, MA: Harvard University Press, 1988), 1–11.
41. Charles E. Hunter, "The Delaware Nativist Revival of the Mid-Eighteenth Century," *Ethnohistory* 18, no. 1 (Winter 1971): 39–49; Alfred A. Cave, "The Delaware Prophet Neolin: A Reappraisal," *Ethnohistory* 46, no. 2 (Spring 1999): 265–90; Gregory Evans Dowd, *War Under Heaven: Pontiac, the Indian Nations, and the British Empire* (Baltimore: Johns Hopkins University Press, 2002), 101–105.
42. Schutt, *Peoples of the River Valleys*, 114–23, 126–49.
43. C. A. Weslager, *The Delaware Indians: A History* (New Brunswick, NJ: Rutgers University Press, 1989), 196–220 (the author's language about Natives is outdated); Jane T. Merritt, *At*

the Crossroads: Indians and Empires on a Mid-Atlantic Frontier, 1700–1763 (Williamsburg, VA, and Chapel Hill, NC: Omohundro Institute of Early American History and Culture, 2003), 235–63.
44. Anthony F. C. Wallace, *King of the Delawares: Teedyuscung, 1700–1763* (Philadelphia: University of Pennsylvania Press, 1949), 157, 176–89, 216, 222, 228–30, 236; Susan Kalter, ed., *Benjamin Franklin, Pennsylvania, and the First Nations: The Treaties of 1736–62* (Urbana, IL: University of Illinois Press, 2006), 208, 265–66, 269–71; Jean R. Soderlund, *Lenape Country: Delaware Valley Society before William Penn* (Philadelphia: University of Pennsylvania Press, 2015), 188–95. Teedyuscung probably referred to "his" land to conform to whites' ideas about personal property and political leadership.
45. Merritt, *At the Crossroads*, 256–63.
46. Kalter, ed., *First Nations Treaties*, 197.
47. Keyword search of "Treaty of Friendship," 1736, in electronic edition of Kalter, ed., *First Nations Treaties* (Project Muse); the triad of concepts is cited on pp. 53, 58, 59.
48. Kalter, ed., *First Nations Treaties*, 221 (denunciation of the Walking Purchase); James H. Merrell, "'I Desire All That I Have Said . . . May Be Taken Down Aright': Revisiting Teedyuscung's 1756 Treaty Council Speeches," *WMQ* 63, no. 4 (October 2006): 777–826, esp. 803.
49. Peace Medal (1757) by Edward Duffield for the Friendly Association, http://digitalpaxton.org/works/digital-paxton/media/peace-medal-1757?path=art; keyword searches in Kalter, ed., *First Nations Treaties*, electronic edition (Project Muse), with results that omit use of these terms to describe ordinary things and in the scholarly notes; see also the Appendix in Kalter's book.
50. Kalter, ed., *First Nations Treaties*, 292, 328–31.
51. Kalter, ed., *First Nations Treaties*, 380, 386, 387, 398 (quotation), 400, 402.
52. Kalter, ed., *First Nations Treaties*, 381–82.
53. Wallace, *King of the Delawares*, 258–61.
54. Jon William Parmenter, "Pontiac's War: Forging New Links in the Anglo-Iroquois Covenant Chain, 1758–1766," *Ethnohistory* 44, no. 4 (Autumn 1997): 617–54, esp. 624–26, 627–28; Michael Witgen, *An Infinity of Nations: How the Native New World Shaped Early North America* (Philadelphia: University of Pennsylvania Press, 2011), 218–19.
55. Robert A. Williams Jr., *The American Indian in Western Legal Thought: The Discourses of Conquest* (New York: Oxford University Press, 1990), 235–56.
56. [BF], *A Narrative of the Late Massacres, in Lancaster County . . .* (Philadelphia, 1764), in *PBF*, 11: 47-69, quotation p. 67; Dowd, *War Under Heaven*, 22–53 and passim; Peter Rhoads Silver, *Our Savage Neighbors: How Indian War Transformed Early America* (New York: W. W. Norton, 2008), 191–260; Schutt, *Peoples of the River Valleys*, 119–20; Edward G. Gray, *Mason-Dixon: Crucible of the Nation* (Cambridge, MA: Harvard University Press, 2023), 109–34.
57. BF et al., Petition to the King [June? 1769], in *PBF*, 16: 166–69; Williams, *American Indian in Western Legal Thought*, 256–65.

6. Fueled by Fossils

1. BF to Deborah Franklin, February 19, 1758, in *PBF*, 7: 380–83.
2. T. H. Breen, *The Marketplace of Revolution: How Consumer Politics Shaped American Independence* (New York: Oxford University Press, 2004); Gary B. Nash, "Franklin and Slavery," *Proceedings of the American Philosophical Society* 150 (2006): 618–35.
3. BF to John Lining, April 14, 1757, in *PBF*, 7: 185–88; Franklin sent a copy of this to Ezra Stiles, promising to buy Stiles a thermometer in London (delivered when he returned in 1762), 2, 1757, in *PBF*, 7: 233.
4. BF to John Lining, April 14, 1757, 188.

5. BF to Deborah Franklin, February 19, 1758, in *PBF*, 7: 380.
6. BF to James Bowdoin, December 2, 1758, in *PBF*, 8: 195.
7. BF to Bowdoin, December 2, 1758, 195–96.
8. BF to Bowdoin, December 2, 1758, 196.
9. H. M. Colvin, ed., *The History of the King's Works*: vol. 5, *1660–1782* (London: HM Stationery Office, 1976), 404–405.
10. BF to Bowdoin, December 2, 1758, 196–97; Jan Woudstra, "The Use of Flowering Plants in Late Seventeenth- and Early Eighteenth-Century Interiors," *Garden History* 28, no. 2 (Winter 2000): 194–208.
11. BF to Bowdoin, December 2, 1758, 197–98.
12. BF to Mary Stevenson, [November? 1760], in *PBF*, 9: 250–51.
13. BF to William Cullen, October 21, 1761, and BF to Henry Home, Lord Kames, October 21, 1761, in *PBF*, 9: 373, 376.
14. Douglas McKie and Niels H. de V. Heathcote, *The Discovery of Specific and Latent Heats* (London: Edward Arnold, 1935); John B. West, "Joseph Black, Carbon Dioxide, Latent Heat, and the Beginnings of the Discovery of the Respiratory Gases," *American Journal of Physiology. Lung Cellular and Molecular Physiology* 306 (2014): L1057–L1063.
15. Howard S. Reed, "Jan Ingenhousz, Plant Physiologist: With a History of the Discovery of Photosynthesis," *Chronica Botanica* 11 (1949): 285–396; Henry Guerlac, *Lavoisier—The Crucial Year: The Background and Origin of His First Experiments on Combustion in 1772* (Ithaca, NY: Cornell University Press, 1961); Jean-Pierre Poirier, *Lavoisier: Chemist, Biologist, Economist*, trans. Rebecca Balinski (Philadelphia: University of Pennsylvania Press, 1996), 72–83; Keiko Kawashima, "Madame Lavoisier et la traduction française de l'Essay on Phlogiston de Kirwan," *Revue d'histoire des sciences* 53, no. 2 (2000): 235–63; John W. Severinghaus, "Fire-Air and Dephlogistication. Revisionisms of Oxygen's Discovery," *Advances in Experimental Medicine and Biology* 543 (February 2003): 7–19; Ian Van Wye, "Châtelet, Lavoisier, Charrière: Negotiating the Borderlands of the Republic of Letters," in *Encounters in the Arts, Literature, and Philosophy: Chance and Choice*, ed. Jérôme Brillaud and Virginie Greene (New York: Bloomsbury Academic, 2021), 107–16.
16. Robert E. Schofield, *The Lunar Society of Birmingham: A Social History of Provincial Science and Industry in Eighteenth-Century England* (Oxford: Clarendon Press, 1963), 28; Everett Mendelsohn, *Heat and Life: The Development of the Theory of Animal Heat* (Cambridge, MA: Harvard University Press, 1964), 108–65.
17. John Canton to BF, January 21, 1762, in *PBF*, 10: 23–26; BF to Ebenezer Kinnersley, February 20, 1762, in *PBF*, 10: 38–39, 48–49; Canton to BF, June 29, 1764, in *PBF*, 11: 244.
18. Adolph B. Benson, ed., *Peter Kalm's Travels in North America*, vol. 2 (New York: Wilson-Erickson, 1937), 644.
19. BF to Sir Alexander Dick, January 21, 1762, in *PBF*, 10: 14–15.
20. BF to Dick, January 21, 1762, and BF to David Hume, [January] 21, 1762, in *PBF*, 10: 15–16, 17.
21. A[lexander] C[armichael] and J[ohn] B[rownlie], *The Edinburgh Smoke Doctor* . . . (Edinburgh, 1757).
22. BF to Henry Home, Lord Kames, October 21, 1761, in *PBF*, 9: 376; BF to Kames, January 27, 1762, in *PBF*, 10: 28–29.
23. Schofield, *Lunar Society*, esp. pp. 23–32, 35–38.
24. Larry Stewart, *The Rise of Public Science: Rhetoric, Technology, and Natural Philosophy in Newtonian Britain, 1660–1750* (New York: Cambridge University Press, 1992), 147–51, 362–69.
25. Richard L. Hills, "How James Watt Invented the Separate Condenser," *Bulletin of the Scientific Instrument Society*, part 1, 57 (1998): 26–29, and part 2, ibid., 58 (1998): 6–10; Jim Andrew, "Boulton, Watt and Wilkinson: The Birth of the Improved Steam Engine," in

Matthew Boulton: Enterprising Industrialist of the Enlightenment, ed. Kenneth Quickenden, Sally Baggott, and Malcolm Dick (Farnham UK: Ashgate, 2013), 85–100.

26. Matthew Boulton to BF, [February 22, 1766], in *PBF*, 13: 166–68; Schofield, *Lunar Society*, 60–62; Andrew, "Boulton, Watt and Wilkinson," 93–96.
27. H. W. Dickinson, *Matthew Boulton* (Cambridge: Cambridge University Press, 1937), ix, 64–74, 75–87, 113–14, 108; Stewart, *Public Science*, 230–54. That Watt consulted chemically indifferent BF questions new emphasis on steam technology's essentially chemical nature; see David Philip Miller, "A New Perspective on the Natural Philosophy of Steams and Its Relation to the Steam Engine," *Technology and Culture* 61, no. 4 (October 2020): 1129–48.
28. *Examination of Doctor Benjamin Franklin, before an August Assembly, Relating to the Repeal of the Stamp Act* . . . [Philadelphia: Hall and Sellers, 1766], in *PBF*, 13: 124–62.
29. BF to Matthew Boulton, March 19, 1766, in *PBF*, 13: 196–97.
30. BF to Boulton, March 19, 1766, 197.
31. Peter Brimblecombe, *The Big Smoke: A History of Air Pollution in London since Medieval Times* (London: Methuen, 1987), 63–89, 90–95; William M. Cavert, *The Smoke of London: Energy and Environment in the Early Modern City* (Cambridge: Cambridge University Press, 2016), 195–212.
32. *Examination of Doctor Benjamin Franklin*, 136, 137, 139.
33. *Examination of Doctor Benjamin Franklin*, 140–41, 148, 159; Boulton to BF, [February 22, 1766], 166.
34. J. T. Desaguliers, *A Course of Experimental Philosophy* . . . , vol. 1 (London, 1734), viii, 559–60; [Nicolas] Gauger, *Fires Improv'd* . . . , trans. J. T. Desaguliers (London, [1715]), iv–vi.
35. Alexander Small to BF, December 1, 1764, in *PBF*, 11: 481–82; Christine MacLeod, *Inventing the Industrial Revolution: The English Patent System, 1660–1800* (Cambridge: Cambridge University Press, 1988), esp. 58–78, 115–57, 201–22; Joel Mokyr, *The Lever of Riches: Technological Creativity and Economic Progress* (New York: Oxford University Press, 1990), 82–83; Liliane Hilaire-Pérez, "Technical Invention and Institutional Credit in France and Britain in the 18th Century," *History and Technology* 16, no. 3 (2000): 285–306; Hilaire-Pérez, "Technology, Curiosity and Utility in France and in England in the Eighteenth Century," *Science and Spectacle in the European Enlightenment*, ed. Bernadette Bensaude-Vincent and Christine Blondel (Aldershot, UK: Ashgate, 2008), 25–42.
36. Harry Emerson Wildes, *Twin Rivers: The Raritan and the Passaic* (New York: Farrar and Rinehart, 1943), 81–82, 98; Elizabeth Martling, "Arent Schuyler and His Copper Mine," *Proceedings of the New Jersey Historical Society* 65 (1947): 126; Veront M. Satchell, "Early Use of Steam Power in the Jamaican Sugar Industry, 1768–1810," *Transactions of the Newcomen Society* 67 (1995): 221–31.
37. BF to Deborah Franklin, June 4, 1765, in *PBF*, 12: 167, 168; Vivian Bruce Conger, "'There Is Graite Odds between A Mans being At Home And A Broad': Deborah Read Franklin and the Eighteenth-Century Home," *Gender and History* 21, no. 3 (October 2009): 592–607.
38. Deborah Franklin to BF, January 8, 1765, in *PBF*, 12: 13–14; entries for 8 pence for delivering ½ cord wood into yard, 9 pence for splitting wood, £2 and 14 shillings for 3 cords wood, £3 and 6 shillings "for splitting Wood and clearing away Snow," memoranda of expenses, LXVII, 81b, Miscellaneous BF Collections (Misc. BF Collections), APS. These expenses are undated but, given the presence of both BF and daughter Sarah Franklin, and Sarah's being given money for some expenses in these memoranda, it's likeliest that this was in the interval from 1762–1764, when Franklin had briefly returned from London and before Sarah married in 1767. See also John L. Cotter, Daniel G. Roberts, and Michael Parrington, *The Buried Past: An Archaeological History of Philadelphia* (Philadelphia: University of Pennsylvania Press, 1992), 88; Conger, "Deborah Read Franklin and the Eighteenth-Century House," 596.

39. Deborah Franklin to BF, February 10, 1765, in *PBF*, 12: 43–44.
40. HSP, "Collection 212: Forges and Furnaces Collection, 1727–1921," 18, https://www.hsp.org/sites/default/files/legacy_files/migrated/findingaid212forgesandfurnaces.pdf; Henry C. Mercer, *The Bible in Iron* . . . , ed. Joseph E. Sandford and Horace M. Mann, 3rd ed. (Doylestown, PA: Bucks County Historical Society, 1961), 81–88, 97–98; Samuel Y. Edgerton Jr., "Heating Stoves in Eighteenth Century Philadelphia," *Bulletin of the Association for Preservation Technology* 3 (1971): 15–104, esp. 23–29, 41; Morrison H. Heckscher, *American Rococo, 1750–1775: Elegance in Ornament* (New York: Metropolitan Museum of Art; [Los Angeles]: Los Angeles County Museum of Art, 1992), 223–28.
41. BF to Hugh Roberts, August 9, 1765, and Roberts to BF, November 27, 1765, in *PBF*, 12: 236, 386–87.
42. Mercer, *Bible in Iron*, 241. The stove was reassembled and cleaned in late 2023, when the Mercer Museum determined that the firebox's back plate had (for decades) been upside down—its slot should have been positioned at the bottom, not the top, thus closer to Franklin's original design. Communication and photographs from Clint Flack, Exhibit Specialist and Preparator, Mercer Museum and Fonthill Castle, December 12, 2023.
43. John Bartram to BF, November 24, 1770, in *PBF*, 17: 291; September 29, 1747, Warwick Furnace, daybook, 1747–1748, vol. 920, p. 121, coll. 212, HSP; Mercer, *Bible in Iron*, 110; Cotter, Roberts, and Parrington, *Buried Past*, 280–81; Franklin stove, c. 1780–1810, object number 1960.0774, Winterthur Museum.
44. John E. Crowley, *The Invention of Comfort: Sensibilities and Design in Early Modern Britain and Early America* (Baltimore: Johns Hopkins University Press, 2001), 184–85; Marie Thébaud-Sorger, "Changing Scale to Master Nature: Promoting Small-Scale Inventions in Eighteenth-Century France and Britain," *Technology and Culture* 61, no. 4 (October 2020): 1076–1107, esp. 1080–82; Mary R. M. Goodwin, *Use of Coal and Fire Grates in Eighteenth-Century Virginia: A Research Report* (Williamsburg, VA: Colonial Williamsburg Foundation, 1963); *The Pennsylvania Gazette* (Philadelphia), July 18, 1765.
45. *The Pennsylvania Gazette*, September 5, 1751, August 20, 1752, January 24, 1765 (quotation).
46. Boston Non-Importation Agreement, August 1, 1768, in David Adler, ed. *Milestone Documents in American History: Exploring the Primary Sources That Shaped America: 1619–1831* (Dallas, TX: Schlager, 2020), 95–101.
47. BF to Anthony Todd, October 29, 1769 [1768], in *PBF*, 15: 246–48; Joyce E. Chaplin, *The First Scientific American: Benjamin Franklin and the Pursuit of Genius* (New York: Basic Books, 2006), 196–200.
48. Editorial notes, *PBF*, 4: 376n, 377n; BF, *Experiments and Observations on Electricity . . . to Which are Added, Letters and Papers on Philosophical Subjects*, 4th ed. (London, 1769), 284–318.
49. BF, *Experiments and Observations on Electricity* (1769), 461–63.
50. Nash, "Franklin and Slavery," 619–20, 622–29; Kevin J. Hayes, "New Light on Peter and King, the Two Slaves Benjamin Franklin Brought to England," *Notes and Queries* 60, no. 2 (June 2013): 205–209.
51. BF, "The Cravenstreet Gazette," September 22, 1770, in *PBF*, 17: 220, 225.
52. BF, "Cravenstreet Gazette," 220–26; on Ann Hardy, editorial note, *PBF*, 10: 334n.
53. Tim Clayton, *James Gillray: A Revolution in Satire* (New Haven: Yale University Press, 2022), 140.
54. Kames to BF, February 18, 1768, in *PBF*, 15: 50-51; BF to Kames, February 28, 1768, in *PBF*, 15: 60–62.
55. [William Smellie], ed., *Encyclopaedia Britannica* . . . , vol. 3 (Edinburgh, 1771), 607, s.v. "smoke"; [James Anderson], *A Practical Treatise on Chimneys: Containing Full Directions for Preventing or Removing Smoke in Houses* (Edinburgh, 1776); Crowley, *Invention of Comfort*, 175–76.
56. Joseph Priestley, *Directions for Impregnating Water with Fixed Air* . . . (London, 1772), 3; Henry Guerlac, "Joseph Black and Fixed Air: Part II," *Isis*, 48, no. 4 (December 1957):

433–56; Laurence Brockliss and Colin Jones, *The Medical World of Early Modern France* (Oxford: Clarendon Press, 1997), 462–65 (concern over air quality from 1750s onward); Robert E. Schofield, *The Enlightened Joseph Priestley: A Study of His Life and Work from 1773 to 1804* (University Park, PA: Penn State University Press, 2004).

57. Joseph Priestley, *Experiments and Observations on Different Kinds of Air* (London, 1774), 36, 50–54, 86–94. On animals as actors in science, see Whitney Barlow Robles, *Curious Species: How Animals Made Natural History* (New Haven: Yale University Press, 2023).

58. BF to Joseph Priestley, [July 1772?], in *PBF*, 19: 215–16 (admiring the defense of vegetation); BF to Priestley, June 7, 1782, in *PBF*, 37: 445 (defending the mice); Priestley, *Different Kinds of Air*, 94–95 (citing BF's letter of 1772).

59. Stephen Hales, *A Treatise on Ventilators* . . . (London, 1758); Marie Thébaud-Sorger, "Capturing the Invisible: Heat, Steam and Gases in France and Great Britain, 1750–1800," in *Compound Histories: Materials, Governance, and Production, 1760–1840*, ed. Lissa L. Roberts and Simon Werrett (Leiden: Brill, 2017), 85–105; Vladimir Janković, *Confronting the Climate: British Airs and the Making of Environmental Medicine* (New York: Palgrave Macmillan, 2010), 41–91; Sasha Handley, *Sleep in Early Modern England* (New Haven: Yale University Press, 2016), 121–31.

60. BF to Samuel Cooper, July 7, 1773, in *PBF*, 20: 269–70.

61. Board of Works to Sir John Cust (speaker of the House of Commons), June 7, 1769, Memorials, p. 212, Work 6/18, National Archives (London); https://www.measuringworth.com/index.php, accessed September 22, 2024. See also Minutes, June 30, 1769, Office of Works, Work 4/14, National Archives (London), where the cost was shaved down to £220.

62. Alexander Small, "Of Ventilation," [before March 15, 1777?], in *PBF*, 23: 488–89.

63. Francis Jennings, "Iroquois Alliances in American History," *The History and Culture of Iroquois Diplomacy: An Interdisciplinary Guide to the Treaties of the Six Nations and Their League*, ed. Francis Jennings (Syracuse, NY: Syracuse University Press, 1985), 49; Francis Jennings, *Empire of Fortune: Crowns, Colonies, and Tribes in the Seven Years War in America* (New York: W. W. Norton, 1988), 18–19, 27–28; Paul Semonin, *American Monster: How the Nation's First Prehistoric Creature Became a Symbol of National Identity* (New York: New York University Press, 2000), 84–110.

64. BF to Jared Eliot, July 16, 1747, in *PBF*, 3: 149; George Croghan, "A List of the Teeth and Bones . . . ," February 7, 1767, in *PBF*, 14: 25–29; BF to George Croghan, August 5, 1767, in *PBF*, 14: 221–22; Mark V. Barrow, Jr., *Nature's Ghosts: Confronting Extinction from the Age of Jefferson to the Age of Ecology* (Chicago: University of Chicago Press, 2009), 16–46.

65. Adrienne Mayor, *Fossil Legends of the First Americans* (Princeton, NJ: Princeton University Press, 2005), 32–72.

66. William Hunter, "Observations on the Bones, Commonly Supposed to be Elephants Bones, Which Have Been Found Near the River Ohio in America," *Philosophical Transactions* 58 (1767): 34–45.

67. The French translation of Franklin's 1751 London edition had appeared in 1756.

68. [Jacques] Barbeu-Dubourg to BF, October 28, 1772, in *PBF*, 19: 347; BF to Barbeu-Dubourg [November 12–16, 1772], in *PBF*, 19: 368; Barbeu-Dubourg, ed., *Oeuvres de M. Franklin* . . . , vol. 2 (Paris, 1773), 199.

69. Account of "Mrs. Stevenson" with Macey Life, December 14, 1764–May 21, 1766, LXVI, 141, BF Papers, Part 13, APS; trade cards of Tuckwell and Cooper, and John Fry and Co., London, Misc. BF Collections, LXVIII, 80-a-d, APS. See Maxine Berg and Helen Clifford, "Selling Consumption in the Eighteenth Century: Advertising and the Trade Card in Britain and France," *Cultural and Social History* 4, no. 2 (2007): 145–70.

70. BF to John Winthrop, July 2, 1768, in *PBF*, 15: 170–71. BF refers to visiting Germany the previous year, though he went in 1766; *Benjamin Franklin's Autobiography*, ed. Joyce E. Chaplin (New York: W. W. Norton, 2012), 398.

71. BF to Winthrop, July 2, 1768, 171; BF, notes and diagram, July 1768, James S. and Frances M. Bradford Collection of Franklin Papers, B F85.bra, 7, APS.
72. A. J. Pacey, "Some Early Heat Engine Concepts and the Conservation of Heat," *British Journal for the History of Science* 7 (1974): 135–45.
73. Bill of Josiah Day, June 25, 1774, Misc. BF Collections, LXVII, 40, APS.
74. BF Journal, London, 1764–1774, p. 35, Ms. Account Books, BF Papers, APS; George King, mason, bill to "Mrs. Stevenson," June 18, 1772, LXVII, BF Papers, Part 13, APS; Bill of Josiah Day, June 25, 1774; receipt for James Buttall, LXVIII, 81a-c, Misc. BF Collections, APS.
75. BF to Barbeu-Dubourg, January 22, 1773, *Oeuvres de M. Franklin*, 2: 118.
76. BF to Barbeu-Dubourg, June 29, 1773, in *PBF*, 20: pp. 248–51, incl. plate. Franklin had paid a London engraver, Matthew Darly, £1 and 12 shillings for unspecified work done in May 1770. If this was for the stove, then the second wooden model of 1772, might not have been for the same device, or represented a subsequent modification. A June 1773 payment to another artist, "Bonner for drawings," £9, 16 shillings, and 6 pence, is on the high end, but not impossible as work related to the stove. See BF Journal, 1764–1774, pp. 29, 49.
77. Lord Le Despencer to BF, May 3, 1774, in *PBF*, 21: 208; excerpt from [Baron Le Despencer], *Abridgement of the Book of Common Prayer* (1773), in *PBF*, 20: 346.
78. [BF], "An Edict by the King of Prussia," September 22, 1773, in *PBF*, 20: 416; John J. McCusker and Russell R. Menard, *The Economy of British America, 1607–1789* (Williamsburg, VA, and Chapel Hill, NC: Institute of Early American History and Culture, 1985), 326.
79. A. N. Davenport, *James Watt and the Patent System* (London: British Library, 1989), 13–37; MacLeod, *Inventing the Industrial Revolution*, 102; Irina Gouzévitch, "Enlightened Entrepreneurs versus 'Philosophical Pirate' (1788–1809): Two Faces of the Enlightenment," in *Matthew Boulton*, 215–30; https://www.measuringworth.com/index.php.
80. R. L. Hills and A. J. Pacey, "The Measurement of Power in Early Steam-Driven Textile Mills," *Technology and Culture* 13, no. 1 (January 1972): 25–43; Stephen Mullen, "James Watt and Slavery in Scotland," *History Workshop*, August 17, 2020, https://www.historyworkshop.org.uk/james-watt-and-slavery-in-scotland; *OED Online*, s.v. "manpower."
81. "A Conversation on Slavery," *Public Advertiser* (London), January 26, 1770, in *PBF*, 17: 37–44; P.E.H. Hair, "Slavery and Liberty: The Case of the Scottish Colliers," *Slavery and Abolition* 21 (2000): 136–51; Nash, "Franklin and Slavery," 629.
82. *Poor Richard, 1742. An Almanack . . .* , in *PBF*, 2: 338.

7. The American Revolution?

1. BF to John Hancock, April 13, 1776, in *PBF*, 22: 400; "Franklin's Description of His Ailments," October 17, 1777, *PBF*, 25: 78–79; BF to Richard and Sarah Bache, May 10, 1785, *The Writings of Benjamin Franklin*, ed. Albert Henry Smyth, vol. 9 (New York: Macmillan, 1905–), 327, 696; Joyce E. Chaplin, *The First Scientific American: Benjamin Franklin and the Pursuit of Genius* (New York: Basic Books, 2006), 250–54.
2. Hannah Griffitts, "Inscription, on a Curious Chamberstove . . . ," November 1776, Hannah Griffitts Papers, LCP in HSP222, Box 1, Folder 7422.F.49–51, HSP.
3. BF to Sarah Bache, June 3, 1779, in *PBF*, 29: 613; Stacy Schiff, *A Great Improvisation: Franklin, France, and the Birth of America* (New York: Henry Holt, 2005); Chaplin, *First Scientific American*, 250–66.
4. Nikolai N. Bolkhovitinov, *Russia and the American Revolution*, trans. and ed. C. Jay Smith (Tallahassee, FL: The Diplomatic Press, 1976), 39, 137; *La révolution américaine et l'Europe, Colloques internationaux du CNRS*, no. 577 (Paris: CNRS, 1979), 369–419; David Armitage, *The Declaration of Independence: A Global History* (Cambridge, MA: Harvard University Press, 2008), esp. 63–102.

5. BF, "List of Household Improvements and Memorandum" [July 12, 1785], in *PBF*, 44: 364; Cash Book, entries for January 1 and July 1, 1780, BF825f6.18, BF Papers, APS; bill for wood, March 21, 1782, item LXIII, 32.5b, BF Papers, APS; budget memorandum, 1783–1784, LXIII, 37, BF Papers, APS. On historical values of currency, see https://www.historical statistics.org/Currencyconverter.html.
6. Hugh Williamson, "An Attempt to Account for the Change of Climate, Which Has Been Observed in the Middle Colonies in North-America," *Transactions of the American Philosophical Society* 1 (January 1, 1769–January 1, 1771): 272–80; Anya Zilberstein, *A Temperate Empire: Making Climate Change in Early America* (New York: Oxford University Press, 2016), 166–67, 171.
7. Donald Jackson, ed., *The Diaries of George Washington*, vol. 3, *1 January 1771–5 November 1781* (Charlottesville, VA: University Press of Virginia, 1978), 340–41; *Pennsylvania Journal and Weekly Advertiser*, May 10, 1780; Brian Fagan, *The Little Ice Age: How Climate Made History, 1300–1850* (New York: Basic Books, 2000), 99–100, 160–66.
8. Jacques Roger, *Buffon: A Life in Natural History*, trans. Sarah Lucille Bonnefoi, ed. L. Pearce Williams (Ithaca, NY: Cornell University Press, 1997), 421–23; George-Louis LeClerc, le Comte de Buffon, *The Epochs of Nature*, trans. and ed. Jan Zalasiewicz, Anne-Sophie Milon, and Mateusz Zalasiewicz (Chicago: University of Chicago Press, 2018), 39–40, 52–53, 113, 126–28.
9. Buffon, *Epochs of Nature*, 177–78, n. 4.
10. Buffon, *Epochs of Nature*, 57–60, 127–28, 137–42, n. 9.
11. Zilberstein, *Temperate Empire*, 148–73; Paul Warde, Libby Robin, and Sverker Sörlin, *The Environment: A History of the Idea* (Baltimore: Johns Hopkins University Press, 2021), 37–38, 103.
12. Jean-François-Clément Morand to BF, August 10, 1777, in *PBF*, 24 : 404–405; Morand to BF, October 5, 1778, in *PBF*, 27: 506–507, 507n; Morand, *L'Art d'exploiter les mines de charbon de terre*, vol. 3 ([Paris], 1779), plate 58, pp. 1585–86; Jean-François Belhoste, "Une silviculture pour les forges, XVIe–XXIXe siècles," in *Forges et forêts: Recherches sur la consommation proto-industrielle de bois*, ed. Denis Woronoff (Paris: Écoles des hautes études en science sociales, 1990), 241–46; Daniel Roche, *A History of Everyday Things: The Birth of Consumption in France, 1600–1800*, trans. Brian Pearce (Cambridge: Cambridge University Press, 2000), 130–34.
13. Chaplin, *First Scientific American*, 250–66.
14. BF, *Des Herrn D. Benjamin Franklin's ... sämmtliche Werke*, ed. G. T. Wenzel [Wenceslaus], 3 vols. (Dresden, 1780), 2: 108–61, 220–27 (heating), 3: 1–28 (population). For his help in translating this, I thank Hans Martin Pech.
15. Johann Jacob Schübler, *Nützliche Vorstellung und deutlicher Unterricht, von zierlichen, bequemen und Holtz ersparenden Stuben-Oefen [wood-saving room stoves]* ... (Nürnberg, 1728), esp. 13–17, appendix; Rolf-Jürgen Gleitsmann, "Rohstoffmangel und Lösungsstrategien: Das Problem vorindustrieller Holzknappheit," *Technologie und Politik*, 16 (1980): 104–54, esp. 108–21 (rising fuel demands), 121–24 (stoves, including their prohibitive cost), 124–35 (conservation); Rolf Peter Sieferle, *The Subterranean Forest: Energy Systems and the Industrial Revolution* (Cambridge: White Horse Press, 2001), 47–61, 138–80. German works read with Google Translate.
16. Ian Inkster, "Technological and Industrial Change: A Comparative Essay," *The Cambridge History of Science*, vol. 4: *Eighteenth-Century Science*, ed. Roy Porter (Cambridge: Cambridge University Press, 2003), 858–66; Michael Williams, *Deforesting the Earth: From Prehistory to Global Crisis* (Chicago: University of Chicago Press, 2003), 179, table 7.2, 285–91; Douglas R. Weiner, "The Predatory Tribute-Taking State: A Framework for Understanding Russian Environmental History," in *The Environment and World History*, ed. Edmund Burke III and Kenneth Pomeranz (Berkeley: University of California Press, 2009), 281–82.

17. *Собрание разных сочинений Вениамина Франклина: Перевод с французскаго / Collected Miscellaneous Writings of Benjamin Franklin: Translation from French* (Moscow, 1803); A. M. Kuchumov, *The Pavlovsk Palace and Park* (Leningrad: Aurora Art Publishers, 1975), plates 11, 44, 57, 67 (stoves), 19, 47, 81, 92, 101, 139, 145, 149, 154, 169, 176, 184, 191, 197, 208, 213, 217 (fireplaces); Bolkhovitinov, *Russia and the American Revolution*, esp. 31–75, 116–17; R. E. F. Smith, "The Russian Stove," *Oxford Slavonic Papers*, New Series, 18 (1985): 83–101; Vidar Lerum, *Sustainable Building Design: Learning from Nineteenth-Century Innovations* (London: Routledge, 2016), 28. For help with the Russian materials in this and the following two notes, I wish to thank Leora Eisenberg.
18. Eufrosina Dvoichenko-Markoff, "Benjamin Franklin, the American Philosophical Society, and the Russian Academy of Science," *Proceedings of the American Philosophical Society* 91, no. 3 (1947): 250–57; Bolkhovitinov, *Russia and the American Revolution*, 115–19; Jonathan R. Dull, *A Diplomatic History of the American Revolution* (New Haven: Yale University Press, 1985), 128–33; Sue Ann Prince, ed., *The Princess and the Patriot: Ekaterina Dashkova, Benjamin Franklin, and the Age of Enlightenment* (Philadelphia: American Philosophical Society, 2006); Russian National Library catalog https://primo.nlr.ru; Russian State Library catalog https://www.rsl.ru/ru; catalog of E. R. Dashkova, Alupka Palace and Park Museum Reserve, French list, p. 182, English list, pp. 19, 145. I thank Alexandr Balinchenko for providing photographs of the pages of the Dashkova catalog, sent on November, 22, 2021.
19. *Nova acta Academiae scientiarum imperialis petropolitanae* (1783); *Mémoires de l'Académie impériale des sciences de St. Pétersbourg* (1803–1806); searches (yielding no Franklin-related findings) done for the following items on https://search.worldcat.org in January 2022 by Ms. Leora Eisenberg: Sergei Nikiforovich Livotov, *Оглавление сочинений заключающихся в 63 x [sic] книгах под названием Труды Вольнаго экономическаго общества / Table of Contents of the Works Contained in 63 Books Entitled Proceedings of the Free Economic Society* (St. Petersburg, 1812); Nikolai Novikov, ed., *Утренний свет / Morning Light* (1777–80); Novikov, ed., *Московское ежемесячное издание / Moscow Monthly Edition* (1781). See also Eufrosina Dvoichenko-Markoff, "The American Philosophical Society and Early Russian-American Relations," *Proceedings of the American Philosophical Society* 94 (1950): 549–610.
20. Fernand Braudel, *The Structures of Everyday Life: The Limits of the Possible*, trans. Siân Reynolds (New York: Harper and Row, 1981), 366; Dull, *Diplomatic History of the American Revolution*, 69–70, 107–13; Ramon Archidona Casteràs, *Franklin i Catalunya/Franklin and Catalonia* (Barcelona: Generalitat de Catalunya, Comissió Amèrica, 1992), 71; Gabriel Paquette and Gonzalo M. Quintero Saravia, "Introduction," in *Spain and the American Revolution: New Approaches and Perspectives*, ed. Paquette and Quintero Saravia (London: Routledge, 2020), 11–18, 26; Gregg French, "Spain and the Birth of the American Republic: Establishing Lasting Bonds of Kinship in the Revolutionary Era," *Spain and the American Revolution*, 186–88.
21. James Sharp, *An Account of the Principle and Effects of the Air Stove-Grates, (which warm rooms, &c. by a continual introduction and exchange of dry fresh air,) Commonly Known by the Name of American Stoves*, 9th ed. ([London]: J. Sharp, [1781?]), 6; James Sharp, *Exposé du principe et des effets des grilles à feu pensilvanienes . . .* (London: J. Sharp, [1781?]).
22. Sharp, *Air Stove-Grates*, 3 ("These Stoves"), 9 ("very strong"), 11.
23. Sharp, *Air Stove-Grates*, 11; James Makittrick Adair, *Medical Cautions . . .* (Bath, 1786), 46–47.
24. Étienne-François Turgot, Marquis de Soumont, to BF, April 25, 1781, in *PBF*, 34: 574; BF to Turgot, May 1, 1781, in *PBF*, 35: 8–9.
25. Jan Ingenhousz to BF, October 2, 1782, in *PBF*, 38: 177; [Henri Justel], "An Account of an Engine that Consumes Smoak . . . ," *Philosophical Transactions* 16 (1686): p. 78 and preceding plate.
26. [Antoine Le Camus and Pierre Abraham Pajon de Moncets], *Poêle hydraulique, oeconomique et de santé* (Paris, 1772); Carl Johan Cronstedt, *Beskrifning på ry inrättning af kakelugnar til weds*

besparning... (Stockholm, 1775); Carl Sparre, *Beskrifning öfwer ätskilliga på flera ställen werkställde inrättningar af eldstäder*... (Stockholm, 1775), APS, cataloged as from library of BF.
27. *Observations sur la physique*... (Paris, 1777), translated from Giuseppe Vernazza, ed., *Lettera del Conte Cisalpino... in cui si descrive un cammino, e stufa du nuova invenzione* [Torino, 1776], 6–7, 13–14. The first edition opens with praise of material improvements informed by science that later editions omitted. On the journal *Observations sur la physique*, see James E. McClellan, "The Scientific Press in Transition: Rozier's Journal and the Scientific Societies in the 1770s," *Annals of Science* 36 (1979): 425–49.
28. Williams, *Deforesting the Earth*, 183–84; Steven L. Kaplan, *The Bakers of Paris and the Bread Question, 1700–1775* (Durham, NC: Duke University Press, 1996), 53–58, 76–77; Jean-Pierre Poirier, *Lavoisier: Chemist, Biologist, Economist*, trans. Rebecca Balinski (Philadelphia: University of Pennsylvania Press, 1996), 135–40.
29. Michel Devèze, "La crise forestière en France dans la première moitié du XVIIIe siècle, et les suggestions de Vauban, Réaumur, Buffon," *Actes du 88e Congrès National des Sociétés Savantes, Clérmont-Ferrand, 1963* (Paris: Imprimerie nationale, 1964), 595–616; Ernest Labrousse and Fernand Braudel, eds, *Histoire économique et sociale de la France*, vol. 2 (Paris: Presses Universitaires de France, 1970–79), 2: 399; Jean Boissière, "La consommation parisienne de bois et les sidérurgies périphériques: essai de mise en parallèle (milieu xve–milieu xixe siècles)," in *Forges et forêts*, 29–52; Roche, *History of Everyday Things*, 107–108, 124, 125–30; Hamish Graham, "Fleurs-de-Lis in the Forest: 'Absolute' Monarchy and Attempts at Resource Management in Eighteenth-Century France," *French History* 23, no. 3 (September 2009): 311–35; William H. Sewell Jr., *Capitalism and the Emergence of Civic Equality in Eighteenth-Century France* (Chicago: University of Chicago Press, 2021).
30. André Vaquier, "Un philanthrope méconnu: Cadet de Vaux (1743–1828)," *Paris et Île-de-France. Mémoires, IX* (Paris: CTHS, 1958).
31. Jean-Charles-Pierre Lenoir to BF, December 4, 1783, *PBF*, 41: 257–58 and illustration facing 257.
32. Lenoir to BF, December 4, 1783, 256–58; BF to Cadet de Vaux, February 5, 1784, in *PBF*, 41: 535–37. BF would give Cadet de Vaux a longer essay on maize c. April 28, 1785, in *PBF*, 44: 106–108.
33. Jan Ingenhousz to BF, October 2, 1782, in *PBF*, 38: 177; Kieko Matteson, *Forests in Revolutionary France: Conservation, Community, and Conflict, 1669–1848* (New York: Cambridge University Press, 2015), 41–42, 71–72, 91–92.
34. [Joseph Jary] to BF [c. June 1779], in *PBF*, 44: 576; Paul Walden Bamford, "French Forest Legislation and Administration, 1660–1789," *Agricultural History* 29, no. 3 (1955): 97–107; Arlette Brosselin, Andrée Corvol, and François Vion-Delphin, "Les doléances contre l'industrie," *Forges et forêts*, 11–28; Peter Sahlins, *Forest Rites: The War of the Demoiselles in Nineteenth-Century France* (Cambridge, MA: Harvard University Press, 1994); Roche, *History of Everyday Things*, 130–34; Matteson, *Forests in Revolutionary France*, esp. 38–153.
35. Charles-Louis-François Fossé to BF, March 23, 1783, in *PBF*, 39: 372–73; Fossé to BF, June 27, 1785, in *PBF*, 44: 306–307.
36. Charles-Louis-François Fossé, *Cheminée économique: a laquelle on a adapté la mécanique de M. Franklin* (Paris, 1786), 1–37; he also specifies the number of rooms he experimented with in Fossé to BF, June 27, 1785, in *PBF*, 44: 306. And see Jan Woudstra, "The Use of Flowering Plants in Late Seventeenth- and Early Eighteenth-Century Interiors," *Garden History* 28, no. 2 (Winter 2000): 194–208; Catherine Horwood, *Potted History: The Story of Plants in the Home* (London: Frances Lincoln Limited, 2007), 11–82; Penny Sparke, *Nature Inside: Plants and Flowers in the Modern Interior* (New Haven: Yale University Press, 2021), 15–22.
37. Joseph-François Desarnod, *Mémoires sur les foyers économiques et salubres de M. le docteur Franklin et du Sr. Desarnod, architecte à Lyon* (Lyon et Paris, 1789), 5, 7, 24, 27, the affidavits are on pp. 35–58, instructions pp. 98–110; on French *privilèges exclusifs* to manufacture

inventions, see Liliane Hilaire-Pérez, "Technical Invention and Institutional Credit in France and Britain in the 18th Century," *History and Technology* 16, no. 3 (2000): 285–306, esp. 294–95; Marie Thébaud-Sorger, "Changing Scale to Master Nature: Promoting Small-Scale Inventions in Eighteenth-Century France and Britain," *Technology and Culture* 61, no. 4 (October 2020): 1076–1107, esp. 1081–82.

38. Jean-Baptiste Le Roy to BF [March 1784?], in *PBF*, 42: 109–10. Franklin's design for burning coal was evidently a dinner-table topic, as a visiting Scotsman noted in May 1784—see extracts of Henry Mackenzie's Journal, May 2, 1784, in *PBF*, 42: 203.
39. Antonio Pace, *Benjamin Franklin and Italy* (Philadelphia: American Philosophical Society, 1958), 76–77.
40. Carlo Giuseppe Campi to BF, July 24, 1774, in *PBF*, 21: 250–51; *Benjamin Franklin's Autobiography*, ed. Joyce E. Chaplin (New York: W. W. Norton, 2012), 94–95, 146; Pace, *Benjamin Franklin and Italy*, 1–13.
41. [Campi, ed.], *Scelta di lettere e di opuscoli del Signor Beniamino Franklin tradotti dall' inglese* (Milan, 1774); BF, "Descrizione della stufa di Pensilvania . . . ," in *Scelta di opuscoli interessanti tradotti da varie lingue*, [ed. Campi], 2nd ed. (Milan, 1775), 1: 15–82; on the full publication history of the journal, see Partizia Delpiano, "Lire les sciences dans l'Italie du xviiie siècle," *Archives internationales d'histoire des sciences* 63 (2013): 288, note 5; Giuseppe Vernazza, *Lettera del Conte Cisalpino . . . in cui si descrive un cammino, e stufa du nuova invenzione* [Torino, 1776], 6–7, 13–14.
42. Pace, *Benjamin Franklin and Italy*, 77–78.
43. The Inquisitor's affidavit appears in Graziosi's third edition of 1791.
44. Search for Graziosi, Antonio, in the Venice consortium catalog, Polo VEA-SBN, https://polovea.sebina.it/SebinaOpac/query/ED:%22Graziosi,%20Antonio%22?locale=eng&context=catalogo, accessed September 22, 2024.
45. Pace, *Franklin and Italy*, 78–79. The copy with uncut pages is a 1788 edition: Box 1788, Massachusetts Historical Society, Boston, MA; the diagram, with notes in Venetian, is in a 1791 edition of *Stufa di Pensilvania*: D791.F832d, John Carter Brown Library, Providence, RI.
46. Karl Appuhn, *A Forest on the Sea: Environmental Expertise in Renaissance Venice* (Baltimore: Johns Hopkins University Press, 2009), 5–11, 30–33, 94–143, 144–94.
47. Appuhn, *Forest on the Sea*, 26–30; Gianmario Nicoletti, *I carbonai: immagini di un mestiere scomparso* (Pordenone, Italy: Savioprint, 1988), 77–91; Mauro Agnoletti, "From the Dolomites to Venice: Rafts and River Driving along the Piave River in Italy (13th to 20th Centuries)," *IA, Journal of the Society for Industrial Archeology* 21, no. 1 (1995): 15–32.
48. Carlo Amoretti and Francesco Soave, eds., *Opuscoli scelti sulle scienze e sulle arti . . .* (Milan, 1778–1803), vol. 4, 410–18, vol. 5, 97–106; Pace, *Franklin and Italy*, 78–80; Giuse Scalva, "L'invenzione di Benjamin Franklin e l'attività manifatturiera di Castellamonte," *I quaderni di Terramia* 8 (2010): 44–50.
49. Scalva, "L'invenzione di Benjamin Franklin," 44–50; Paolo Malanima, "Energy Consumption in England and Italy, 1560–1913. Two Pathways toward Energy Transition," *Economic History Review* 69, no. 1 (2016): 78–103. Screenshot of eBay.it listing available from author; the stove, offered by ernestino-lgb, wasn't complete, alas, having only the side, back, and front plates, not the top, nor the stand for it.
50. Pace, *Franklin and Italy*, 293. The surviving *Poor Richard, 1732. An Almanack . . .* , is in the Rosenbach Library (Philadelphia); search for *Descrizione della stufa di Pensilvania*, https://search.worldcat.org, accessed July 16, 2024.
51. [BF], "Father Abraham's Speech," in *PBF*, 7: 326–40; Chaplin, *First Scientific American*, 157–58; Sophus A. Reinert, "The Way to Wealth around the World: Benjamin Franklin and the Globalization of American Capitalism," *American Historical Review* 120, no. 2 (Winter 2015): 61–97.

8. Novus Ordo Seclorum

1. Treaty of Peace between the United States and Great Britain, September 3, 1784, in *PBF*, 40: 571.
2. On BF's multiplying maladies, see Stanley Finger, *Doctor Franklin's Medicine* (Philadelphia: University of Pennsylvania Press, 2006), 267–75, 280–92, 300–306.
3. BF, "Suppositions and Conjectures on the Aurora Borealis, c. December 7, 1778, in *PBF*, 28: 190–95.
4. BF, "Aurora Borealis," 191–92; BF to Joseph Priestley, February 8, 1780, in *PBF*, 31: 456. In 1772, BF helped a mariner, Isaac L. Winn, publish in the *Philosophical Transactions* of the Royal Society a hypothesis about the aurora borealis's effects on wind and storms, also offering his own brief comment. See *PBF*, 19: 236–39.
5. BF to Joseph Banks, [September 2], 1783, in *PBF*, 40: 543–52; Jacques-Étienne Montgolfier to BF, September 11, [1783], in *PBF*, 40: 609–10; BF to Richard Price, September 16, 1783, in *PBF*, 41: 6–8.
6. BF to Joseph Banks, November 21, 1783, in *PBF*, 41: 216–20. See also Joyce E. Chaplin, *The First Scientific American: Benjamin Franklin and the Pursuit of Genius* (New York: Basic Books, 2006), 293–302; Charles Coulson Gillispie, *The Montgolfier Brothers and the Invention of Aviation 1783-1784: With a Word on the Importance of Ballooning for the Science of Heat and the Art of Building Railroads* (Princeton, NJ: Princeton University Press, 2014).
7. Charles Blagden to BF, June 22, 1783, July 10, [1783], in *PBF*, 40: 211, 293; John Jeffries, *A Narrative of the Two Aerial voyages of Doctor Jeffries with Mons. Blanchard . . .* (London, 1786), which explains that, to reduce weight, the men had jettisoned a "packet" of letters (carefully noting the trajectory of its descent) but kept at least one, the only one known to have survived: William Franklin to William Temple Franklin, December 16, 1784 (written half past midnight), in *PBF*, 43: 362–63; Richard P. Hallion, *Taking Flight: Inventing the Aerial Age from Antiquity through the First World War* (New York: Oxford University Press, 2003), 59–60.
8. Joseph Priestley to BF, September 27, 1779, in *PBF*, 30: 407–408; Jan Ingenhousz to BF, December 5, 1780, in *PBF*, 34: 121; BF to Benjamin Vaughan, April 29, 1784, in *PBF*, 42: 200.
9. BF "Loose Thoughts on a Universal Fluid," [June 25, 1784], in *PBF*, 42: 359; with minor changes, the essay was published after BF's death as "A New and Curious Theory of Light and Heat," *Transactions of the American Philosophical Society* 3 (1793): 5–8. See also BF to David Rittenhouse, June 20, 1788, https://franklinpapers.org.
10. BF, "Loose Thoughts on a Universal Fluid," 359, 361.
11. BF, "Loose Thoughts on a Universal Fluid," 361.
12. BF, "Meteorological Imaginations and Conjectures," in *PBF*, 42: 289–94. See also Oliver Hochadel, "'In Nebula Nebulorum': The Dry Fog of the Summer of 1783 and the Introduction of Lightning Rods in the German Empire," *Transactions of the American Philosophical Society* 99 (2009): 45–70; David McCallam, "Vers une histoire de la résilience: les réponses socio-écologiques en France à l'éruption volcanique du Laki en 1783," *Dix-huitième siècle* 54 (2022): 101–14.
13. *Memoirs of the Literary and Philosophical Society of Manchester* 11 (1785): 357–61.
14. BF, "Meteorological Imaginations and Conjectures," 290–91.
15. *OED Online*, s.v. "albedo," 2.a.
16. BF, "Meteorological Imaginations and Conjectures," 291–92.
17. BF, "Meteorological Imaginations and Conjectures," 292–93. See also discussion of meteors and comets (Halley's comet, specifically) in *Poor Richard Improved, 1757*, in *PBF*, 7: 90; David Rittenhouse to BF, December 31, 1780, in *PBF*, 34: 224–25 (on a meteor entering "our Atmosphere"); Richard B. Stothers, "The Great Dry Fog of 1783," *Climatic Change* 32 (January 1996): 79–89; Thorvaldur Thordarson and Stephen Self, "Atmospheric and Environmental Effects of the 1783–1784 Laki Eruption: A Review and Reassessment," *Journal*

of Geophysical Research: Atmospheres 108, 4011 (2003): AAC 10–1-FFR 9–6; Rudolf Brázdil et al., "European Floods during the Winter 1783/1784: Scenarios of an Extreme Event during the 'Little Ice Age,'" *Theoretical and Applied Climatology* 100 (2010): 163–89; Katrin Kleemann, *A Mist Connection: An Environmental History of the Laki Eruption of 1783 and Its Legacy* (Berlin and Boston: De Gruyter, 2023).

18. BF to John Pringle, January 6, 1758, in *PBF*, 7: 357; BF to Samuel Mather, July 7, 1773, in *PBF*, 20: 288; BF to Jean-Louis Giraud Soulavie, September 22, 1782, in *PBF*, 38: 126.
19. Chaplin, *First Scientific American*, 240–41, 289–91, 310–11, 320–24.
20. BF, "A Letter from Dr. B. Franklin, to Dr. Ingenhausz, Physician to the Emperor, at Vienna, on the Causes and Cure of Smokey Chimneys," *Transactions of the American Philosophical Society* 2 (1786); second volume of *Nouvelles experiences et observations sur divers objets de physique* (Paris, 1789); *Opuscoli scelti sulle scienze e sulle art*... (Milan, 1778–1803): 9: 410–18; 10: 97–106; Beatrice Marguerite Victory, *Benjamin Franklin and Germany* ([Philadelphia]: University of Pennsylvania Press, 1915), 160. The essay would also appear in *Philosophical and Miscellaneous Papers Lately Written by B. Franklin*... (London, 1787).
21. BF, "Smokey Chimneys," 10.
22. BF, "Smokey Chimneys," 11–13.
23. BF, "Smokey Chimneys," 14.
24. BF, "Smokey Chimneys," 27–28; L. H. Butterfield, Leonard C. Faber, and Wendell D. Garrett, eds., *Diary and Autobiography of John Adams*, vol. 3 (Cambridge, MA: Harvard University Press, 1961), 418.
25. BF, "Smokey Chimneys," 15–17.
26. BF, "Smokey Chimneys," 16–17.
27. BF, "Smokey Chimneys," 17.
28. BF, "Smokey Chimneys," 17–20, 33. BF was responding to Jose Mac Packe [anagram of James Peacock], *Oikidia, or Nutshells: Being Ichnographic Distributions for Small Villas*... (London, 1785), in which Peacock defined architectural principles, including for chimneys and fireplaces (pp. 16–17), based on proportions. That "Mac Packe" styled himself a "Bricklayers Labourer" may especially have irked Franklin, a former artisan himself.
29. BF, "Smokey Chimneys," 22–24.
30. BF, "Smokey Chimneys," 21–25.
31. BF, "Smokey Chimneys," 28–29.
32. BF, "Smokey Chimneys," 29–30, 40. The unnamed London "landlord" was most likely for the first place BF lived, in Little Britain near Smithfield; the two named London proprietors from whom he rented rooms later were women. *Benjamin Franklin's Autobiography*, ed. Joyce E. Chaplin (New York: W. W. Norton, 2012), 42, 47, 156n. The luxurious French "cheminée à double foyer" (double-hearth fireplace) is further described in "Art du fumiste," *Encyclopédie méthodique: arts et métiers méchaniques*..., 8 vols. (Paris, 1782–1791), 3: 122, 128.
33. BF, "Smokey Chimneys," 30–31, 33–37.
34. BF, "Smokey Chimneys," 25.
35. BF, "Smokey Chimneys," 31–32.
36. BF, "Smokey Chimneys," 25–26, 37–38, 41.
37. BF, "Description of a New Stove for Burning of Pitcoal, and Consuming All Its Smoke, by Dr. Benjamin Franklin," *Transactions of the American Philosophical Society* 2 (1786): 43–60, quotation 57.
38. BF, "Description of a New Stove," 57–60.
39. BF, "Description of a New Stove," 68–70. The Craven Street room thought to be his bedchamber has a fireplace and measures 360 x 440 cm (11.8 x 14.4 ft) while the study/parlor (with fireplace) is 430 x 510 cm (14.1 x 16.7 ft). The bedchamber's dressing room has no fireplace. Email correspondence with Ms. Caitlin Hoffman, Benjamin Franklin House (London), June 21, 2021.

40. BF, "Description of a New Stove," 64, 66–67, 70.
41. BF, "Description of a New Stove," 70–71.
42. BF, "Description of a New Stove," 72–73; ledger, 1764–1775, penultimate page, Miscellaneous BF Collections, APS—BF estimated £150 of damages to "Bathing House & Fences at Pasture destroy'd."
43. BF, "Description of a New Stove," 73–74.
44. Francis Hopkinson to BF, October 18, 1782, in *PBF*, 38: 229–30.
45. "Silvester," *The Pennsylvania Gazette* (Philadelphia), August 21, 1782.
46. "Silvester," *The Pennsylvania Gazette*, August 21, 1782.
47. The law was repealed on August 29, a week after Hopkinson's piece had appeared: *The Pennsylvania Gazette*, September 4, 1782.
48. BF to Francis Hopkinson, December 24, 1782, in *PBF*, 38: 490.
49. Mark Reinberger and Elizabeth McLean, *The Philadelphia Country House: Architecture and Landscape in Colonial America* (Baltimore: Johns Hopkins University Press, 2015), 234–36.
50. Robert I. Alotta, *Street Names of Philadelphia* (Philadelphia: Temple University Press, 1975), 13, 34, 39.
51. Bureau of the Census, *Heads of Families at the First Census of the United States Taken in 1790: Pennsylvania* (Washington, DC: Government Printing Office, 1908), 8.
52. Francis Jennings, "Iroquois Alliances in American History," in *The History and Culture of Iroquois Diplomacy: An Interdisciplinary Guide to the Treaties of the Six Nations and Their League*, ed. Francis Jennings, et al. (Syracuse, NY: Syracuse University Press, 1985), 58.
53. David Zeisberger, entry for June 18, 1777, *The Moravian Mission Diaries of David Zeisberger, 1772–1781*, eds. Hermann Wellenreuther and Carola Wessel, trans. Julie Weber (University Park, PA: Penn State University Press, 2005), 365.
54. Charles J. Kappler, ed., *Indian Affairs: Laws and Treaties*, vol. 1 (Washington, DC: Government Printing Office, 1904), 3, 4–5.
55. Zeisberger, entry for September 30, 1778, *Diaries of David Zeisberger*, 470; Randolph C. Downes, *Council Fires on the Upper Ohio: A Narrative of Indian Affairs in the Upper Ohio Valley until 1795* (Pittsburgh: University of Pittsburgh Press, 1940), 215–17; C. A. Weslager, *The Delaware Indians: A History* (New Brunswick, NJ: Rutgers University Press, 1989), 282–328.
56. Amy C. Schutt, *Peoples of the River Valleys: The Odyssey of the Delaware Indians* (Philadelphia: University of Pennsylvania Press, 2007), 175–87; Michael John Witgen, *Seeing Red: Indigenous Land, American Expansion, and the Political Economy of Plunder in North America* (Williamsburg, VA, and Chapel Hill, NC: Omohundro Institute of Early American History and Culture, 2021), 121–25, 131–35.
57. BF, "Information to Those Who Would Remove to America," [1784], in *PBF*, 41: 603.
58. Adam Smith, *The Wealth of Nations* (1776), ed. Edwin Cannan (New York: Modern Library, 2000), 80–83; James Bonar, ed., *A Catalogue of the Library of Adam Smith* (London: Macmillan, 1894), 72, 148; Hiroshi Mizuta, *Adam Smith's Library: A Catalogue* (Oxford: Oxford University Press, 2000), 96; Joyce E. Chaplin, *Benjamin Franklin's Political Arithmetic: A Materialist View of Humanity* (Washington, DC: Smithsonian Institution Libraries, 2006); J. A. Leo Lemay, *The Life of Benjamin Franklin*, vol. 3: *Soldier, Scientist, and Politician, 1748–1757* (Philadelphia: University of Pennsylvania Press, 2014), 605–609.
59. Smith, *Wealth of Nations*, 89–90, 715.
60. Smith, *Wealth of Nations*, 417–20.
61. BF, sketches of a fire-grate, in *PBF*, 41: xxx.
62. John Bezís-Selfa, *Forging America: Ironworkers, Adventurers, and the Industrious Revolution* (Ithaca, NY: Cornell University Press, 2004), 180–81; "Valley Forge: Washington's Headquarters," National Park Service, https://www.nps.gov/vafo/learn/historyculture/washingtons_headquarters.htm.

63. BF to Jane Franklin Mecom, September 21, 1786, https://franklinpapers.org.
64. Johann David Schoepf, *Travels in the Confederation [1783–1784]*, trans. and ed. Alfred J. Morrison, vol. 1 (Philadelphia: William J. Campbell, 1911), 36–37.
65. Samuel Gustaf Hermelin, *Report about the Mines in the United States of America, 1783*, trans. Amandus Johnson (Philadelphia: John Morton Memorial Museum, 1931), 11–13, 16–17, 40, 43, 45, 46–47, 50, 54, 64, 72, 74, 76; Bezís-Selfa, *Forging America*, 22, 24, 26; Benjamin Latrobe to Erick Bollmann, July 1, 1809, *The Correspondence and Miscellaneous Papers of Benjamin Henry Latrobe*, ed. John C. Van Horne et al., vol. 2 (New Haven: Yale University Press, 1984–1988), 743.
66. Clarence J. Glacken, *Traces on the Rhodian Shore: Nature and Culture in Western Thought from Ancient Times to the End of the Eighteenth Century* (Berkeley: University of California Press, 1976), 654–81, 685–93; Conevery Bolton Valenčius, *The Health of the Country: How American Settlers Understood Themselves and Their Land* (New York: Basic Books, 2004), 214–19.

9. Keystone State

1. Henry Adams, *The Education of Henry Adams: An Autobiography* (Boston: Houghton Mifflin, 1918), 334; Gary B. Nash, *First City: Philadelphia and the Forging of Historical Memory* (Philadelphia: University of Pennsylvania Press, 2002), 144–75.
2. Their first surviving letter is Michel-Guillaume St. John de Crèvecœur to BF, August 27, 1781, in *PBF*, 35: 415–17, on US prisoners of war; on steamboats, see Crèvecœur to BF, January 30, 1788, and BF to Crèvecœur, February 16, 1788, https://franklinpapers.org.
3. John Fitch to BF, October 12, 1785, https://franklinpapers.org; BF to Francis Hopkinson, March 27, 1786, https://franklinpapers.org; BF to Jean-Baptiste Le Roy, February 17, 1788, https://franklinpapers.org; BF to Benjamin Vaughan, May 14, 1788, https://franklinpapers.org; BF to Jean-Baptiste Le Roy, October 25, 1788, *The Writings of Benjamin Franklin*, 10 vols., ed. Albert Henry Smyth (New York: Macmillan, 1905–), 9: 679; Brooke Hindle, *The Pursuit of Science in Revolutionary America, 1735–1789* (Williamsburg, VA, and Chapel Hill, NC: Institute of Early American History and Culture, 1956), 373–77.
4. Christine MacLeod, *Inventing the Industrial Revolution: The English Patent System, 1660–1800* (Cambridge: Cambridge University Press, 1988), table 9.3, p. 172, and p. 175; Ian Inkster, "Technological and Industrial Change: A Comparative Essay," *The Cambridge History of Science*, vol. 4: *Eighteenth-Century Science*, ed. Roy Porter (Cambridge: Cambridge University Press, 2003), 858.
5. James Rumsey to BF, August 10, 1788, https://franklinpapers.org; Ben Marsden, *Watt's Perfect Engine: Steam and the Age of Invention* (New York: Columbia University Press, 2002), 175–76.
6. *Benjamin Franklin's Autobiography*, ed. Joyce E. Chaplin (New York: W. W. Norton, 2012), 89, 111.
7. Chaplin, *Benjamin Franklin's Autobiography*, 197, 198.
8. Chaplin, *Benjamin Franklin's Autobiography*, 110–11.
9. Chaplin, *Benjamin Franklin's Autobiography*, 110–11; Colin T. Ramsey, "Stealing Benjamin Franklin's Stove: A New Identification for the 'Ironmonger in London,'" *ANQ* 20, no. 2 (Spring 2007): 25–30.
10. I. Bernard Cohen, *Science and the Founding Fathers: Science in the Political Thought of Thomas Jefferson, Benjamin Franklin, John Adams, and James Madison* (New York: W. W. Norton, 1995), 238–43.
11. Granville Sharp to BF, [May 9?, 1785], in *PBF*, 44: 157–58; Sharp to BF, June 17, 1785, in *PBF*, 44: 270–74; BF to Sharp, July 5, 1785, in *PBF*, 44: 334–36; Joyce E. Chaplin, *An Anxious Pursuit: Agricultural Innovation and Modernity in the Lower South, 1730–1815*

(Williamsburg, VA, and Chapel Hill, NC: Institute of Early American History and Culture, 1993), 114–16, 259–61, 309–11; Walter Johnson, *River of Dark Dreams: Slavery and Empire in the Cotton Kingdom* (Cambridge, MA: Harvard University Press, 2013), 73–96.

12. John Bezís-Selfa, *Forging America: Ironworkers, Adventurers, and the Industrious Revolution* (Ithaca, NY: Cornell University Press, 2004), 180–90; Erica Armstrong Dunbar, *A Fragile Freedom: African American Women and Emancipation in the Antebellum City* (New Haven: Yale University Press, 2008); Cory James Young, "A Just and True Return: Pennsylvania's Surviving County Slave Registries, 1780–1826," *Journal of Slavery and Data Preservation*, 3 (2022): https://doi.org/10.25971/QS08-YE30.

13. Gary B. Nash and Jean R. Soderlund, *Freedom by Degrees: Emancipation in Pennsylvania and Its Aftermath* (New York: Oxford University Press, 1991), 4, 8–9, 74–166; Erica Armstrong Dunbar, *Never Caught: The Washingtons' Relentless Pursuit of Their Runaway Slave, Ona Judge* (New York: 37ink, 2017).

14. Richard Allen, *The Life Experience and Gospel Labors of the Rt. Rev. Richard Allen...* (Nashville: Abington Press, 1960), 17–19; Richard S. Newman, *Freedom's Prophet: Bishop Richard Allen, the AME Church, and the Black Founding Fathers* (New York: NYU Press, 2008), 14–77.

15. Chaplin, *Benjamin Franklin's Autobiography*, 114–15.

16. Three unpublished BF letters (https://franklinpapers.org) of June 30, 1787: to the Cherokee Indian Queen, to "Indian Chiefs," to "The Cornstalk." See earlier letter from "Tobocah and a Chichasaw Captain," June 19, 1787, for the probable identity of the "Indian Chiefs."

17. Chaplin, *Benjamin Franklin's Autobiography*, 115; Alfred P. James, "Benjamin Franklin's Ohio Valley Lands," *Proceedings of the American Philosophical Society* 98, no. 4 (1954): 255–65.

18. Gregory Evans Dowd, *War Under Heaven: Pontiac, the Indian Nations, and the British Empire* (Baltimore: Johns Hopkins University Press, 2004), 271–73; Colin G. Calloway, *The Shawnees and the War for America* (New York: Viking Penguin, 2007), 145 (quotation), 165 (steamboats); Gregory Hitch and Marcus Grignon, "A Forest of Energy: Settler Colonialism, Knowledge Production, and Maple Syrup Kinship in the Menominee Community," *American Quarterly* 75, no. 2 (2023): 251–77.

19. *Opere filosofiche di Beniamino Franklin...* (Padova, 1783); Jan Golinski, *British Weather and the Climate of Enlightenment* (Chicago: University of Chicago Press, 2007), 210–11; Richard Kirwan, "Of the Variations of the Atmosphere," *Transactions of the Royal Irish Academy* 8 (1802): 269–507, esp. 384, 397 (Kirwan didn't accept Franklin's hypothesis of volcanic climate forcing, however, pp. 482–87).

20. Alexander von Humboldt, *Personal Narrative of Travels to the Equinoctial Regions of the New Continent during the Years 1799–1804*, trans. Helen Maria Williams, vol. 1 (London, 1814), 29, 48; Alexander von Humboldt, *Cosmos*, trans. E. C. Otté, vol. 1 (Baltimore: Johns Hopkins University Press, 1997), 308–309. On Humboldt's conceptualization of nature as a set of interconnected environments, see Aaron Sachs, *The Humboldt Current: Nineteenth-Century Exploration and the Roots of American Environmentalism* (New York: Viking, 2006), esp. 74–88.

21. John K. Alexander, *Render Them Submissive: Responses to Poverty in Philadelphia, 1760–1800* (Amherst: University of Massachusetts Press, 1980), 124–25; Billy G. Smith, "The Material Lives of Laboring Philadelphians, 1750 to 1800," *WMQ* 38, no. 2 (April 1981): 169–202.

22. May 1, 1795, *Early Proceedings of the American Philosophical Society for the Promotion of Useful Knowledge* ([Philadelphia]: McCalla and Stavely, 1884), 231; A. William Hoglund, "Forest Conservation and Stove Inventors, 1789–1850," *Forest History* 5, no. 4 (Winter 1962): 2–8.

23. "A Letter from Mr. C. W. Peale to the Editor of the Weekly Magazine," *Weekly Magazine*, 1 (Philadelphia), March 31, 1798, 265–69; Lillian B. Miller et al., eds., *The Selected Papers of Charles Willson Peale and His Family*, vol. 2 (New Haven: Yale University Press,

c1983 –), 140–41; C. W. Peale and Raphaelle Peale, "Description of Some Improvements in the Common Fire-Place . . . ," *Transactions of the American Philosophical Society* 5 (1802): 320–24; Samuel Y. Edgerton Jr., "Heating Stoves in Eighteenth-Century Philadelphia," *Bulletin of the Association for Preservation Technology* 3, no. 2/3 (1971): 15–104, esp. 24–26; Laura Turner Igoe, "'Processes of Nature and Art': The Ecology of Charles Willson Peale's Smoke-Eaters and Stoves," *A Greene Country Towne: Philadelphia's Ecology in the Cultural Imagination*, ed. Alan C. Braddock and Laura Turner Igoe (University Park, PA: Penn State University Press, 2016), 37–39.

24. *Peale Papers*, 1: 218–20. The design specifies a stack of eleven bricks. See also *Weekly Magazine*, 2, July 21, 1798, 353; Igoe, "Peale's Smoke-Eaters and Stoves," 34–36, 40–43.

25. Edgerton Jr., "Heating Stoves," 82–83; Franklin Stove, c. 1795, Metropolitan Museum of Art, https://www.metmuseum.org/art/collection/search/3833.

26. Benjamin Thompson, "New Experiments upon Heat," *Philosophical Transactions* 76 (1786): 273–304, esp. 293, 297–98, 302; [Benjamin Thompson], *Count Rumford's Experimental Essays, Political, Economical, and Philosophical. Essay IV. Of Chimney Fire-Places . . .* (London, 1796), 297–301, 308–22, 324–53.

27. Thompson, *Experimental Essays . . . IV*, 347n.

28. Jane Austen, *Northanger Abbey* (1817), ed. Anne Henry Ehrenpreis (Harmondsworth, UK: Penguin, 1972), 168; *New York Commercial Advertiser*, November 3, 1798; BF, *Observations on Smoky Chimneys . . .* (London: I. and J. Taylor, 1793), Franklin Collection, 381 Ob9 1793, Beinecke Library; Franklin's work shares a publisher with two of the others it's bound with.

29. James Gillray, engraver, "Advantages of wearing muslin dresses!" (Great Britain, 1802), https://www.loc.gov/resource/cph.3g08774; E. Claire Cage, "The Sartorial Self: Neoclassical Fashion and Gender Identity in France, 1797–1804," *Eighteenth-Century Studies* 42 (2009): 193–215; Alison Matthews David, "Blazing Ballet Girls and Flannelette Shrouds: Fabric, Fire, and Fear in the Long Nineteenth Century," *Textile: The Journal of Cloth and Culture* 14 (2016): 244–67; Anne Higonnet, *Liberty, Equality, Fashion: The Women Who Styled the French Revolution* (New York: W. W. Norton, 2024), 8–10, 45–49, 97–107, 165–73, 186–93.

30. James Watt, Patent #1485, submitted June 14, 1785, enrolled July 9, 1785, British Library. I thank David Armitage for retrieving text of the patent in July 2021 when I was unable to travel to London.

31. Thompson, *Experimental Essays . . . IV*, 364n, 365; *Encyclopaedia Britannica*, s.v., "smoke," 3rd ed. (Edinburgh, 1797), 17: 547–56; Peter Brimblecombe, *The Big Smoke: A History of Air Pollution in London since Medieval Times* (London: Methuen, 1987), 93–95.

32. William Stanley Jevons, *The Coal Question: An Enquiry Concerning the Progress of the Nation, and the Probable Exhaustion of Our Coal Mines* (London: Macmillan, 1865); Steve Sorrell, "Jevons' Paradox Revisited: The Evidence for Backfire from Improved Energy Efficiency," *Energy Policy* 37, no. 4 (April 2009): 1456–69; Astrid Kander, Paolo Malanima, and Paul Warde, *Power to the People: Energy in Europe over the Last Five Centuries* (Princeton, NJ: Princeton University Press, 2013), 99, 111, 133; Mark Reinberger and Elizabeth McLean, *The Philadelphia Country House: Architecture and Landscape in Colonial America* (Baltimore: Johns Hopkins University Press, 2015), 131.

33. [Isabella Batchelder] James, *Memorial of Thomas Potts, Junior* (Cambridge [MA]: private press, 1874), 54–55; Henry C. Mercer, *The Bible in Iron . . .* , ed. Joseph E. Sandford and Horace M. Mann, 3rd ed. (Doylestown, PA: Bucks County Historical Society, 1961), 110, 243–44, illus. 336–44; Milo M. Naeve, *John Lewis Krimmel: An Artist in Federal America* (Newark: University of Delaware Press, 1987), 164–65, figure 5.8; Bernard L. Herman, *Town House: Architecture and Material Life in the Early American City, 1780–1830* (Williamsburg, VA, and Chapel Hill, NC: Omohundro Institute of Early American History and

Culture, 2005), 77–79, 84–85, 93; Howell John Harris, "'The Stove Trade Needs Change Continually': Designing the First Mass-Market Consumer Durable, ca. 1810–1930," *Winterthur Portfolio* 43:4 (Winter 2009): 365–406.

34. BF to Jane Mecom, May 30, 1787, https://franklinpapers.org; BF, final will and codicil, June 23, 1789, http://franklinpapers.org.
35. Adams cited in Joseph J. Ellis, *Passionate Sage: The Character and Legacy of John Adams* (New York: W. W. Norton, 1993), 66; https://librarycompany.org/2018/05/21/shareholder-spotlight-mr-robert-grace/; Franklin Estate Inventory, April 26, 1790, Miscellaneous BF Collections, APS.
36. *General Advertiser* (Philadelphia), May 22, 1792; John L. Cotter, Daniel G. Roberts, and Michael Parrington, *The Buried Past: An Archaeological History of Philadelphia* (Philadelphia: University of Pennsylvania Press, 1993), 91.
37. BF, final will and codicil, June 23, 1789; value of codicil calculated using https://measuringworth.com, accessed September 22, 2024.
38. Harold C. Syrett, ed., *The Papers of Alexander Hamilton*, vol. 10: *December 1791–January 1792* (New York: Columbia University Press, 1966), 230–340 (list of economic desiderata, p. 249; quotations, pp. 259 and 260; debts to Smith in notes on pp. 127–29).
39. Alfred D. Chandler, "Anthracite Coal and the Beginnings of the Industrial Revolution in the United States," *Business History Review* 46, no. 2 (1972): 141–81; Frederick Moore Binder, *Coal Age Empire: Pennsylvania Coal and Its Utilization to 1860* (Harrisburg: Pennsylvania Historical and Museum Commission, 1974); H. Benjamin Powell, *Philadelphia's First Fuel Crisis: Jacob Cist and the Developing Market for Pennsylvania Anthracite* (University Park, PA: Penn State University Press, 1978); Bezís-Selfa, *Forging America*, 26; Christopher F. Jones, "A Landscape of Energy Abundance: Anthracite Coal Canals and the Roots of American Fossil Fuel Dependence, 1820–1860," *Environmental History* 15, no. 3 (July 2010): 449–84.
40. R. V. Reynolds and Albert H. Pierson, *Fuel Wood Used in the United States, 1630–1930* (Washington, DC: United States Department of Agriculture, 1942), 4, 11; Brooke Hindle, "Introduction: The Span of the Wooden Age," in *America's Wooden Age: Aspects of Its Early Technology*, ed. Brooke Hindle (Tarrytown, NY: Sleepy Hollow Restorations, 1975), esp. 10, 12, 36; Charles F. Carroll, "The Forest Society of New England," in *America's Wooden Age*, 13–36; Michael Williams, *Americans and Their Forests: A Historical Geography* (Cambridge: Cambridge University Press, 1989), 134–39; Priscilla J. Brewer, *From Fireplace to Cookstove: Technology and the Domestic Ideal in America* (Syracuse, NY: Syracuse University Press, 2000), ch. 4; Jones, "Landscape of Energy Abundance," 461–63.
41. Charles Dickens, *American Notes for General Circulation*, ed. John S. Whitley and Arnold Goldman (London: Penguin Books, 1985), 99, 306, 311; Codman Hislop, *Eliphalet Nott* (Middletown, CT: Wesleyan University Press, 1971), 255–71; Rolf Peter Sieferle, *The Subterranean Forest: Energy Systems and the Industrial Revolution* (Cambridge: White Horse Press, 2001), 183–84.
42. Nathaniel Hawthorne, "Fire-Worship," in *Mosses from an Old Manse*, ed. William Charvat et al. (Columbus, OH: Ohio State University Press, 1974), 138, 143, 146; Sam H. Schurr and Bruce C. Netschert, *Energy in the American Economy, 1850–1975: An Economic Study of Its History and Prospects* (Baltimore: Johns Hopkins University Press, 1960), 36–37; B. R. Mitchell, *Economic Development of the British Coal Industry, 1800–1914* (Cambridge: Cambridge University Press, 1984), 1; Paolo Malanima, "Energy Consumption in England and Italy, 1560–1913: Two Pathways toward Energy Transition," *Economic History Review* 69 no. 1 (2016): 78–103, esp. 87–90.
43. Richard Slotkin, *Regeneration Through Violence: The Mythology of the American Frontier, 1600–1860* (Middletown, CT: Wesleyan University Press, 1973), 466–516; Philip J. Deloria, *Playing Indian* (New Haven: Yale University Press, 1998), 13–20, 27, 45–58; Jean M. O'Brien,

Firsting and Lasting: Writing Indians out of Existence in New England (Minneapolis: University of Minnesota Press, 2010), 105–43.

44. Andrew M. Schocket, *Founding Corporate Power in Early National Philadelphia* (DeKalb, IL: Northern Illinois University Press, 2007); Albert J. Churella, *The Pennsylvania Railroad*, vol. 1: *Building an Empire, 1846–1917* (Philadelphia: University of Pennsylvania Press, 2013); Richard Sylla and Robert E. Wright, "Corporation Formation in the Antebellum United States in Comparative Context," *Business History* 55, no. 4 (2013): 653–69. See also Philip J. Stern, *Empire, Incorporated: The Corporations That Built British Colonialism* (Cambridge, MA: Harvard University Press, 2023).

45. Thomas J. Misa, *A Nation of Steel: The Making of Modern America, 1865–1925* (Baltimore: Johns Hopkins University Press, 1995), 164–77; "Mercer Museum and Fonthill Castle," https://www.mercermuseum.org/about/mercer-museum/; "About H. F. du Pont," Winterthur Museum, Garden and Library, https://www.winterthur.org/about-h-f-du-pont.

46. Thomas Piketty, *Capital in the Twenty-First Century*, trans. Arthur Goldhammer (Cambridge, MA: Harvard University Press, 2014), 72–109.

47. Karl Marx, *Capital: A Critique of Political Economy*, trans. Ben Fowkes, vol. 1 (London: Penguin/New Left Review, 1976), 142, *n18*; Bruce H. Yenawine, *Benjamin Franklin and the Invention of Microfinance*, ed. Michele R. Costello (London: Pickering and Chatto, 2010), 47–54, 62–63, 115. See also John R. Aiken, "Benjamin Franklin, Karl Marx, and the Labor Theory of Value," *Pennsylvania Magazine of History and Biography* 90, no. 3 (1966): 378-84; Gregory Slack, "Marx's Argument for the Labor Theory of Value," *Review of Radical Political Economics* 53, no. 1 (March 2021): 143–56.

48. Thomas Paine, "The Construction of Iron Bridges" (1803), in *The Writings of Thomas Paine*, ed. Moncure Daniel Conway, vol. 4 (New York: G. P. Putnam's Sons, 1896), 445–49; Eric Foner, *Free Soil, Free Labor, Free Men: The Ideology of the Republican Party Before the Civil War* (New York: Oxford University Press, 1970); Nash and Soderlund, *Freedom by Degrees*, 167–204; Christopher M. Osborne, "Invisible Hands: Slaves, Bound Laborers, and the Development of Western Pennsylvania," *Pennsylvania History* 72, no. 1 (2005): 77–99; Edward G. Gray, *Tom Paine's Iron Bridge: Building a United States* (New York: W. W. Norton, 2016), which doesn't mention the slavery, either.

49. John Majewski, *A House Dividing: Economic Development in Pennsylvania and Virginia before the Civil War* (New York: Cambridge University Press, 2000), 1–11, 38–53, 59–84, 102–109, 128–35; Nash, *First City*, 152–61; Sean Patrick Adams, "A Comparison between the Richmond Coal Basin and Pennsylvania's Anthracite Fields: Slave Labour, Free Labour and the Political Economy," in *Towards a Comparative History of Coalfield Societies*, ed. Stefan Berger, Andy Croll, and Norman LaPorte (Aldershot, UK: Routledge, 2005), 113–26; Jeremy Zallen, *American Lucifers: The Dark History of Artificial Light, 1750–1865* (Chapel Hill: University of North Carolina Press, 2019), 214–55.

50. Robert Sutcliff, *Travels in Some Parts of North America . . .* (York, England, 1811), 184.

51. Paul A. Gilje and Howard B. Rock, "'Sweep O! Sweep O!': African-American Chimney Sweeps and Citizenship in the New Nation," *WMQ* 51, no. 3 (July 1994): 507–38; Newman, *Freedom's Prophet*, 22, 56–58.

52. Gilje and Rock, "African-American Chimney Sweeps," 507–38; Anneliese Harding, *John Lewis Krimmel: Genre Artist of the Early Republic* (Winterthur, DE: Winterthur Museum, 1994), 28–29; Igoe, "Peale's Smoke-Eaters and Stoves," 41–45; Edward G. Gray, *Mason-Dixon: Crucible of the Nation* (Cambridge, MA: Harvard University Press, 2023), 192–96, 237–40.

53. Abraham Lincoln, "Address at Gettysburg, Pennsylvania," November 19, 1863, in *Speeches and Writings*, vol. 2: *1859–1865, Speeches, Letters, and Miscellaneous Writings, Presidential Messages and Proclamations*, [ed. Don E. Fehrenbacher] (New York: Library of America, 1989), 536 (quotations); Nash, *First City*, 144–75.

54. John F. Kasson, *Civilizing the Machine: Technology and Republican Values in America 1776–1900* (New York: Viking, 1976), 162–67; Adam Goodheart, "The Machine of the Myth," *Design Quarterly* 155 (Spring 1992): 24–28.
55. Misa, *Nation of Steel*, 1–44; Andrew Garn, *Bethlehem Steel* (New York: Princeton Architectural Press, 1999), 9–46.
56. Misa, *Nation of Steel*, 45–131, 211–51; Garn, *Bethlehem Steel*, 3–46.
57. Audra Simpson, *Mohawk Interruptus: Political Life across the Borders of Settler States* (Durham, NC: Duke University Press, 2014), 2–5, 7, 65, 168–69, 197.
58. John Bodnar, *Steelton: Immigration and Industrialization, 1870–1940* (Pittsburgh: University of Pittsburgh Press, 1977); Dennis C. Dickerson, *Out of the Crucible: Black Steelworkers in Western Pennsylvania, 1875–1980* (Albany: State University of New York Press, 1986); Paul Kahan, *The Homestead Strike: Labor, Violence, and American Industry* (New York: Routledge, 2014); W.E.B. Du Bois, "The Princess Steel," introduction by Adrienne Brown and Britt Rusert, *PMLA* 130, no. 3 (2015): 819–29; Thomas Dublin and Walter Licht, *The Face of Decline: The Pennsylvania Anthracite Region in the Twentieth Century* (Ithaca, NY: Cornell University Press, 2016).
59. Rebecca Onion, "The Other NRA (or How the Philadelphia Eagles Got Their Name)," *Slate*, May 22, 2013; *Wikipedia*, s.v. "Proposed Acquisition of U.S. Steel by Nippon Steel," accessed September 18, 2024.
60. Brian Black, *Petrolia: The Landscape of America's First Oil Boom* (Baltimore: Johns Hopkins University Press, 2000), 21–26.
61. Christopher Jones, "The Carbon-Consuming Home: Residential Markets and Energy Transitions," *Enterprise and Society* 12, no. 4 (December 2011): 790–823; Vaclav Smil, *Energy and Civilization: A History* (Cambridge, MA: MIT Press, 2017), 245–92; Odinn Melsted and Irene Pallua, "The Historical Transition from Coal to Hydrocarbons: Previous Explanations and the Need for an Integrative Perspective," *Canadian Journal of History* 53, no. 3 (2018): 395–422.
62. Clay McShane, *Down the Asphalt Path: The Automobile and the American City* (New York: Columbia University Press, 1994).
63. Yenawine, *Franklin and the Invention of Microfinance*, 88–96.
64. Yenawine, *Franklin and the Invention of Microfinance*, 75–76, 84–85, 105–108, 141.
65. Paul Semonin, *American Monster: How the Nation's First Prehistoric Creature Became a Symbol of National Identity* (New York: New York University Press, 2000), 111–35, 214–34.
66. Semonin, *American Monster*, 300–14; Mark V. Barrow Jr., *Nature's Ghosts: Confronting Extinction from the Age of Jefferson to the Age of Ecology* (Chicago: University of Chicago Press, 2009), esp. 28, 33–43.
67. Henry Clepper, "Rise of the Forest Conservation Movement in Pennsylvania," *Pennsylvania History* 12, no. 3 (1945): 200–16; Peter Linehan, "The Teacher and the Forest: The Pennsylvania Forestry Association, George Perkins Marsh, and the Origins of Conservation Education," *Pennsylvania History* 79, no. 4 (2012): 520–36. On later state and US forest preservation efforts, see Williams, *Americans and Their Forests*, 393–424.
68. Clepper, "Rise of the Forest Conservation Movement in Pennsylvania," 201; Peter Dunwiddie et al., "Old-Growth Forests of Southern New England, New York, and Pennsylvania," *Eastern Old-Growth Forests: Prospects for Rediscovery and Recovery*, ed. Mary Byrd Davis, foreword by John Davis (Washington, DC: Island Press, 1996), 127.
69. Richard H. Grove, *Green Imperialism: Colonial Expansion, Tropical Island Edens and the Origins of Environmentalism, 1600–1860* (New York: Cambridge University Press, 1996); Mark David Spence, *Dispossessing the Wilderness: Indian Removal and the Making of the National Parks* (New York: Oxford University Press, 2000); Delaware Nation v. Commonwealth of Pennsylvania et al. (2006), https://sct.narf.org/documents/deleware_nation/brief_in_opposition_binney_smith.pdf.

70. Martin J. S. Rudwick, *Worlds Before Adam: The Reconstruction of Geohistory in the Age of Reform* (Chicago: University of Chicago Press, 2008), 35–46.
71. Henry Shaler Williams, "Correlation Papers: Devonian and Carboniferous," *United States Geological Survey Bulletin* 80 (Washington, DC: Government Printing Office, 1891), esp. 115, 118–19, 135; Amadeus W. Grabau, *A Textbook of Geology*, vol. 2 (Boston: D. C. Heath, c. 1920–1921), 475; Paul Lucier, *Scientists and Swindlers: Consulting on Coal and Oil in America, 1820–1890* (Baltimore: Johns Hopkins University Press, 2008), 106, 351, *n157*.
72. Maura Shapiro, "Eunice Newton Foote's Nearly Forgotten Discovery," *Physics Today*, August 23, 2021; Raymond Zhong and Keith Collins, "See How 2023 Shattered Records to Become the Hottest Year," *The New York Times*, January 12, 2023; "June 2024 Marks 12th Month of Global Temperatures at 1.5°C Above Pre-industrial Levels," Copernicus Climate Change Service, July 10, 2024, https://climate.copernicus.eu/june-2024-marks-12th-month-global-temperatures-15degc-above-pre-industrial-levels; "Summer 2024—Hottest on Record Globally and for Europe," Copernicus Climate Change Service, September 6, 2024, https://climate.copernicus.eu/copernicus-summer-2024-hottest-record-globally-and-europe.

Coda: God Helps Them Who Help Themselves

1. June, *Poor Richard, 1736. An Almanack...*, in *PBF*, 2: 140; Andrew Kirk, "Appropriating Technology: The Whole Earth Catalog and Counterculture Environmental Politics, *Environmental History* 6, no. 3 (July 2001): 374–94.
2. BF, final will and codicil, June 23, 1789, https://franklinpapers.org.
3. Raymond Arsenault, "The End of the Long Hot Summer: The Air Conditioner and Southern Culture," *Journal of Southern History* 50, no. 4 (November 1984): 597–628; Max Roser, "Energy Poverty and Indoor Air Pollution," *Our World in Data*, July 5, 2021; "Sources of Greenhouse Gas Emissions," EPA, https://www.epa.gov/ghgemissions/sources-greenhouse-gas-emissions.
4. On climate science: Kirk, "Appropriating Technology," 385; Naomi Oreskes, "The Scientific Consensus on Climate Change," *Science*, New Series, 306, no. 5702 (December 3, 2004): 1686; Naomi Oreskes and Erik M. Conway, *Merchants of Doubt: How a Handful of Scientists Obscured the Truth on Issues from Tobacco Smoke to Global Warming* (New York: Bloomsbury Press, 2010). On internationalism and science: G. R. de Beer, "The Relations between Fellows of the Royal Society and French Men of Science when France and Britain Were at War," *Notes and Records of the Royal Society* 9 (1952): 244–99; Joyce E. Chaplin, *The First Scientific American: Benjamin Franklin and the Pursuit of Genius* (New York: Basic Books, 2006), 284–85, 306, 355–56; Chaplin, *Round about the Earth: Circumnavigation from Magellan to Orbit* (New York: Simon and Schuster, 2012), 345–47; Benjamin W. Goossen, "A Benchmark for the Environment: Big Science and 'Artificial' Geophysics in the Global 1950s," *Journal of Global History* 15, no. 1 (2020): 149–68.
5. Building on Franklin and Smith and Hamilton (and others), twentieth-century economist Simon Kuznets defined GDP as a nation's aggregate consumption, plus private investments, plus government spending, plus net exports—minus imports. GDP has become a metric of successful governance in industrial economies, despite objections that it correlates more reliably with environmental damage than with human happiness. It's been less remarked that Kuznets accepted (probably via Smith) Franklin's discounting of Indigenous people. Swift "growth of population," he said, had been a global trend since 1750, with a pattern set initially in North America, with its "relatively 'empty' areas," a fiction he repeated: "I should be inclined to stress the importance of territorial expansion to the young and 'empty' countries that were the overseas offshoots of Europe"—though they were not, in fact, "empty." Simon Kuznets, *Modern Economic Growth: Rate, Structure, and Spread* (New Haven: Yale

University Press, 1966), 34, 36, 38, 39, 336–39, 341–42; Simon Kuznets, *Economic Growth of Nations: Total Output and Production Structure* (Cambridge, MA: Harvard University Press, 1971), table 1, pp. 13–14, p. 18. See also Vibha Kapuria-Foreman and Mark Perlman, "An Economic Historian's Economist: Remembering Simon Kuznets," *The Economic Journal* 105 (November 1995): 1524–47; Robert W. Fogel, "Simon S. Kuznets, April 30, 1901–July 9, 1985," *Biographical Memoirs: National Academy of Sciences* 79 (2001): 203–30; Peter Vanham, "A Brief History of GDP—and What Could Come Next," *World Economic Forum*, December 13, 2001, 576; Joseph Stiglitz, Amartya Sen, and Jean-Paul Fitoussi, *Mismeasuring Our Lives: Why GDP Doesn't Add Up* (New York: New Press, 2010); Daniel Susskind, *Growth: A History and a Reckoning* (Cambridge, MA: Harvard University Press, 2024).

6. "Federally Recognized Indian Tribes and Resources for Native Americans," https://www.usa.gov/tribes; website for the Lenape Nation of Pennsylvania, https://www.lenape-nation.org; Nick Estes (Kul Wicasa), *Our History Is the Future: Standing Rock Versus the Dakota Access Pipeline, and the Long Tradition of Indigenous Resistance* (New York: Verso, 2019).

7. *A Matter of Interpretation: Federal Courts and the Law*, ed. Amy Gutmann (Princeton, NJ: Princeton University Press, 1997); Dennis J. Goldford, *The American Constitution and the Debate over Originalism* (New York: Cambridge University Press, 2005); Jonathan Gienapp, "Written Constitutionalism, Past and Present," *Law and History Review* 39, no. 2 (2021): 321–60.

8. J. R. McNeill and Peter Engelke, *The Great Acceleration: An Environmental History of the Anthropocene Since 1945* (Cambridge, MA: Harvard University Press, 2014).

9. "Global Warming of 1.5° C," IPCC, https://www.ipcc.ch/sr15; "Health Topics: Climate Change," World Health Organization, https://www.who.int/health-topics/climate-change#tab=tab_1; Jos Lelieveld et al., "Air Pollution Deaths Attributable to Fossil Fuels: Observational and Modelling Study," *BMJ* 383 (2023): e077784, https://doi:10.1136/bmj-2023-077784; "There Could Be 1.2 Billion Climate Refugees by 2050. Here's What You Need to Know," *Zurich Magazine*, January 17, 2024, https://www.zurich.com/en/media/magazine/2022/there-could-be-1-2-billion-climate-refugees-by-2050-here-s-what-you-need-to-know; Adrien Bilal and Diego R. Känzig, *The Macroeconomic Impact of Climate Change: Global vs. Local Temperature*, NBER (National Bureau of Economic Research) Working Paper 32450, 2024.

ACKNOWLEDGMENTS

Everything has a history, and I think of this book's history in terms of the people I want and need to thank, in the order in which they made me thankful—except for the very last person, who's out of order.

When Sheila Jasanoff invited me to speak at the Science, Technology, and Society (STS) Circle at Harvard meeting, I was grateful for the opportunity, but didn't have an STS project underway I could talk about. I wondered if I might fashion a slim article, at the least, out of my recent realization that Benjamin Franklin had invented his famous stove during a very cold winter in the Little Ice Age, so I concocted a presentation out of that. Its warm reception (thanks especially to Lisa Randall, Neil Maher, Cindy Ott, Aaron Sachs, and the late James McCarthy) was very encouraging. Later versions of that talk benefited from comments by participants at the Benjamin Franklin Initiative (University of Wisconsin), the Cambridge Talks (Harvard Graduate School of Design), the Seminar on Early American History and Culture at Columbia University, and a Radcliffe Exploratory Seminar on "Climate and Colonization" that I organized with my colleague Matthew Liebmann (thank you, Matt).

As my slim article threatened to become a book, and while I was writing my way out of a proliferation of projects on Thomas Robert Malthus, conversations with Eric Slauter and the late Julie Saville at the University of Chicago were extremely helpful. When Cameron Strang and Christopher Parsons invited me to contribute to a special environmental history issue of *Early American Studies*, I offered them an essay on the Industrial Revolution and depictions of early American material life in pop culture (hello, *Assassin's Creed*). I'm not sure this was what Chris and Cameron were expecting, but they said they thought I was on to something—I'm deeply grateful to both of them. Back in 2017, before most people (including academics) wanted to talk about the climate crisis, the gang at BackStory Radio did a whole program on climate history, invited me to be on it, and asked all the right questions. Much gratitude to Brian Balogh, Nathan Connolly, Joanne Freeman, and Peter Onuf.

While the project grew, I learned a great deal from the colleagues who sized up various versions of the shaggy beast: the Franklin Humanities Institute at Duke University; the Department of History at Yale University; the Seminar in the Department of History at the Johns Hopkins University; the Department of History at Wesleyan University, where I was invited to give the Meigs Lecture in United States History (thank you, especially, Demetrius Eudell); the Center for InterAmerican Studies, Bielefeld University (danke, Eleonora Rohland); the American Origins Seminar at the University of Southern California–Huntington Library Early Modern Studies Institute; the Cogut Institute for the Humanities at Brown University; the Bard Graduate Center; the Seminar on Environmental and Agricultural History at MIT; the Early American Seminar at Harvard's Mahindra Humanities Center; and the Western Society for Eighteenth-Century Studies, which invited me to give their plenary lecture in February 2020, right before the COVID-19 shutdown—thank you, especially, Ann Little, for a sustaining infusion of collegiality as we sank into the emergency.

During and just after the shutdown, I gave Zoom presentations about the project to the Wolfenbütteler Arbeitskreis Frühneuzeitforschung/Wolfenbüttel Early Modern Research Group (thank you, Andreas Mahler and Cornel Zwierlein), to the Rachel Carson Center for Environment and Society in Munich, to my colleagues in Harvard's Department of History (especially Ann Blair and Josh Meija), and to the James Harvey Young Seminar at Emory University. An inspiring Dumbarton Oaks conference on climate and resources, long deferred because of the pandemic, finally took place—thank you, Thaïsa Way, for your persistence and for including me in that meeting. And then, as the manuscript took its final form, I received excellent advice from attendees at seminars at the Shelby Cullom Davis Center for Historical Studies at Princeton University (thank you, David Bell and Anthony Grafton), the Colloquium for the Study of Early and Native Americans at Harvard, and a second round at the Cambridge Talks of the Harvard Graduate School of Design.

Financial support from several sources let me travel to archives before the COVID shutdown and gave me time to write afterward. I'm honored to have been awarded a Guggenheim Foundation fellowship, a National Endowment for the Humanities Fellowship at the Library Company of Philadelphia, a grant from the Kenan Institute for Ethics at Duke University (thank you, Norman Wirzba), and sabbatical support from the Mahindra Humanities Center, Harvard University (thank you, Dean Robin Kelsey). Along the way, Anthony Grafton, Steven Shapin, Karin Wulf, and Robert Ritchie have generously written many letters of reference for me.

I'm delighted to have the opportunity to thank the archivists at the places where I did my primary research: the John Carter Brown Library, the American Philosophical Society, the Library Company of Philadelphia, the Historical Society of Pennsylvania, the National Archives (London), and the Pottstown Historical Society, with its extraordinary volunteer staff. I owe a special debt, as well, to the team at the American Philosophical Society that digitized the Franklins' Shop Books, which were the only archival sources I could read during the pandemic shutdown, a convenience (and strange consolation) I never thought I'd need. I'm also grateful to staff at several museums: the Mercer Museum (Emma Falcon and Clint Flack, especially), the Winterthur Museum, and the Benjamin Franklin House in London.

Once I was writing in earnest, Timothy Breen, Anthony Grafton, and Saffron Sener gave thoughtful responses to individual chapters of the book, and several heroic colleagues read drafts of the whole thing: Camden Elliott, Philip J. Stern, Steven Shapin, and Shuichi Wanibuchi—I cannot thank you enough. Gabriel Pizzorno offered calm help with a problematic digitized image. For countless pieces of advice, and general bucking up, I'm grateful to Clint and Mona Chaplin, Lauren Kerr, Laura J. Martin, Darrin McMahon, Evander Price, Alide Cagidemetrio, Sonja Asal, and Rebecca Goetz.

Segments of my article "The Franklin Stove: Modern Materiality, Made in Pennsylvania," published in the *William and Mary Quarterly*, vol. 81, no. 2 (April 2024), are reproduced in chapters 2, 3, and 9. I'm grateful to the *WMQ* at the Omohundro Institute of Early American History and Culture for granting permission to reproduce this material, as well as for the bracing peer review and rigorous editing they did—deepest appreciation to Josh Piker and John Bezís Selfa, especially. I'd also like to thank Jennifer Motter, the OIEAHC editorial apprentice I hired to help with proofreading the book.

Benjamin Franklin was a printer as well as an inventor, and it gives me the greatest pleasure, in this book about him, to thank some of the people who continue to make books. Andrew Wylie believed in this project, helped me refine my proposal, and then secured a publisher. Alexander Star took on the project at Farrar, Straus and Giroux—

I am still pinching myself, and deeply grateful to Alex for his enthusiasm for the project and his sharp insights. I'm likewise indebted to my other editor, Ian Van Wye. From his comments on my first draft to his line edits of the final version, Ian's eye for the authorial lapse (logical, grammatical, arithmetical . . .) was rivaled only by his tact or wry incredulity (as needed) in pointing it out. Tremendous thanks, as well, to Nancy Elgin, FSG's production editor, to copy editor Janet Rosenberg, and to proofreaders Winn Foreman and Kory Stamper. Na Kim gave me a beautiful and witty cover, and Patrice Sheridan did the elegant interior design—in Caslon, Franklin's preferred typeface. Gina Guilinger provided the index, and then Lakeside Book Company (founded 1864) printed the ink-and-paper books.

My final debt is actually the first one, the one that's out of order, because it's to the person who, some time ago now, advised my doctoral dissertation. Jack Greene is an expert on the American Revolution, but he always told his graduate students it was "the most overrated event in American history." He thought the Civil War, with its massively unfinished business, was the more important event. I agree, but hope he thinks I may have a point, too, about the Industrial Revolution. Really, his point was that his students didn't have to work on what he worked on, and we were supposed to revolt against all the received historical tropes and topics, even (somewhat ironically) history's celebrated revolutions. I'm not sure I'd have had the nerve to tackle climate history without that early education in dissent—thank you, Jack.

INDEX

Page numbers in *italics* refer to illustrations.

abolitionism, 309–10, 311, 321; *see also* enslavement
Académie royale des sciences, 259, 270
Account of the New Invented Pennsylvanian Fire-Places, An (Franklin), 96–98, 128, 136–37, 162, 215–16, 249
Account of the Principle and Effects of Air Stove-grates, An (Sharp), 249–51, *250*
Adams, Abigail, 7, 303
Adams, Henry, 303
Adams, John, 7, 281, 303, 322
Adirondacks, 339
aeoluses, 282
aerophobia, 281
"age of sail," 23–24
air: "in the air" metaphor, 269–70; carbon dioxide, 10, 199, 220, 273, 339, 344; chemical composition of, 199–200; definitions of, 198–99; devices for circulating, 221–22; heat and expansion of, 99–100; life-giving properties of, 220–21; oxygen, 199, 206, 220–21, 273, 288, 339; pollution, 10, 26, 194–95, 206, 223–24, 319–20; pressure, 142–44; quality of indoor, 223; *see also* atmosphere(s)
Albany Congress, 177
albedo, 276
Allegheny Plateau, 22
Allen, Richard, 311, 331
Allhazar, Shedid, 72–73
almanacs, 89–90; *see also Poor Richard's Almanacks*
American Magazine (periodical), 87
American Philosophical Society, 242, 248, 306, 337, 339
American Revolution, 4, 8, 11, 61, 64, 157, 190, 192, 239–40, 242–43, 247, 257, 268, 295, 304, 312, 348–49
Anderson, James, 220
Anglo-Wabanaki wars, 46, 49, 159
Anishinaabeg Council of the Three Fires, 180, 187
Anthropocene, 244, 269
anti-Catholicism, 39
anti-smoke campaigns, 10, 219–20
Aristotle, 142
artificial intelligence (AI), 5

ash, 76
Asseroni (axe makers), 78
Atlantic storm systems, 10, 169–70, 173, 215, 272, 313–14
atmosphere(s): artificial, 42, 128, 221, 253, 258, 347; Boyle's Law, 142; circulation in, 148, 215, 272–74; clothing as an artificial, 20–21; definitions of, 141–42; electrical, 145; Franklin's experiments and observations on, 10, 142–44, 145, 147–49, 169–70, 175–78, 215, 239, 252–53, 272–74, 276–77, 313; human impacts on, 10–11, 199, 344; indoor, 9–10, 88, 141; natural, 128–29; outdoor, 10; plant life and, 175–77, 273; pollution in, 10, 26, 194–95; pressure, 42
aurora borealis, 270–71
Austen, Jane, 318

Bache, Benjamin Franklin, 322–23
Bacon, Francis, 41
baffles, 76, 103, 105, 106, 108, 253
balloons, hot-air, 271–72
Bank of North America, 327
Barbeu-Dubourg, Jacques, 226–27, 230, 232, 244, 245, 252, 259, 260, 288
barometers, 42–43, 77, 100
Barry Lyndon (film), 6–7
Barton, Benjamin Smith, 338
Bartram, John, 134–36, 137, 171–72, 175, 212–13, 225
Bartram, William, 172
Beaver Wars, 48–49
Ben (enslaved person), 123
Benjamin Franklin Bridge, 333
Berkeley, Norborne, 213
Bethlehem Steel, 333, 334
Biddle family, 67
Big Bone Lick, 224, 338
biopolitics, 164
Birch, William, 316
Bird, Mark, 211
Black, Joseph, 198–99, 220
Black Death, 25, 356*n4*
Black people: autonomy and precarity of enslaved, 122–25; chimney-sweepers, 331; colonists' fears of resembling, 118; emancipated, 331, 334; enslaved, 11, 12, 13, 28, 60, 121–25, 217; freedom purchased by, 125, 310–11; ironworkers' expertise, 63–64; minority population in Pennsylvania, 295; use of consumer goods, 123–24; *see also* enslavement and enslaved people
Blagden, Charles, 272, 314
body heat, 43, 176, 193–94
Boerhaave, Herman, 42, 76, 91, 106, 276
bonfires, 39–40, 83, 311–12
Bonhomme Richard (ship), 257
Book of Common Prayer, 232–33, 309
Boone, Daniel, 326
Boston Evening-Post (periodical), 133
Boston Fire Society, 73
Boston News-Letter (periodical), 137
Boston University, 337
Boulton, Matthew, 203–208, 229, 233–34, 257, 305–307, 319
Bouquet, Henry, 189
Bowdoin, James, 195–96, 201, 285, 308
Boyle, Robert, 41–42, 142
Bradford, Andrew, 87
Brand, Stewart, 345
brand identity, 96, 130
Branson, William, 92, 93
Brattle Square Church, 222
braziers, 31, 96, 104, 256, 285, 287, 291
bread riots, 254
bricklayers, 106, 107, 195, 283
bridges, 332, 333
British Admiralty, 24, 28
Brooklyn Bridge, 332
Brunswick River, 89
Buffon, comte de, Georges-Louis Leclerc, 243, 338
Burnet, Thomas, 74, 145, 225
business corporations, 327–28
Buttall, James, 230
Buzaglo, Abraham, 213–14

Cadet de Vaux, Antoine-Alexis, 255–56
Callowhill Street, 295
calorimeter, 254
Campbell, Archibald, 143
Campbell, William, 120
Campi, Carlo Giuseppe, 260
Candide (Voltaire), 209
capitalism, 9, 28, 45, 61, 119–20, 205, 267, 327–29, 333–34, 346
carbonated water, 220
carbon dioxide, 8, 10, 199, 220, 273, 339, 344
Carboniferous period, 340–41, *342–43*
Caribbean colonies, 28, 59, 63, 161, 164, 318
Carnegie, Andrew, 328, 334, 336
Carnot, Nicolas Léonard Sadi, 229
Castellamonte Franklin stoves, 264–65

INDEX

Catherine II, 240, 247
Cayuga nation, 37
cedar cultivation, 134–36, 175
Centennial Exposition (1876), 332, 351
central heating, 4, 9, 282, 284
Cesar (enslaved person), 123–24, 125
charcoal, 25, 64–65, 104, 264
Charles II, 52
Charles (II) le Moyne de, Baron de Longueuil, 224
Charlotte, Queen (1744–1818), 219
Chaumont, Jacques-Donatien Le Ray de, 240
Cheminée économique (Fossé), 258
chemistry, 41, 42, 142, 198–200, 206, 220–21, 272–73
chimneys: architectural design principles, 392*n*28; baffles, 76, 103, 105, 106, 108, 253; in draft-resistant houses, 280–82; emergence of, 31, 285; fires in, 116–17; flues, 32, 202, 286; oversized, 282–83; problems with multiple, 283–84; proliferation of, 285–86; registers, 195–96, 201–202; smoky, 220, 280–87; used for cooling, 196–97; wind in, 284
chimney-sweepers, 32, 112, 116, 201, 330–31
Chippendale, Thomas, 153
Chrysler Building, 333
civic corporations, 327
Civil War, 303, 329, 331, 335
Clare, Martin, 74, 76–77, 97, 99, 102, 252
Clarke, William, 163
Clermont, SS, 307
climate change: anthropogenic, 8, 10–11, 46–47, 69, 176, 242, 269, 351; fossils as proof of, 224–25; Franklin stoves as an adaptation to, 3–4, 5, 14–15, 132–33, 348; suffering from, 133; technology, 15–16; white settlement as cause of, 132; as a wicked problem, 15, 16, 346; *see also* Little Ice Age
climate(s): ancient, 277–78; artificial, 118, 141; climatic phenomena, 10; control of indoor, 9–10, 112; global differences in, 27–28, 139; human-generated, 129; microclimates, 10, 294; mitigation, 349
climate science, 27, 74, 277–78, 313–14
clothing: as an artificial atmosphere, 20–21; fire hazards, 318–19; fur hats, 236; for keeping warm, 20–21, 31, 123–24; as status symbol, 124
coal: assumed inexhaustibility of, 243–44, 325–26; British consumption, 206, 218–19; coke, 332; costs, 93; domestic uses, 9, 25–26; efficiency, 320; enslavement and, 234–35; Franklin's embrace of, 192, 288–90, 291–92; Franklin's London consumption, 227; imports into British colonies, 168–69, 214; mining, 26, 214, 227, 234–35, 244–45, 301, 324; Pennsylvania's dependence on, 11–12, 14, 324–25; pitcoal, 15, 287, 288; pollution from, 10, 26, 194–95, 206, 223–24, 319–20; power plants, 335–36; sea, 26, 153, 214; smoke from, 32, 194–95, 206, 286–87, 325; soot produced by, 26, 32, 33, 201
cold, sense of, 41–42
Colden, Cadwallader, 137, 142
colds (illness), 72, 103, 115, 281
Cold War, 349
Colebrookdale Furnace, 64, 65, 122
Coleridge, Samuel Taylor, 172
colonists: anxiety of physical adaptations to environment, 118; expectations of North America, 29–30; ideas about forests, 47–48; lack of outdoor common sense, 77–78; Lenape nation and, 48, 49, 51–54, 78; natural resources assumed to belong to, 136; perspectives on abundance and scarcity, 378*n*2; population growth, 157–58, 167, 207, 295; wood fuel consumption, 36, 78, 97–98, 166–67
colonization, 22, 24, 28; resistance to, 159, 180–83, 187–90, 312–13; as visible from space, 161–62, 244
combustion, 193–94, 200–201, 203, 280
Committee to Alleviate the Miseries of the Poor, 179–80
Compleat Housewife, The (Smith), 67
conservation: Bartram's cedar orchards plan, 134–36; forest, 98, 339–40; forestry as, 23, 24, 25, 246, 263, 301, 339–40; Franklin's stove project as means of, 84, 86, 98, 255; growth vs., 54–55, 156–57; imperialism and, 54, 136; Indigenous, 22; preventative, 136; strategies, 168–69
Constitution, US, 309, 311, 321, 351
Constitutional Convention, 305, 306, 311
consumer economy, 29–30, 33–34, 45, 65, 156, 162
consumerism, 29, 157, 192, 254–55, 352
Continental Army, 300
Continental Congress, 238, 296–97
convection heating, 10, 77, 105–106, 110, 144, 270, 280

cooling of homes, 196–97, 347
Cooper, James Fenimore, 327
Cooper, Samuel, 222
copyright, 309, 311
Cornwall Furnace, 315
corporations, 327–28
Cosmos (Humboldt), 314
council fires, 37–38, *38*, 40, 48, 51, 80–81, 82–83, 133, 160, 182–87, *189*, 311–12, 350
Course of Experimental Philosophy (Desaguliers), 203, 208
Covenant Chain, 79, 81, 177, 180, 183–84
Coventry Forge, 64, 93–95, 105, 119–23, 371*n*21
"Cravenstreet Gazette, The" (Franklin), 217–18
creolization, 35
Crèvecœur, J. Hector St. John de, 305
Crocker Building, 333
Croghan, George, 224–25, 244
Cullen, William, 198
Cuvier, Georges, 338

Dalesme, André, 76, 252, 259, 287, 291
Dampier, William, 170
Daniell (Native man), 119
Darly, Matthew, 386*n*76
Darwin, Charles, 162
Darwin, Erasmus, 199–200, 203
Dashwood, Francis, Baron Le Despencer, 232–33, 287, 290, 309
Davis, John, 120
Day, Josiah, 229–30, 232
De Brahm, William Gerard, 272
Declaration of Independence, 11, 239–40, 292
Declaratory Act (1766), 205
Defoe, Daniel, 30, 40–41
deforestation: cause of climate change, 10, 175, 184; cause of flooding, 339–40; colonization and, 28–29, 166–67; European, 23–24; farming as cause of 134–35; iron manufacture and, 65–66, 301; New England, 301; Pennsylvanian fireplace as solution to, 15, 246, 248; Russia, 247; *see also* conservation
de Grasse, François Joseph Paul, 257
Delaware Nation of Oklahoma, 340, 350
Delaware River, 87, 89, 333
Denny, William, 82
Desaguliers, J. T., 74–76, 97, 114, 115, 153, 196, 203, 208–209

Desarnod, Joseph-François, 258–59
Descartes, René, 33
Description of a New-Invented Stove-Grate (Durno), 151–52
Descrizione della Stufa di Pensilvania, 260, 262, *263*, 266
desertification, 28
Dewees, William, Jr., 129
Dick (enslaved person), 122
Dick, Alexander, 201
Dickens, Charles, 326
dietetic theories, 57–58, 194
Dolfin, Daniel, 261
drafts, 77, 100, 102–103, 111–12, 114, 115, 195–97, 220, 279–82, 322
Du Bois, W.E.B., 334
dung, as fuel source, 25
du Pont, Henry Francis, 328
Durham Furnace, 64, 67, 81
Durno, James, 151–53, 308

Early American Imprints, Series I, 69
Echo-Hawk, Roger C. (Pawnee), 21
ecology, 13, 54, 349–50
economics: conservation vs. growth, 54–55, 156–57; growth, 12, 13, 65, 155–56, 299, 349–50; labor as main source of value, 58–59; Marxist, 329; origins of modern, 12–13; population growth as key to, 58
economy(ies): of abundance, 266–67; consumer, 29–30, 33–34, 45, 65, 156, 162, 192; frontier, 119; household management as, 12; inflation, 314; mixed, 59, 65, 68, 84; nonconservationist, 299; organic, 158–59, 168–69, 191, 226; political, 12–13, 44, 54; thriving vs. stationary, 299
"Edict by the King of Prussia" (Franklin), 233
Edinburgh Philosophical Society, 198
Edinburgh Smoke Doctor, The, 202
electricity: atmospheric, 145, 147–49; electrical wheel inventions, 145–47; energy source, 335–36; Franklin's experiments with, 144–51; heat generated by, 200; kite experiment, 148; lightning, 147–48
Elementa Chemiae (Boerhaave), 91
Elizabeth Furnace, 65
Ellis, Robert, 120
emancipation, 13, 124, 235, 330, 331
Empire State Building, 333

Encyclopaedia Britannica, 219–20, 320
energy autonomy, 240
energy sources: access to modern, 347; carbon-based, 8; food, 23; labor as, 12, 58–59, 120, 121, 160–61; oil, 335; peat, 25, 138; power plants, 335–36; Sweden, 140–41; *see also* coal; fossil fuels; labor; wood and wood fuel
enslavement and enslaved people: ability to keep warm, 90–91; abolitionist efforts, 309–10; of Black people, 11, 12, 13, 28, 60; Caribbean colonies, 59, 63, 161, 164; in Chesapeake, 63; coal and, 234–35; as economically inferior to free labor, 160–61, 299; enshrined in Constitution, 321; freedom purchased by, 310–11; of Indigenous peoples, 28, 60; iron industry use of, 12, 63–64, 121–27, *126*, 165, 329–30; iron restraints usage, 64; Lenape opposition to, 60, 164, 296; in Pennsylvania, 63; in Philadelphia, 164–65, 179; property ownership by, 122; Quaker opposition to, 60, 310; resistance to, 13–14; sugar industry use of, 28, 161; in Virginia, 11, 63
Enters, Margaret, 120
environmental management, 22–23
Époques de la nature (Buffon), 243
"Essay for the improvement of estates, An" (Bartram), 134–36
Essay upon Projects (Defoe), 40–41
Evans, Lewis, 97, 113, 137, 139, 172–75, 224
Evelyn, John, 25, 26, 84, 168
Experiments and Observations on Electricity (Franklin), 148, 152, 162, 215, 226, 272
extinctions, 174–75, 337–39

Fahrenheit, Daniel Gabriel, 42–43, 45
famine, 27, 88, 277
Faraday, Michael, 147
farming, deforestation performed for, 134–35
Fawkes, Guy, 39
Federation of Seven Fires, 48
fermentation, 194
festina lente, 16–17, *16*, 353
film and television, premodern depictions of domestic heating in, 6–8
firebacks, 91, 104–105, 125–27, *126*, 130, 211
firefighting, 73–74, 149, 293
fireplaces: architectural design principles, 392*n28*; chimney design, 76; "cis-Alpine," 253; colonial, 36; convection heating, 77; decorations, 33; decorative, 210–11; early designs, 101–105; film depictions of, 6–8; Franklin's experiments with, 99–105; as luxury goods, 316–17; mentioned in *Poor Richard*, 71; origins, 31; oversized, 282–83; registers for, 195–96; Rumford, 317–19; smoke from, 76, 89; Staffordshire Fire-Place, 285; *see also* Pennsylvanian fireplace; stoves
fire(s): bonfires, 39–40, 83; chimney, 116–17; colonial use of, 97; cultural uses of, 13, 20; Franklin's Sphere of Fire, 273–74; Great Fire of London, 55, 73; Indigenous uses of, 13, 20, 22, 37–39, *38*, 40, 51, 80–81, 82–83, 160, 182–87, 311–12; for keeping warm, 20–21, 31; Lenape uses of, 51, 182, 184; presided over by men vs. women, 31, 37, 186; safety, 73–74; urban, 72–74; women's clothing hazards, 318–19
Fires Improv'd (Desaguliers), 75–76, 97, 98
firewood, *see* wood and wood fuel
"Fire-Worship" (Hawthorne), 326
First Nations (Canada), 188
Fitch, John, 305–307
five-plate stoves, 91
Flatiron Building, 333
Fleet, Thomas, 133
flues, 31, 32, 202, 286
fog, 254, 274–78, 319
Folger, John, 34
Folger, Timothy, 215
Foote, Eunice, 344
forced-air heat, 103
Ford, Thomas, 120
forestry, 25, 246, 263, 301, 339–40
forests: atmospheric cooling by, 175–77; clearcutting of for ironworks, 64–65; colonists' ideas about, 47–48; conservation of, 98, 339–40; cultivation of, 134–35; European treatment of, 22–23, 55; in *Increase of Mankind*, 161–62; Indigenous treatment of, 23; Ireland, 24, 54; lumbering laws, 55; North American, 28–29; old-growth acreage, 22, 339; regrowth of, 356*n4*; Sweden, 140–41; terraforming, 161–62, 176–77, 243–44; *see also* deforestation
Fort Allen, 325
Fort Augusta, 159
Fort Detroit, 297
Fort Pitt, 169, 185, 188, 325
Foscolo, Ugo, 264
Fossé, Charles-Louis François, 258

fossil fuels: and climate change, 8–9; dependence on, 304; Franklin's study of, 192–93; pollution from, 206; transition from organic to, 12, 234; *see also* coal; petroleum

fossils, 170, 171, 173–75, *174*, 224–26, 313, 338, 341

France, 245, 254–60, 274–75

Franklin (miniseries), 7

Franklin, Benjamin: abolitionist efforts, 13, 309–10, 321; anti-smoke campaign, 10, 219–20; as ambassador to France, 236–37, 239–40, 245, 268–69; arrival in Philadelphia, 50, 56–57, 58; Craven Street rooms, 193, 195–96, 201, 217–18, 227, 229–30, 288–89, 392*n*32, 392*n*39; death, 322; dietary beliefs, 57–58, 71–72; estate, 322–23; fire insurance scheme of, 149; health concerns, 236, 269, 321, 322; land speculation, 84, 312; in London, 57–58, 191–93, 210–11, 284–85, 392*n*32, 392*n*39; Market Street house, 95–96, 99, 371*n*21; as militia colonel, 177–78; as a Modern Prometheus, 19, 134, 150, 238; as namesake of Victor Frankenstein, 151; Paris house, 240–41; philanthropy, 303–304, 323–24, 337; Poor Richard alter ego, 3, 16–17, 71, 156; portrait, *151*, *265*; as postmaster, 77, 93; as president of American Philosophical Society, 306; as printer, 40, 51, 66–67, 69–70, 77, 79, 86, 89, 96; racism, 161–62, 163; rejection of pacifism, 177; as "universal Smoke Doctor," 10, 219, 277; US commission to Canada, 236; use and acceptance of enslaved labor, 127, 164, 191–92, 217–18; views on economic value of humans, 13; virtues of, 58

Franklin, Benjamin, inventions of: electrical wheel, 145–47; fireplace registers, 195–96, 201, 227; legacy, 314; lightning rod, 147, 150, 238; vase stove, 229–33, *231*, 244, 287–90, 322; *see also* Franklin stoves; Pennsylvanian fireplace

Franklin, Benjamin, scientific experiments of: atmospheric circulation, 215, 272–74; atmospheric heating and cooling, 175–78, 239, 252–53, 313; atmospheric pressure, 142–44; aurora borealis, 270–71; color and thermal energy, 91–92, 197–98; convection, 10, 77, 105–106, 144, 169–70; electricity, 144–51; Gulf Stream, 10, 215, 272, 278, 313; heat engines, 228–30; hot-air balloons, 271–72; kite experiment, 148; legacy, 313–14; physics of heat, 200–203; pulse glasses, 227–29; thermal variation, 10, 144, 197

Franklin, Benjamin, writings of: *Account of the New Invented Pennsylvanian Fire-Places, An,* 96–98, 128, 136–37, 162, 215–16, 249; "Cravenstreet Gazette, The," 217–18; *"Description of a New Stove for Burning of Pitcoal, and Consuming All its Smoke,"* 290–91, *291* "Edict by the King of Prussia," 233; *Experiments and Observations on Electricity,* 148, 152, 162, 215, 226, 272; "Information to Those Who Would Remove to America," 298, 300; "Loose Thoughts on a Universal Fluid," 272; memoirs, 307–309; "Meteorological Imaginations and Conjectures," 275; *Narrative of the Late Massacres,* 188; *Observations Concerning the Increase of Mankind,* 155–56, 157–63, 192, 207, 245–46, 298, 338; *The Pennsylvania Gazette,* 69–70, 72, 79, 83, 86, 88–89, 93, 96, 147, 148; *Plain Truth,* 177; *Poor Richard. An Almanack,* 67, 69–72, 86–87, 88, 96, 132, 175, 178, 345–46; *Poor Richard Improved,* 134, 266; "Smokey Chimneys," 279; "Speech of Father Abraham," 248, 266; *Way to Wealth, The,* 266–67

Franklin, Deborah Read Rogers, 65, 95–96, 139, 194, 210–11, 217, 234, 309, 323

Franklin, Francis Folger, 95–96

Franklin, James, 39, 40, 46

Franklin, John, 129, 136

Franklin, Josiah, 33, 34

Franklin, Peter, 129, 130, 136

Franklin, Sarah, 95–96, 383*n*38

Franklin, William, 93, 391*n*7

Franklin, William Temple, 241, 272, 391*n*7

Franklin Fund, 323–24, 329, 336–37

Franklin Institute, 337

Franklin stoves: as an adaptation to climate change, 14–15, 256–57; as a collaboration with French authorities, 255–57; design modifications, 4, 255, 308, 356*n*6; London sliding register, 195–96, 285; mentioned in memoirs, 307–309; as a project, 15–16; purpose of, 6*n*; rotating stove, 290–92, *291*; vase stove, 229–30, *231*, 238, 244, 255, 258, 287–90, *291*, 322; *see also* Pennsylvanian fireplace

Franklin Trade School, 336

Frederick II (1712–1786), 233
French and Indian War, 83, 155, 163, 177; *see also* Seven Years' War
French Revolution, 255, 257
Friendly Association for Regaining and Preserving Peace with the Indians by Pacific Measures, 184
Fulton, Robert, 307
Fumifugium (Evelyn), 26
fur hats, 236
fur trade, 48–49

Galileo Galilei, 42
Gauger, Nicolas, 74–76, 90, 97, 98, 103, 108, 153
General Magazine, The (periodical), 87
Gentleman and Cabinetmaker's Director, The (Chippendale), 153
Gentleman's Magazine (periodical), 163
geologic time, 170, 172–75, 277, 340–41, *342–43*, 350
George (enslaved person), 122
George III, 188, 219
geothermal air-conditioning, 197
Germany, 104, 245–46, 279, 301
Gettysburg Address, 331
Gibbons, John, 120
Gillray, James, 318, 319
Gnadenhütten massacre, 297
Golden Gate Bridge, 332
Goldoni, Carlo, 262
Grace, Rebecca, Savage Nutt, 67, 95, 211, 322
Grace, Robert, 67, 94–97, 105, 108, 121, 123, 125, 129, 136, 143, 157, 211, 308, 321, 371*n*21, 375*n*2
Grand Ohio Company, 190
Gravier, Charles, comte de Vergennes, 239, 255
Graziosi, Antonio, 261–62
Great Famine (1845–1852), 88
Great Fire of London (1666), 55, 73
Great Plague of London (1665–1666), 55
Great Snow (1717), 34
Great Treaty (1722), 78–82
greenhouse gases, 199, 347, 352
Griffitts, Hannah, 237–38, 243, 322
Gronovius, Johann, 137, 139, 140
gross domestic product (GDP), 352, 400*n*5
growth economics, 12, 13, 65, 155–56, 299, 349–50
Guericke, Otto von, 42

Gulf Stream, 10, 215, 272, 278, 313
Gunpowder Treason Day, 39–40, 312

Hackett, Jonathan, 120
Hadley, George, 170
Hales, Stephen, 175–76, 221–22
Hamilton, Alexander, 209, 324
Hardy, Ann, 217
Harrison, John, 209
Harvard University, 337
Haudenosaunee Confederation, 21, 37–38, 48, 49, 51, 52, 78–82, 159, 171–73, 180, 182–83, 186, 190, 295–96, *297*
Hawthorne, Nathaniel, 326
haze, 254, 274–78, 319
hearths, 31
heat: absorption or reflection of, 42, 91, 197–98, 275–76; artificial vs. natural, 144; Franklin's experiments with the physics of, 200–203; generated by electricity, 200; index, 42; latent, 199*n*; measuring, 41–43; metabolic role in producing, 57–58, 254; methods of generating, 194; produced by bodies, 43, 176, 193–94; thermal energy, 91–92, 197–98
heating of homes: advice from almanacs, 89–90; ancient Rome, 30; associated with women, 216–17; central heating, 4, 7, 9, 282, 284; consumer expectations for, 20; costs, 90–91; discussed in *Poor Richard*, 70–71; economic effects, 33–34; feelings of warmth, 99–101; film depictions, 6–8; forced-air, 103; Franklin's experiments with, 85–86, 92–93, 99–105; as integral to a good life, 2, 45, 86, 219; as purpose of Franklin stoves, 6*n*; recommended temperatures, 251; whole-house heating, 320–21; *see also* fireplaces; stoves
Hellfire Club, 232
Hermelin, Samuel Gustaf, 301
Hildebrand, Barbara, 120
Holy Inquisitor (Venice), 261
homesteads, 83
Hopewell Furnace, 211
Hopkinson, Francis, 292–94
horsepower, 234
hot-air balloons, 271–72
House of Commons, 196, 223
Humboldt, Alexander von, 314
Hume, David, 202
humidity, 41–42, 253, 258
Hunter, William, 225–26
Huron nation, 172

hydrogen, 271
hydrostatics, 200
Hyme, William, 94
hydropower, 247

Iceland, 276–77
illnesses: Black Death, 25, 356*n4*; caused by drafts, 102–103, 114; colds, 72, 103, 115, 281; contagion, 281; Great Plague of London, 55
Imperial Academy of Sciences (Russia), 247
imperialism, 136, 247
improvement: age of, 12–13; eighteenth-century rage for, 5; seventeenth-century ideas of, 40–41; today, 349
indentured servants, 120, 168–69, 373*n70*
Independence Hall, *see* Pennsylvania State House
Indigenous American nations: alliances among, 180, 187, 189; attacks on, 187; colonists' fears of resembling, 118; conservation efforts, 55; council fires, 37–38, *38*, 40, 48, 51, 80–81, 82–83, 133, 160, 182–87, *189*, 311–12, 350; diplomatic practices, 13, 37–38, 40; displacement of, 119, 159; enslavement of, 28, 60; environmental caretaking by, 21–22, 29; ethos of sharing resources, 160; extinction narrative about, 174–75; fossil discoveries and, 224–26; in the frontier economy, 119; iron industry, 50; peacemaking power of women, 51, 52; recognition of, 350; removal of, 298–99, 301–302, 312–13, 340; settlement patterns, 171–72; sovereignty of, 187–88, 347; sweathouses, 51; treaties with, 13, 38–39, 52–54, 78–82; uses of fire, 13, 20, 22, 37–39; winter habits of, 37; *see also individual tribes and nations*
indoor climate control, 9–10, 112, 346–47
industrialism, 9
Industrial Revolution, 4, 8, 9–14, 43, 157, 203, 344, 351
"Information to Those Who Would Remove to America" (Franklin), 298, 300
Ingenhousz, Jan, 199, 252, 257, 273, 279, 285
inglenooks, 101
Inquiry into the Nature and Causes of the Wealth of Nations, An (Smith), 298–99
Intergovernmental Panel on Climate Change, 8
internal combustion engine, 336
International Geophysical Year, 349

"in the air" metaphor, 269–70
intoxication, 57–58
inventions: expectations for, 5; inability of to solve wicked problems, 15; patenting of, 130, 208–209, 233–34, 259, 306–309, 315, 318, 319; ventilators, 221–22; *see also* Franklin, Benjamin, inventions of
Ireland, 24, 54–55
Iron Act (1750), 157, 162, 165
iron industry: blast furnaces, 61–63; bloomery forges, 61; Chesapeake, 61, 63; company stores, 119–21, 123; enslaved labor, 12, 63–64, 121–27, *126*, 165, 329–30; expertise of Black ironworkers, 63–64; fuel needs of, 64–65, 367*n36*; global, 60–61; growth of, 61, 62; indentured servants, 373*n70*; Ireland, 24, 54; labor forces, 118–25; Pennsylvania, 12, 50–51, 60, 62–66; prisoner of war labor, 300; refineries, 61–63; in Russia, 247; in Sweden, 140–41; waged and contract labor, 119–21; white investment in, 64–65; women in, 65, 67, 124
Iroquois Confederacy, *see* Haudenosaunee Confederacy
Ishmael (enslaved person), 122, 124
Italy, 75, 102, 253–54, 259–66, 286

Jackson, Richard, 245
Jefferson, Thomas, 302
Jevons, William Stanley, 320
Jevons Paradox, 320
John Adams (miniseries), 7
John Carter Brown Library, 262
Johnston, Jacob, 121
Johnstown Flood, 339–40
joint-stock companies, 327
Jones, Absalom, 311
Jones, David, 120
Josiah (Native man), 119
Journal de physique (periodical), 270
Journal des sçavans (periodical), 76, 252
Junto, 73, 91–92, 94

Kalm, Pehr, 139–40, 201
Kames, Henry Home, Lord, 202, 219
Kanickhungo (Seneca), 80
Kant, Immanuel, 19, 134, 150, 238, 345
keeping warm: as allowing for culturally specific activities, 33, 133; with clothing, 20–21, 31, 123–24; costs of, 90–91; by dancing, *30*; with diet, 57–58; feelings of warmth, 99–101; with fire, 20–21,

INDEX

31; with fireplaces and flues, 32; *see also* heating of homes
Keith, William, 67
Keith's Furnace, 67
Kent, John, 119
kerosene, 336
King (enslaved person), 217, 218, 234
King, Thomas (Oneida), 185
Kingelo, Martin, 120
King Philip's War, 45–46
Kirwan, Richard, 314
"Kubla Khan" (Coleridge), 172
Kubrick, Stanley, 6–7, 33
Kuznets, Simon, 400*n*5

labor: economic value of, 58–59; as an energy source, 12, 58, 120, 121; exploitation of, 329, 334; indentured servants, 120, 168–69; inferiority of enslaved to free, 160–61, 299; in iron industry, 118–25; prisoners of war used as, 300; serfs, 247; strikes, 334; unionized, 334; waged and contract, 119–21; *see also* enslavement and enslaved people
Lack, John, 123
Lafayette, Marquis de, 242
Laki Fissure event, 276–77
La Mecanique du feu (Gauger), 75–76
land and land use: agricultural, 59–60; appropriation of Indigenous, 59–60, 77, 81–84; colonial expansion, 169; English property law, 54, 347; Franklin's western speculation in, 84, 312; Indigenous ideas of, 13, 21, 53; by ironworks, 64
L'art d'exploiter les mines de charbon de terre (Morand), 244–45
Last Glacial Period (LGP), 20–22
Last of the Mohicans, The (Cooper), 327
Latrobe, Benjamin Henry, 301
Lavoisier, Antoine, 199, 254
Lavoisier, Marie-Anne Pierrette Paulze, 199, 317
Lenapehoking, 51–52
Lenape nation: alliance with Haudenosaunee, 52, 159, 182; alliance with US, 296–97; colonists and, 48, 49, 51–54, 78; council fires, 38, 182–87; cultural revival, 180–82; declaration as a nation of women, 51, 52, 182; erosion of land rights of, 83–84; fictionalized, 327; homeland, 51–52; inland exile, 82, 297–98; iron industry and, 51; opposition to enslavement, 60, 164, 296; pacifism, 181, 296; Penn's treaty with, 52–54; Treaty of Fort Pitt, 296–97; use of fire, 51; views on nature, 13, 21, 53, 60; Wyoming Valley settlement, 181–82
Lenoir, Jean-Charles-Pierre, 239, 255–57
Leopold, Grand Duke of Tuscany, later Holy Roman Emperor Leopold II, 260, 264
Lepanto, Battle of, 24
Lettera del Conte Cisalpino (Vernazza), 261
Leutmann, Johann Georg, 287, 291
Library Company of Philadelphia, 64–65, 67, 97, 316
lightning, 147–48, 200
lightning rods, 147, 150, 202, 238
Lincoln, Abraham, 331
Lindbergh Charles, 333
Linnaeus, Carl, 134, 139
Little George (enslaved person), 122
Little Ice Age, 3, 5–6, 9, 20, 27, 34, 47, 68–69, 90, 179, 237, 277, 313, 347, 356*n*4
Logan, James, 67, 73, 80, 81, 90, 91, 94, 97, 103, 168
longhouses, 51
"Loose Thoughts on a Universal Fluid" (Franklin), 273
Louis XIV, 131
Louis XVI, 239
Lunar Society, 203

Mac Packe, Jose, 392*n*28
Maironi da Ponte, Giovanni, 262
maize, 256
Malthus, Thomas Robert, 162
Malthusian trap, 163–64
manpower, 234
Marelli, Giuseppe, 260–61
Marx, Karl, 329
Maryland: iron industry in, 61; winter conditions, 88–89
Massachusetts: old-growth acreage, 339; population growth, 295, winter conditions, 46–48, 88–90, 133, 300–301; *see also* Anglo-Wabanaki Wars
Massachusetts Historical Society, 262
material states, balancing of, 57–58
Mather, Cotton, 34–35, 91
Maunder Minimum, 69
Mazzei, Filippo, 260
Mecom, Jane Franklin, 300–301
Medieval Climate Anomaly, 23
Memoirs of the Literary and Philosophical Society of Manchester, 275
Mercer, Henry, 328

Mercer Museum, 3, 166, 212, 384*n42*
Meridith, Samuel, 120
Metacomet, "Philip" (Wampanoag), 46
"Meteorological Imaginations and Conjectures" (Franklin), 275
Miami nation, 64
Michaux, François André, 339
microclimates, 10, 294
Middle Ages, 23, 26, 31–32, 91, 142, 246, 262, 277
military-industrial complex, 333
Mills, John, 121
mineral resources in America, 60, 174, 337
mining industry, 60, 64, 209–10; *see also* coal: mining
Mississippian period, 341
Mitchell, John, 143–44
Mittelberger, Gottlieb, 179
Mohawk nation, 37, 48, 79
Montgolfier, Jacques-Étienne, 271
Morand, Jean-François-Clément, 244–45, 255, 287
Moravian missionaries, 159–60, 297
Morgan, J. Pierpont, 328
Morgan Library, 328
Morris, Anthony, Jr., 93
Mother Bethel Church (Philadelphia), 311
Motion of Fluids, Natural and Artificial, The (Clare), 76–77
Mount Hekla, 276
Mount Joy Forge, 64, 300
Mount Pleasant Furnace, 94

Nairne, Edward, 228
Napoleonic Wars, 4, 314, 324
Narragansett nation, 36
Narrative of the Late Massacres (Franklin), 188
National Recovery Administration, 334
Native Americans, *see* Indigenous peoples; *specific tribes and nations*
natural resources: assumed to belong to colonists, 136; cultural uses of, 21–23; depletion of, 4, 132; development of, 12; finite nature of, 91; hazards of clearing, 86–87; Indigenous ethos of sharing, 160; mineral, 60, 174; and population growth, 158–59; protection of urban, 294–95; sovereignty over, 53; *see also* conservation, war
Nenockeman (Native man), 119
Neolin (Lenape), 180–81, 187, 312
Netherlands, 103–104, 137–39

Newcomen, Thomas, 203
Newcomen engine, 43, 203–204, 210
New Deal, 334
New-England Courant, The (periodical), 46, 49
New Experiments and Observations Touching Cold (Boyle), 41
Newton, Isaac, 42, 75, 91, 276
New-York Weekly Post-Boy (periodical), 137
Nippon Steel, 334
Noah's flood, 46–47
North America: colonial terraforming of, 161–62, 176–77; colonists' expectations of, 29–30; forests, 28–29
North American Sylva (Michaux), 339
Northanger Abbey (Austen), 318
Nott, Eliphalet, 325
novus ordo seclorum motto, 268–69, 302
Nutt, Anna, 67, 121, 123, 321
Nutt, Samuel, Sr., 64

Observations Concerning the Increase of Mankind (Franklin), 155–56, 157–63, 192, 207, 245, 298, 338
Observations on the Late and Present Conduct of the French (Clarke), 163
Ohio River, 174
Oneida nation, 37, 160, 297–98
Onondaga nation, 37, 79, 82, 171
Opere filosofiche di Beniamino Franklin, 313
Order of the Friars of St. Francis of Wycombe, 232
Ottawa nation, 180
overwintering, 34
oxygen, discovery of, 199, 206, 220–21

pacifism, 53, 177, 181, 295, 296
Paine, Thomas, 329
Panama Canal, 333
Paris Agreement (2015), 352
Parker, James, 89, 129, 136–37
Parks, William, 129
patents, 43, 208–209, 233–34, 306–309, 311, 315, 319
Patton, John, 125–27
Pavlovsk Palace, 247
Paxton Boys, 188, 295
peacemaking: Lenape pacifism, 181, 296; Quaker pacifism, 53; role of Indigenous women in, 51, 52; Treaty of Paris (1783), 268–69, 297–98, 311; *see also* council fires, treaties
Peacock, James, 392*n28*

INDEX

Peale, Charles Willson, 315–16, 338
peat, 25, 138, 141
Pemberton, Israel, 94
Penn, Hannah Callowhill, 78, 79, 83, 295
Penn, John, 79
Penn, Richard, 79
Penn, Thomas, 79, 80, 183
Penn, William, 50, 327, 341; colony mapping, 55–56; "Holy Experiment" of, 56; land transfer negotiations, 77, 79, 81–82, 83–84; "purchase" of Philadelphia, 53; statue of, 333; treaty with Lenape, 52–54
Pennsylvania: abolition in, 321, 331–32; coal deposits, 324–25; economic growth, 59–60, 65; enslavement in, 60, 63–64, 310; forest conservation, 339–40; forests, 22; history, 11–14; industrialization, 303–304, 332–34; iron industry, 12, 50–51, 60, 62–66, 330; map, *95*; militia, 177–78; mixed economy, 59, 65, 68, 84; naming of, 52; as part of Appalachia, 301, 335; petroleum, 335; population growth, 59, 83, 295; present-day lack of recognized tribes, 350; rust belt, 335; treaty with Haudenosaunee, 78–82; treaty with Lenape, 52–54, 81; *see also* Philadelphia
Pennsylvania Abolition Society, 310, 321
Pennsylvania Act for the Gradual Abolition of Slavery, 310
Pennsylvania Assembly, 293
Pennsylvania Gazette, The (periodical), 69–70, 72, 79, 83, 86, 88–89, 93, 96, 147, 148, 168, 178, 293, 331
Pennsylvania Hospital, 327
Pensilvanian Air-grates (Sharp), 249–50
Pennsylvanian fireplace: advantages of, 114–18; advertisements for, 93–94, 129, 137; airbox technology, 110–12; assembly and installation, 106–10, *107*, *109*; audience for, 132–34, 164, 165–66; Durno's appropriation of, 151–53, 308; Dutch marketing efforts, 137–39; in England, 249–51; in Europe, 239; firebox technology, 110–11; in France, 251–59; fuel needs, 113, 115–16, 166–67; invention of, 85–86; in Italy, 259–66, *263*; labor used in forging, 118–25; market competition, 130–31; mentioned in memoirs, 307–309; modifications to, 212, 315, 386n76; ornamental design, 130–31, 212–13;
production, 129, 157, 211–12; sales, 129–30, 136–37, 165, 371n21, 375n2; solar analogy, 131–32, 141; as solution to deforestation, 15, 246; surviving fragments of, 212–13; ventilation, 111–13
Pennsylvanian Period, 340–41, *342–43*
Pennsylvania Railroad, 328, 334
Pennsylvania Rock Oil Company, 335
Pennsylvania State House, 73, 293, 315, 316, *317*
Pennsylvania Station, 328
Pennzoil, 335
Peter (enslaved person), 191, 217, 218
petroleum, 12, 335, 336
Petty, William, 54, 224–25
Philadelphia: enslavement bans, 60; enslavement in, 164–65, 179; firewood shortage, 167; Franklin's arrival in, 50, 56–57, 58; fuel costs, 179; layout of, 55–56; prevalence of stoves and fireplaces, 316; "purchase" of, 53; street names, 55–56, 295; trees and microclimates, 293
Philadelphia Contributionship for Insuring Houses from Loss by Fire, 149
Philadelphia Eagles, 334
Philadelphia Gas Works, 329
Philosophical Transactions (periodical), 41, 72, 76, 143, 147–48, 252, 272
photosynthesis, 199, 273, 279
Pierre-Simon, Marquis de Laplace, 254
Pilgrim's Progress, The (Bunyan), 67, 105
Pittsburgh Steelers, 334
Plain Truth (Franklin), 177
plantation, meanings of term, 63
political arithmetic, 44, 54, 157–59, 164, 167
political economy, 12–13, 44, 54, 298–99, 324
pollution, 10, 26, 194–95, 206, 223–24, 319–20, 352
Pontiac (Ottawa), 187–88
Pontiac's War, 188–89
Poor Richard, 3, 16–17, 71
Poor Richard. An Almanack, 67, 69–72, 86–87, 88, 96, 132, 175, 178, 266, 345–46
Poor Richard Improved, 134, 266
Pope Day, 39–40, 312
population growth: colonial, 157–58, 167, 207, 295; and demand for wood, 25; Franklin's survey of, 157–62, 323–24; historical force of, 157; as key to economic growth, 58; sustainability of, 163–64; theories of, 45
Potts, John, 64, 93, 108, 122, 124, 129, 212–13

Potts, Stephen, 94
Potts, Thomas, 65, 93, 108, 119, 121, 129, 212–13
Potts family, 64–65, 94–95
power plants, 335–36
Practical Treatise on Chimneys (Anderson), 220
Priestley, Joseph, 199, 220–21, 248, 273, 344
"Princess Steel, The" (Du Bois), 334
Principio Iron Works, 67
printing, 40, 67, 69, 93, 96–97, 127, 241
prisoners of war, 300
privacy, 33, 285–86, 289–90
Proclamation Act (1763), 205, 302
Projecting Age, The, 40–41
Prometheus, 19, 72, 134, 150, 238
Ptolemy, 142
pulse glasses, 227–28

Quakers: engagement with Indigenous peoples, 180–81, 184; Great Meeting House, 80; iron industry investment, 64; opposition to enslavement, 60, 310; pacifism, 53; use of enslaved labor, 165
Quaker State, 335

railroads, 332, 333
Reading Furnace, 92, 93, 95
Réaumur, René Antoine Ferchault de, 45
recycling, 3, 40
Reflections on the Motive Power of Fire (Carnot), 229
refraction, 289
registers, 195–96, 201, 316
Religious Society of Friends, *see* Quakers
"Report on the Subject of Manufactures" (Hamilton), 324
respiration, 43, 176, 199, 200, 203, 220–21, 273
Rittenhouse, Benjamin, 315
Rittenhouse, David, 273, 305, 315
Roberts, Hugo, 212
Robinson, Jonathan, 122–23
Roman Warm Period, 22
Roosevelt, Franklin Delano, 334
Rothrock, Joseph T., 339
Royal Society of London, 41, 72, 76, 143–44, 147–48, 252, 272
Rumford fireplace, 317–19; *see also* Thompson, Benjamin
Rumsey, James, 305–307
Russia, 23, 60, 115, 240, 246–49
Russian National Library, 248

Russian State Library, 248
Rutter, Thomas, 64, 65

Saratoga, Battle of, 268
Saunders, Richard, *see* Poor Richard
Sawantaeny (Seneca), 78, 81
Scelta di opuscoli interessanti tradotti da varie lingue (Marelli), 260
Scheele, Carl Wilhelm, 199
Schöpf, Johann David, 301
Schuyler, Arent, 209
Schuyler, Elizabeth, 209
scientific investigation and discovery: atmospheric phenomena, 313–14; atmospheric pressure, 142–44; basal metabolism, 254; carbon dioxide, 199, 220, 273, 344; chemistry, 41, 142, 198–200; convection, 10, 77, 105–106, 144, 169–70; electricity, 144–51; fossils, 10, 171, 173–75, 224–26; geologic time, 170, 172–75; oxygen, 220–21; photosynthesis and transpiration, 175–77, 273; smoke circulation, 284; thermal variation, 91, 144, 170, 197; *see also* Franklin, Benjamin, scientific experiments of
Scots Magazine, The (periodical), 163
sea coal, 26, 153, 214
Second Hundred Years' War, 4, 24
Seneca nation, 37, 78
Seneca Oil, 335
serfs, 247
Seven Years' War, 4, 155, 163, 177, 179, 182–87, 218, 257, 311, 349
Shamokin settlement, 159–60, 171
Sharp, Granville, 309
Sharp, James, 249–51, *250*, 259, 308, 309
Shawnee nation, 159, 312, 313
shipbuilding, 23–24
Six Nations, *see* Haudenosaunee Confederation
skyscrapers, 333, 334
slavery, *see* enslavement and enslaved people
Sleepy Hollow (film), 7–8
Small, Alexander, 209
Small, William, 203
smallpox, 72, 80, 189–90
Smith, Adam, 162, 298–99, 324
Smith, Eliza, 67
smog, 26, 194–95, 325
smoke: anti-smoke campaigns, 10, 219–20; in chimneys, 92, 280–87; from coal, 32, 194–95, 206, 286–87, 325; as different

from heat and air, 100, 279; dispersal of, 202–203; *Encyclopaedia Britannica* entry on, 220, 320; from fireplaces, 76, 89; Franklin's attempts to manipulate, 285–88
smokeless furnace, 252
"Smokey Chimneys" (Franklin), 279
Snellius, Willebrord, 142
social sciences, 44, 163–64, 350
Société royale de médecine, 259
Society of Arts, 209
soot, 26, 32, 33, 76, 200–201, 287
Spain, 43, 249
"Speech of Father Abraham" (Franklin), 248, 266
Spirit of St. Louis (airplane), 333
Sputnik I, 349
Stamp Act (1765), 205, 207–208, 223, 227
steamboats, 305–307, 309–10, 321
steam engines, 203–208, 209–10, 228, 233–34, 257, 305, 309, 319, 332, 349, 383n27
steam power, 9, 14, 43, *44*, 305, 309, 326
steel, 12, 24, 64, 92, 115, 140, 157, 202, 204, 207, 229, 258, 325, 328, 330, 332–34
Stenton (James Logan Home), 90, 103
Stevenson, Margaret, 193, 217
Stevenson, Mary, 197–98, 216, 283
Stiles, Ezra, 176, 381n3
Stock Exchange, 333
stoves: advertisements for, 93–94; chest-of-drawer models, 241; Dutch, 103–104, 152; five-plate, 91; German, 104, 105, 152, 246; meanings of, 32; patents, 315; production of, 94–95; Russian, 247; six-plate, 211; ten-plate, 211; *see also* Franklin stoves
Streaphon (enslaved person), 123–24, 125
sugar industry, 28, 29, 161, 299
Susquehanna Company, 181, 186
Susquehanna River, 79, 82, 159, 171
Susquehannock people, 52, 188
sweathouses, 51
Sweden, 139–41, 350
Sylva (Evelyn), 25, 168

Tamanend (Lenape), 52, 53–54, 327
Tammany Hall, 327
Tammany societies, 327
technology: climate-crisis, 15–16; colonial, 35–36; energy-transition, 353; expectations of, 5; faith in, 5, 346; inability of to solve wicked problems, 15; scientific discovery and, 43–45; sustainable, 197; techno-escape fantasies, 346
techno-optimism, 5, 17
Tecumseh (Shawnee), 312
Teedyuscung (Lenape), 180–86, 188, 295, 325
Telluris Theoria Sacra, or Sacred Theory of the Earth (Burnet), 74
temperance, 57, 58
temperature: measuring, 41–43, 227–28, 282; recommended indoor, 251; rising of Earth's, 341, 344, 352
Tenskwatawa (Shawnee), 312
Terence, 66
terraforming, 28, 161–62, 176–77, 243–44
thermometers, 42, 45, 75, 113–14, 251, 282, 381n3
Thomas, George, 308
Thompson, Benjamin, Count Rumford, 316–18, 319–20
thunderstorms, 144, 147–48, 150, 173
Torricelli, Evangelista, 42
Townshend Duties, 214
Trafalgar, Battle of, 24
Transactions of the American Philosophical Society (periodical), 243, 248, 279
Travels (Bartram), 172
treaties, 13, 38–39, 52–54, 78–82, 183–86, 296–97, 311–12
Treaty of Fort Pitt, 296–97
Treaty of Fort Stanwix, 190
Treaty of Paris, 268–69, 297–98, 311
trees, 293–95; *see also* forests; wood and wood fuel
Tryon, Thomas, 57
Tuckwell and Cooper, 227
Tufts University, 337
Turgot, Étienne-François, 251–52

UN Climate Change Conference, 352
Union Fire Company, 73–74, 148
Union Iron Works, 167
University of Pennsylvania, 337, 338
University of St. Andrews, 198
US Constitution, 11
US Patent Office, 315
US Steel, 328, 334

Valley Forge, 242, 300, 301
vasistas, 282
Vaughan, Benjamin, 245–46
vegetarian diets, 57

Venice, 98, 102, 261–66, 286, 390n45
ventilation, 75, 90, 111–13, 152–53, 196, 220–23, 253, 279–85
Vernazza, Giuseppe, 261
Verri, Alessandro, 264
Verri, Pietro, 264
Virginia, 11; iron industry, 60–61, 330; population growth, 295
virtues, Franklin's list of, 58
volcanic activity, 10, 276–77, 356n4
Volta, Alessandro, 260
Voltaire, 53–54, 188, 209
Vorontsova, Yekaterina Romanovna, 248
Vulcanus Famulans (Leutmann), 287

Wabanaki Confederacy, 21, 37, 38, 45–46, 48, 133
Walking Purchase, 81–82, 154, 159, 168, 172, 180, 181, 325, 340
Wallace, Alfred Russel, 162
wampum, 13, 186, 187, *297*, 298
war, 4, 23–24, 46, 60, 62, 163, 170–71, 177–78, 187–88, 237–40, 257, 300; *see also names of wars*
warmth, *see* keeping warm
War of Jenkins' Ear, 83
War of the League of Augsburg, 4
warships, 23–24
Warwick Furnace, 67, *68*, 94–95, 96, 105, 119–25, 129, 211, 321, 371n21
Washington, George, 240, 242, 295, 300, 310, 316
Watt, James, 203–208, 228, 233–34, 257, 305–307, 319, 383n27
Way to Wealth, The (Franklin), 266–67
wealth, 13, 266–67, 328–29, 346
weather: accounts of, 88; complaints about, 34; concerns about, 69; data collection, 242; forecasting, 70; observations about, 26–27; theories of, 74; thought to cause colds, 72; *see also* climate change; winter
Wedgwood, Josiah, 130, 203
Wence, Jacob, 120
West, Benjamin, 189, *189*
Whiston, William, 145
Whole Earth Catalog (periodical), 345, 346

wicked problems, 15, 16–17, 346
Will (enslaved person), 122
Williams, Henry Shaler, 341
Williams, Roger, 36
Williamson, Hugh, 242, 243
Willson, Israel, 120
wind chill, 42
winter: colonists' concerns about, 69; customs of Indigenous peoples, 34–35, 179; effects of summer haze on, 254, 274–78; Franklin's reporting on, 69–70, 85–89; Mather's views on, 34–35, 91; Native testimony about, 133; overwintering, 34; rising firewood costs during, 178–79; at Valley Forge, 242
Winter Meditations (Mather), 34–35, 91
Winter Palace, 247
Winterthur Museum, 126, 213, 328
Witch, The (film), 7–8
witches and witchcraft, 8, 27, 31
women: Black chimney-sweepers, 331; clothing fire hazards, 318–19; heating of homes associated with, 216–17; and household labor, 217–18; in the iron industry, 65, 67, 120, 121, 124; Lenape nation's declaration as a nation of, 51, 52, 182; peacemaking power of Indigenous, 51, 52; in science, 216; as tenders of the hearth, 31, 37, 186
wood and wood fuel: alternatives to, 25–26, 138; colonial consumption of, 36, 78, 97–98, 166–67; commercial uses of, 28–29; conservation efforts, 25; costs of, 90, 178–79; film depictions of, 6–8; industrial uses, 24; iron industry use of, 64–65, 367n36; population growth and demand for, 25; resource scarcity, 237, 249, 254–55, 300–301; in Russia, 247; in shipbuilding, 23–24; shortages of, 4, 23; subsidies for the poor, 87–88, 314, 348; UK consumption of, 325–26; US consumption of, 325–26; Venetian shortages, 262–64; *see also* deforestation
World's Fair (1893), 332
Wyoming Valley settlement, 181–82, 324–25